FIBONACCI AND CATALAN NUMBERS

FIBONACCI AND CATALAN NUMBERS

AN INTRODUCTION

Ralph P. Grimaldi

Rose-Hulman Institute of Technology

WILEY

A JOHN WILEY & SONS, INC., PUBLICATION

Published by John Wiley & Sons, Inc., Hoboken, New Jersey
Published simultaneously in Canada

For general information on our other products and services or for technical support, please contact our Customer Care Department within the United States at (800) 762-2974, outside the United States at (317) 572-3993 or fax (317) 572-4002.

Wiley also publishes its books in a variety of electronic formats. Some content that appears in print may not be available in electronic formats. For more information about Wiley products, visit our web site at www.wiley.com.

Library of Congress Cataloging-in-Publication Data:

Grimaldi, Ralph P.
  Fibonacci and catalan numbers : an introduction / Ralph P. Grimaldi.
    p. cm.
  Includes bibliographical references and index.
    ISBN 978-0-470-63157-7
  1. Fibonacci numbers.   2. Recurrent sequences (Mathematics)   3. Catalan numbers (Mathematics)   4. Combinatorial analysis.   I. Title.
    QA241.G725 2012
    512.7'2–dc23

                                                                    2011043338

10   9   8   7   6   5   4   3   2   1

Dedicated to the Memory of

Josephine and Joseph

and

Mildred and John

and

Madge

◼◼◼ CONTENTS

In January of 1992, I presented a minicourse at the joint national mathematics meetings held that year in Baltimore, Maryland. The minicourse had been approved by a committee of the Mathematical Association of America—the mission of that committee being the evaluation of proposed minicourses. In this case, the minicourse was especially promoted by Professor Fred Hoffman of Florida Atlantic University. Presented in two two-hour sessions, the first session of the minicourse touched upon examples, properties, and applications of the sequence of Fibonacci numbers. The second part investigated comparable ideas for the sequence of Catalan numbers. The audience was comprised primarily of college and university mathematics professors, along with a substantial number of graduate students and undergraduate students, as well as some mathematics teachers from high schools in the Baltimore and Washington, D.C. areas.

Since its first presentation, the coverage in this minicourse has expanded over the past 19 years, as I delivered the material nine additional times at later joint national mathematics meetings—the latest being the meetings held in January of 2010 in San Francisco. In addition, the topics have also been presented completely, or in part, at more than a dozen state sectional meetings of the Mathematical Association of America and at several workshops, where, on occasion, some high school students were in attendance. Evaluations provided by those who attended the lectures directed me to further relevant material and also helped to improve the presentations.

At all times, the presentations were developed so that everyone in the audience would be able to understand at least some, if not a substantial amount, of the material. Consequently, this resulting book, which has grown out of these experiences, should be looked upon as an *introduction* to the many interesting properties, examples, and applications that arise in the study of two of the most fascinating sequences of numbers. As we progress through the various chapters, we should soon come to understand why these sequences are often referred to as *ubiquitous*, especially in courses in discrete mathematics and combinatorics, where they appear so very often. For the Fibonacci numbers, we shall find applications in such diverse areas as set theory, the compositions of integers, graph theory, matrix theory, trigonometry, botany, chemistry, physics, probability, and computational complexity. We shall find the Catalan numbers arise in situations dealing with lattice paths, graph theory, geometry, partial orders, sequences, pattern avoidance, partitions, computer science, and even sporting events.

FEATURES

Following are brief descriptions of four of the major features of this book.

1. **Useful Resources**

    The book can be used in a variety of ways:

    (i) As a textbook for an introductory course on the Fibonacci numbers and/or the Catalan numbers.

    (ii) As a supplement for a course in discrete mathematics or combinatorics.

    (iii) As a source for students seeking a topic for a research paper or some other type of project in a mathematical area they have not covered, or only briefly covered, in a formal mathematics course.

    (iv) As a source for independent study.

2. **Organization**

    The book is divided into 36 chapters. The first 17 chapters constitute Part One of the book and deal with the Fibonacci numbers. Chapters 18 through 36 comprise Part Two, which covers the material on the Catalan numbers. The two parts can be covered in either order. In Part Two, some references are made to material in Part One. These are usually only comparisons. Should the need arise, one can readily find the material from Part One that is mentioned in conjunction with something covered in Part Two.

    Furthermore, each of Parts One and Two ends with a bibliography. These references should prove useful for the reader interested in learning even more about either of these two rather amazing number sequences.

3. **Detailed Explanations**

    Since this book is to be regarded as an introduction, examples and, especially, proofs are presented with detailed explanations. Such examples and proofs are designed to be careful and thorough. Throughout the book, the presentation is focused primarily on improving understanding for the reader who is seeing most, if not all, of this material for the first time.

    In addition, every attempt has been made to provide any necessary background material, whenever needed.

4. **Exercises**

    There are over 300 exercises throughout the book. These exercises are primarily designed to review the basic ideas provided in a given chapter and to introduce additional properties and examples. In some cases, the exercises also extend what is covered in one or more of the chapters. Answers for all the odd-numbered exercises are provided at the back of the book.

ANCILLARY

There is an *Instructor's Solution Manual* that is available for those instructors who adopt this book. The manual can be obtained from the publisher via written request on departmental letterhead. It contains the solutions for all the exercises within both parts of the book.

ACKNOWLEDGMENTS

If space permitted, I should like to thank each of the many participants at the mini-courses, sectional meetings, and workshops, who were so very encouraging over the years. I should also like to acknowledge their many helpful suggestions about the material and the way it was presented.

The work behind this book could not have been possible without the education I received because of the numerous sacrifices made by my parents Carmela and Ralph Grimaldi. Thanks are also due to Helen Calabrese for her constant encouragement. As an undergraduate at the State University of New York at Albany, I was so very fortunate to have professors like Robert C. Luippold, Paul T. Schaefer, and, especially, Violet H. Larney, who first introduced me to the fascinating world of abstract algebra. When I attended New Mexico State University as a graduate student, there I crossed paths, both inside and outside the classroom, with Professors David Arnold, Carol Walker, and Elbert Walker. They had a definite impact on my education. Even more thanks is due to Professor Edward Gaughan and, especially, my ever-patient and encouraging advisor, Professor Ray Mines. Also, it was on a recent sabbatical at New Mexico State University where I was able to put together so much of the material that now makes up this book. Hence, I must thank their mathematics department for providing me with an office, with such a beautiful view, and the resources necessary for researching so much of what is written here.

One cannot attempt to write a book such as this without help and guidance. Consequently, I want to thank John Wiley & Sons, Inc. for publishing this book. On a more individual level, I want to thank Shannon Corliss, Stephen Quigley, and Laurie Rosatone for their initial interest in the book. In particular, I must acknowledge the assistance and guidance provided throughout the development of the project by my editors Susanne Steitz-Filler and Jacqueline Palmieri. Special thanks are due to Dean Gonzalez for all his efforts in developing the figures. Finally, I must gratefully applaud the constant help and encouagment provided by Senior Production Editor Kristen Parrish who managed to get this author over so many hurdles that seemed to pop up.

I also want to acknowledge the helpful comments provided by Charles Anderson and the reviewers Professor Gary Stevens of Hartwick College and Professor Barry Balof of Whitman College. My past and present colleagues in the Mathematics Department at Rose-Hulman Institute of Technology have been very supportive during the duration of the project. In particular, I thank Diane Evans, Al Holder, Leanne Holder, Tanya Jajcay, John J. Kinney, Thomas Langley, Jeffery J. Leader, David Rader, and John Rickert. I must also thank Dean Art Western for approving the sabbatical which provided the time for me to start writing this book.

I thank Larry Alldredge for his help in dealing with the computer science material, Professor Rebecca DeVasher of the Rose-Hulman Institute of Technology for her guidance on the applications in chemistry, and Professor Jerome Wagner of the Rose-Hulman Institute of Technology for his enlightening remarks on the physics applications.

A book of this nature requires the use of many references. The members of the library staff of the Rose-Hulman Institute of Technology were always so helpful when books and articles were needed. Consequently, it is only fitting to recognize the behind-the-scenes efforts of Jan Jerrell and, especially, Amy Harshbarger.

The last note of thanks belongs to Mrs. Mary Lou McCullough, the now-retired one-time secretary of the Rose-Hulman mathematics department. Although not directly involved in this project, her friendship and encouragement in working with me on several editions of another book and numerous research articles had a tremendous effect on my work in writing this new book. I shall remain ever grateful for everything she has done for me.

Unfortunately, any remaining errors, ambiguities, or misleading comments are the sole responsibility of the author.

RALPH P. GRIMALDI
Terre Haute, Indiana

PART ONE

THE FIBONACCI NUMBERS

CHAPTER 1

Historical Background

Born around 1170 into the Bonacci family of Pisa, Leonardo of Pisa was the son of the prosperous merchant Guglielmo, who sought to have his son follow in his footsteps. Therefore, when Guglielmo was appointed the customs collector for the Algerian city of Bugia (now Bejaia), around 1190, he brought Leonardo with him. It was here that the young man studied with a Muslim schoolmaster who introduced him to the Hindu-Arabic system of enumeration along with Hindu-Arabic methods of computation. Then, as he continued his life in the mercantile business, Leonardo found himself traveling to Constantinople, Egypt, France, Greece, Rome, and Syria, where he continued to investigate the various arithmetic systems then being used. Consequently, upon returning home to Pisa around 1200, Leonardo found himself an advocate of the elegant simplicity and practical advantage of the Hindu-Arabic number system—especially, when compared with the Roman numeral system then being used in Italy. As a result, by the time of his death in about 1240, Italian merchants started to recognize the value of the Hindu-Arabic number system and gradually began to use it for business transactions. By the end of the sixteenth century, most of Europe had adjusted to the system.

In 1202, Leonardo published his pioneering masterpiece, the *Liber Abaci* (*The Book of Calculation* or *The Book of the Abacus*). Therein he introduced the Hindu-Arabic number system and arithmetic algorithms to the continent of Europe. Leonardo started his work with the introduction of the Hindu-Arabic numerals: the nine Hindu figures 1, 2, 3, 4, 5, 6, 7, 8, 9, along with the figure 0, which the Arabs called "zephirum" (cipher). Then he addressed the issue of a place value system for the integers. As the text progresses, various types of problems are addressed, including one type on determinate and indeterminate linear systems of equations in more than two unknowns, and another on perfect numbers (that is, a positive integer whose value equals the sum of the values of all of its divisors less than itself—for example, $6 = 1 + 2 + 3$ and $28 = 1 + 2 + 4 + 7 + 14$). Inconspicuously tucked away between these two types of problems lies the one problem that so many students and teachers of mathematics seem to know about—the notorious "Problem of the Rabbits."

Fibonacci and Catalan Numbers: An Introduction, First Edition. Ralph P. Grimaldi.
© 2012 John Wiley & Sons, Inc. Published 2012 by John Wiley & Sons, Inc.

Before continuing at this point, let us mention that although Leonardo is best known for the *Liber Abaci*, he also published three other prominent works. The *Practica Geometriae* (*Practice of Geometry*) was written in 1220. The *Flos* (*Flower* or *Blossom*) was published in 1225, as was the *Liber Quadratorum* (*The Book of Square Numbers*). The latter work established Leonardo as a renowned number theorist.

CHAPTER 2

The Problem of the Rabbits

In the now famous "Problem of the Rabbits," Leonardo introduces us to a person who has a pair of newborn rabbits—one of each gender. We are interested in determining the number of pairs of rabbits that can be bred from (and include) this initial pair in a year if

(1) each newborn pair, a female and a male, matures in one month and then starts to breed;

(2) two months after their birth, and every month thereafter, a now mature pair breed at the beginning of each month. This breeding then results in the birth of one (newborn) pair, a female and a male, at the end of that month; and,

(3) no rabbits die during the course of the year.

If we start to examine this situation on the first day of a calendar year, we find the results in Table 2.1 on p. 6.

We need to remember that at the end of each month, a newborn pair (born at the end of the month) grows to maturity, regardless of the number of days—be it 28, 30, or 31—in the next month. This makes the *new maturity* entry equal to the sum of the *prior maturity* entry plus the *prior newborn* entry. Also, since each mature pair produces a newborn pair at the end of that month, the *newborn* entry for any given month equals the *mature* entry for the prior month. From the third column in Table 2.1, we see that at the end of the year, the person who started with this one pair of newborn rabbits now has a total of 233 pairs of rabbits, including the initial pair.

This sequence of numbers—namely, 1, 1, 2, 3, 5, 8, 13, 21, 34, 55, ... —is often called the *Fibonacci sequence*. The name *Fibonacci* is a contraction of *Filius Bonaccii*, the Latin form for "son of Bonaccio," and the name was given to the sequence in May of 1876 by the renowned French number theorist François Edouard Anatole Lucas (pronounced Lucah) (1842–1891). In reality, Leonardo was not the first to describe the sequence, but he did publish it in the *Liber Abaci*, which introduced it to the West.

Fibonacci and Catalan Numbers: An Introduction, First Edition. Ralph P. Grimaldi.
© 2012 John Wiley & Sons, Inc. Published 2012 by John Wiley & Sons, Inc.

TABLE 2.1

	Number of Pairs of Newborn Rabbits	Number of Pairs of Mature Rabbits	Total Number of Pairs of Rabbits
Start January 1	1	0	1
One Month Later February 1	0	1	1
Two Months Later March 1	1	1	2
Three Months Later April 1	1	2	3
Four Months Later May 1	2	3	5
Five Months Later June 1	3	5	8
Six Months Later July 1	5	8	13
Seven Months Later August 1	8	13	21
Eight Months Later September 1	13	21	34
Nine Months Later October 1	21	34	55
Ten Months Later November 1	34	55	89
Eleven Months Later December 1	55	89	144
One Year Later January 1	89	144	233

The Fibonacci sequence has proved to be one of the most intriguing and ubiquitous number sequences in all of mathematics. Unfortunately, when these numbers arise, far too many students, and even teachers of mathematics, are only aware of the connection between these numbers and the "Problem of the Rabbits." However, as the reader will soon learn, these numbers possess a great number of fascinating properties and arise in so many different areas.

CHAPTER 3

The Recursive Definition

Upon examining the sequence in the middle column of Table 2.1, we see that after the first two entries, each entry is the sum of the two preceding entries. For example, $1 = 1 + 0$, $2 = 1 + 1$, $3 = 2 + 1$, $5 = 3 + 2$, $8 = 5 + 3$, ..., $55 = 34 + 21$. So we are able to determine later numbers in the sequence when we know the values of earlier numbers in the sequence. This property now allows us to define what we shall henceforth consider to be *the* Fibonacci numbers. Consequently, the sequence of Fibonacci numbers is defined, *recursively*, as follows:

For $n \geq 0$, if we let F_n denote the nth Fibonacci number, we have

(1) $F_0 = 0$, $F_1 = 1$ (The Initial Conditions)

(2) $F_n = F_{n-1} + F_{n-2}$, $n \geq 2$ (The Recurrence Relation)

Therefore, the sequence $F_0, F_1, F_2, F_3, \ldots$, which appears in the middle column of Table 2.1, now has a different starting point, namely, F_0, from the sequence F_1, $F_2, F_3, \ldots$, which appears in the third column of Table 2.1. This sequence—$F_0, F_1, F_2, F_3, \ldots$—is now accepted as the standard definition for the sequence of Fibonacci numbers. It is one of the earliest examples of a recursive sequence in mathematics. Many feel that Fibonacci was undoubtedly aware of the recursive nature of these numbers. However, it was not until 1634, when mathematical notation had sufficiently progressed, that the Dutch mathematician Albert Girard (1595–1632) wrote the formula in his posthumously published work *L'Arithmetique de Simon Stevin de Bruges.*

Using the recursive definition above, we find the first 25 Fibonacci numbers in Table 3.1.

TABLE 3.1

$F_0 = 0$	$F_5 = 5$	$F_{10} = 55$	$F_{15} = 610$	$F_{20} = 6765$
$F_1 = 1$	$F_6 = 8$	$F_{11} = 89$	$F_{16} = 987$	$F_{21} = 10,946$
$F_2 = 1$	$F_7 = 13$	$F_{12} = 144$	$F_{17} = 1597$	$F_{22} = 17,711$
$F_3 = 2$	$F_8 = 21$	$F_{13} = 233$	$F_{18} = 2584$	$F_{23} = 28,657$
$F_4 = 3$	$F_9 = 34$	$F_{14} = 377$	$F_{19} = 4181$	$F_{24} = 46,368$

Fibonacci and Catalan Numbers: An Introduction, First Edition. Ralph P. Grimaldi.
© 2012 John Wiley & Sons, Inc. Published 2012 by John Wiley & Sons, Inc.

CHAPTER 4

Properties of the Fibonacci Numbers

As we examine the entries in Table 3.1, we find that the greatest common divisor of $F_5 = 5$ and $F_6 = 8$ is 1. This is due to the fact that the only positive integers that divide $F_5 = 5$ are 1 and 5, and the only positive integers that divide $F_6 = 8$ are 1, 2, 4, and 8. We shall denote this by writing $\gcd(F_5, F_6) = 1$. Likewise, $\gcd(F_9, F_{10}) = 1$, since 1, 2, 17, and 34 are the only positive integers that divide $F_9 = 34$, while the only positive integers that divide $F_{10} = 55$ are 1, 5, 11, and 55. Hopefully we see a pattern developing here, and this leads us to our first general property for the Fibonacci numbers.

Property 4.1: For $n \geq 0$, $\gcd(F_n, F_{n+1}) = 1$.

Proof: We note that $\gcd(F_0, F_1) = \gcd(0, 1) = 1$. Consequently, if the result is false, then there is a first case, say $n = r > 0$, where $\gcd(F_r, F_{r+1}) > 1$. However, $\gcd(F_{r-1}, F_r) = 1$. So there is a positive integer d such that $d > 1$ and d divides F_r and F_{r+1}. But we know that

$$F_{r+1} = F_r + F_{r-1}.$$

So if d divides F_r and F_{r+1}, it follows that d divides F_{r-1}. This then contradicts $\gcd(F_{r-1}, F_r) = 1$. Consequently, $\gcd(F_n, F_{n+1}) = 1$ for $n \geq 0$. $\qquad\square$

Using a similar argument and Property 4.1, the reader can establish our next result.

Property 4.2: For $n \geq 0$, $\gcd(F_n, F_{n+2}) = 1$.

To provide some motivation for the next property, we observe that

$$F_0 + F_1 + F_2 + F_3 + F_4 + F_5 = 0 + 1 + 1 + 2 + 3 + 5 = 12 = 4 \cdot 3$$
$$F_1 + F_2 + F_3 + F_4 + F_5 + F_6 = 1 + 1 + 2 + 3 + 5 + 8 = 20 = 4 \cdot 5$$
$$F_2 + F_3 + F_4 + F_5 + F_6 + F_7 = 1 + 2 + 3 + 5 + 8 + 13 = 32 = 4 \cdot 8.$$

Fibonacci and Catalan Numbers: An Introduction, First Edition. Ralph P. Grimaldi.
© 2012 John Wiley & Sons, Inc. Published 2012 by John Wiley & Sons, Inc.

These results suggest the following:

Property 4.3: The sum of any six consecutive Fibonacci numbers is divisible by 4. Even further, for $n \geq 0$ (with n fixed),

$$\sum_{r=0}^{5} F_{n+r} = F_n + F_{n+1} + F_{n+2} + F_{n+3} + F_{n+4} + F_{n+5} = 4F_{n+4}.$$

Proof: For $n \geq 0$,

$$\sum_{r=0}^{5} F_{n+r} = F_n + F_{n+1} + F_{n+2} + F_{n+3} + F_{n+4} + F_{n+5}$$
$$= (F_n + F_{n+1}) + F_{n+2} + F_{n+3} + F_{n+4} + (F_{n+3} + F_{n+4})$$
$$= 2F_{n+2} + 2F_{n+3} + 2F_{n+4} = 2(F_{n+2} + F_{n+3}) + 2F_{n+4}$$
$$= 4F_{n+4}. \qquad \square$$

In a similar manner, one can likewise verify the following:

Property 4.4: The sum of any ten consecutive Fibonacci numbers is divisible by 11. In fact, for $n \geq 0$ (with n fixed),

$$\sum_{r=0}^{9} F_{n+r} = 11F_{n+6}.$$

Our next property was discovered by Edouard Lucas in 1876. A few observations help suggest the general result:

$$F_0 + F_1 + F_2 = 2 = 3 - 1 = F_4 - 1$$
$$F_0 + F_1 + F_2 + F_3 = 4 = 5 - 1 = F_5 - 1$$
$$F_0 + F_1 + F_2 + F_3 + F_4 = 7 = 8 - 1 = F_6 - 1.$$

Property 4.5: For $n \geq 0$, $\sum_{r=0}^{n} F_r = F_{n+2} - 1$.

Proof: Although this summation formula can be established using the Principle of Mathematical Induction, here we choose to use the recursive definition of the Fibonacci numbers and consider the following:

$$F_0 = F_2 - F_1$$
$$F_1 = F_3 - F_2$$
$$F_2 = F_4 - F_3$$
$$\vdots \quad \vdots \quad \vdots$$
$$F_{n-1} = F_{n+1} - F_n$$
$$F_n = F_{n+2} - F_{n+1}.$$

When we add these $n + 1$ equations, the left-hand side gives us $\sum_{r=0}^{n} F_r$, while the right-hand side provides $(F_2 - F_1) + (F_3 - F_2) + \cdots + (F_{n+1} - F_n) + (F_{n+2} - F_{n+1}) = -F_1 + (F_2 - F_2) + (F_3 - F_3) + \cdots + (F_n - F_n) + (F_{n+1} - F_{n+1}) + F_{n+2} = F_{n+2} - F_1 = F_{n+2} - 1.$ $\square$

Passing from first powers to squares, we find that

$$F_0^2 = 0^2 = 0 = 0 \times 1$$
$$F_0^2 + F_1^2 = 0^2 + 1^2 = 1 = 1 \times 1$$
$$F_0^2 + F_1^2 + F_2^2 = 0^2 + 1^2 + 1^2 = 2 = 1 \times 2$$
$$F_0^2 + F_1^2 + F_2^2 + F_3^2 = 0^2 + 1^2 + 1^2 + 2^2 = 6 = 2 \times 3$$
$$F_0^2 + F_1^2 + F_2^2 + F_3^2 + F_4^2 = 0^2 + 1^2 + 1^2 + 2^2 + 3^2 = 15 = 3 \times 5.$$

From what is suggested in these five results, we conjecture the following:

Property 4.6: For $n \geq 0$, $\sum_{r=0}^{n} F_r^2 = F_n \times F_{n+1}$.

Proof: Here we shall use the Principle of Mathematical Induction. For $n = 0$, we have $\sum_{r=0}^{0} F_r^2 = F_0^2 = 0^2 = 0 = 0 \times 1 = F_0 \times F_1 = F_0 \times F_{0+1}$. This demonstrates that the conjecture is true for this first case and provides the basis step for our inductive proof. So now we assume the conjecture true for some fixed (but arbitrary) $n = k$ (≥ 0). This gives us $\sum_{r=0}^{k} F_r^2 = F_k \times F_{k+1}$. Turning to the case where $n = k + 1$ (≥ 1), we have

$$\sum_{r=0}^{k+1} F_r^2 = \left(\sum_{r=0}^{k} F_r^2 \right) + F_{k+1}^2 = (F_k \times F_{k+1}) + F_{k+1}^2$$
$$= F_{k+1} \times (F_k + F_{k+1}) = F_{k+1} \times F_{k+2}.$$

Consequently, the truth of the case for $n = k + 1$ follows from the case for $n = k$. So our conjecture is true for all $n \geq 0$, by the Principle of Mathematical Induction. $\square$

At this point, let us mention three more properties exhibited by the Fibonacci numbers. There are so many! The reader should find a wealth of such results in References [38, 50]. The first two of these properties are also due to Edouard Lucas from 1876. The third was discovered in 1680 by the Italian-born French astronomer and mathematician Giovanni Domenico (Jean Dominique) Cassini (1625–1712). This result was also discovered independently in 1753 by the Scottish mathematician and landscape artist Robert Simson (1687–1768). We shall leave the proofs of all three results for the reader. However, we shall obtain the result due to Cassini in another way, when we introduce a 2×2 matrix whose components are Fibonacci numbers in Chapter 14.

Property 4.7: For $n \geq 1$, $\sum_{r=1}^{n} F_{2r-1} = F_1 + F_3 + \cdots + F_{2n-1} = F_{2n}$.

Property 4.8: For $n \geq 1$, $\sum_{r=1}^{n} F_{2r} = F_2 + F_4 + \cdots + F_{2n} = F_{2n+1} - 1$.

Property 4.9: For $n \geq 1$, $F_{n-1}F_{n+1} - F_n^2 = (-1)^n$.

At this point, we realize that the Fibonacci numbers do possess some interesting properties. But surely there must be places where these numbers arise—other than the "Problem of the Rabbits." In Chapter 5, we shall encounter some examples where these numbers arise, and start to learn why these numbers are often referred to as ubiquitous.

EXERCISES FOR CHAPTER 4

1. Prove Property 4.2—that is, for $n \geq 0$, $\gcd(F_n, F_{n+2}) = 1$.
2. Provide an example to show that $\gcd(F_n, F_{n+3}) \neq 1$ for some $n \geq 0$.
3. For $n \geq 1$, prove that $F_{2(n+1)} = 2F_{2n} + F_{2n-1}$.
4. For $n \geq 2$, prove that $F_{n+2} + F_{n-2} = 3F_n$.
5. For $n \geq 2$, prove that $F_{n+2} + F_n + F_{n-2} = 4F_n$.
6. For $n \geq 1$, prove that $F_{n+1}^3 = F_n^3 + F_{n-1}^3 + 3F_{n-1}F_nF_{n+1}$.
7. For $n \geq 2$, prove that $F_{3n} = 4F_{3n-3} + F_{3n-6}$.
8. Prove that $\sum_{r=0}^{9} F_{n+r} = 11F_{n+6}$.
9. Use the Principle of Mathematical Induction to prove that for $n \geq 0$, $\sum_{r=0}^{n} F_r = F_{n+2} - 1$.
10. Fix $n \geq 0$. Prove that $\sum_{r=1}^{m} F_{n+r} = F_{m+n+2} - F_{n+2}$.
11. Prove Property 4.7—that is, for $n \geq 1$, $\sum_{r=1}^{n} F_{2r-1} = F_{2n}$.
12. Prove Property 4.8—that is, for $n \geq 1$, $\sum_{r=1}^{n} F_{2r} = F_{2n+1} - 1$.
13. Verify the result due to Giovanni Cassini in Property 4.9 for $n = 3, 4, 5$, and 6.
14. Use the Principle of Mathematical Induction to prove Property 4.9—that is, for $n \geq 1$, $F_{n-1}F_{n+1} - F_n^2 = (-1)^n$.
15. For $n \geq 1$, prove that $F_nF_{n+1}F_{n+2} = F_{n+1}^3 + F^{n+1}(-1)^{n+1}$.
16. Jodi starts to write the Fibonacci numbers on the board in her office, using the recursive definition. She writes the correct values for the numbers F_0, F_1, F_2, ..., F_{n-1}, but then Professor Brooks distracts her and she writes $F_n + 1$ instead of the actual value F_n. (a) If she does not make any further mistakes in using the recursive definition for the Fibonacci numbers, what value does she write next? (b) What does she write instead of the actual value of F_{n+2}? (c) In general, for $r > 0$, what value does she write instead of the actual value of F_{n+r}?
17. Use the Principle of Mathematical Induction to prove that for $n \geq 1$,

$$\sum_{r=1}^{n} rF_r = F_1 + 2F_2 + 3F_3 + \cdots + nF_n = nF_{n+2} - F_{n+3} + 2.$$

(This formula is an example of a *weighted sum* involving the Fibonacci numbers.)

18. For $n \geq 0$, prove that $F_n^2 + F_{n+3}^2 = 2(F_{n+1}^2 + F_{n+2}^2)$. (H. W. Gould, 1963) [24].

19. For $n \geq 1$, prove that $\sum_{i=1}^{n}(-1)^{i+1} F_{i+1} = (-1)^{n-1} F_n$.

20. For $n \geq 1$, prove that $F_{n+1}^2 - F_n^2 = F_{n-1} F_{n+2}$.

21. For $n \geq 0$, prove that $F_n^2 + F_{n+4}^2 = F_{n+1}^2 + F_{n+3}^2 + 4F_{n+2}^2$. (M. N. S. Swamy, 1966) [52].

22. For $n \geq 1$, prove that $\sum_{i=1}^{2n} F_i F_{i+1} = F_{2n+1}^2 - 1$. (K. S. Rao, 1953) [45].

23. For $n \geq 1$, prove that

$$\sum_{i=1}^{n} F_i F_{i+1} = F_{n+1}^2 - \frac{1}{2}\left[1 + (-1)^n\right].$$ (T. Koshy, 1998) [37].

24. (a) For any real numbers a and b, prove that

$$\left[a^2 + b^2 + (a+b)^2\right]^2 = 2\left[a^4 + b^4 + (a+b)^4\right].$$

[This result is known as *Candido's identity*, in honor of the Italian mathematician Giacomo Candido (1871–1941).]

(b) For $n \geq 0$, prove that

$$\left(F_n^2 + F_{n+1}^2 + F_{n+2}^2\right)^2 = 2(F_n^4 + F_{n+1}^4 + F_{n+2}^4).$$

25. For $n \geq 0$, prove that $F_{n+5} - 3F_n$ is divisible by 5. [Alternatively, this can be stated as $F_{n+5} \equiv 3F_n \pmod 5$.]

26. For $n \geq m \geq 1$, prove that $\sum_{r=m}^{n} F_r = F_{n+2} - F_{m+1}$.

27. (a) Verify that $(F_3 + F_4 + F_5 + F_6) + F_4 = F_8$.

(b) For what value of n is it true that $(F_4 + F_5 + F_6 + F_7 + F_8) + F_5 = F_n$?

(c) Fix $n \geq 1$ and $m \geq 1$. What is $(F_n + F_{n+1} + F_{n+2} + \cdots + F_{n+m}) + F_{n+1}$? (This fascinating tidbit was originally recognized by W. H. Huff.)

28. For $n \geq 3$, prove that

$$F_n + F_{n-1} + F_{n-2} + 2F_{n-3} + 2^2 F_{n-4} + 2^3 F_{n-5} + \cdots + 2^{n-4} F_2 + 2^{n-3} F_1 = 2^{n-1}.$$

CHAPTER 5

Some Introductory Examples

As the title indicates, this chapter will provide some examples where the Fibonacci numbers arise. In particular, one such example will show us how to write a Fibonacci number as a sum of binomial coefficients. In addition, even more examples will arise from some of the exercises for the chapter.

Example 5.1: [Irving Kaplansky (1917-2006)] : For $n \geq 1$, let $S_n = \{1, 2, 3, \ldots, n\}$, and let $S_0 = \varnothing$, the *null*, or *empty*, set. Then the number of subsets of S_n is 2^n. But now let us count the number of subsets of S_n with no consecutive integers. So, for $n \geq 0$, we shall let a_n count the number of subsets of S_n that contain no consecutive integers. We consider the situation for $n = 3$, 4, and 5. In each case, we find the empty set $\varnothing$; otherwise, there would have to exist two integers in $\varnothing$ of the form x and $x + 1$. Either such integer contradicts the definition of the null set. So the subsets with no consecutive integers for these three cases are as follows:

$$n = 3: \quad S_3 = \{1, 2, 3\}$$
$$\text{Subsets}: \quad \varnothing, \{1\}, \{2\}, \{3\}, \{1, 3\}$$
$$n = 4: \quad S_4 = \{1, 2, 3, 4\}$$
$$\text{Subsets}: \quad \varnothing, \{1\}, \{2\}, \{3\}, \{4\}, \{1, 3\}, \{1, 4\}, \{2, 4\}$$
$$n = 5: \quad S_5 = \{1, 2, 3, 4, 5\}$$
$$\text{Subsets}: \quad \varnothing, \{1\}, \{2\}, \{3\}, \{4\}, \{1, 3\}, \{1, 4\}, \{2, 4\},$$
$$\{5\}, \{1, 5\}, \{2, 5\}, \{3, 5\}, \{1, 3, 5\}$$

Note that when we consider the case for $n = 5$, only two situations can occur, and they cannot occur simultaneously:

(i) 5 is not in the subset: Here we can use any of the eight subsets for S_4—as we see from the first line of subsets for S_5.

(ii) 5 is in the subset: Then we cannot have 4 in the subset. So we place the integer 5 in each of the five subsets for S_3 and arrive at the five subsets in the second line of subsets for S_5. Consequently, we have $a_5 = a_4 + a_3$.

Fibonacci and Catalan Numbers: An Introduction, First Edition. Ralph P. Grimaldi.
© 2012 John Wiley & Sons, Inc. Published 2012 by John Wiley & Sons, Inc.

The above argument generalizes to give us

$$a_n = a_{n-1} + a_{n-2}, \quad n \geq 2, \quad a_0 = 1, \quad a_1 = 2.$$

The recurrence relation in this case is the same as the one for the Fibonacci numbers, but the initial conditions are different. Here we have $a_0 = 1 = F_2$ and $a_1 = 2 = F_3$. Therefore,

$$a_n = F_{n+2}, \quad n \geq 0.$$

Example 5.2: As in Example 5.1, we shall let $S_n = \{1, 2, 3, \ldots, n\}$, for $n \geq 1$. Then for any nonempty subset A of S_n, we define $A + 1 = \{a + 1 \mid a \in A\}$. So if $n = 4$ and $A = \{1, 2, 4\}$, then $A + 1 = \{2, 3, 5\}$, and we see that $A \cup (A + 1) = S_5$. Consequently, for $n \geq 1$, we shall now let g_n count the number of subsets A of S_n such that $A \cup (A + 1) = S_{n+1}$. Such subsets A of S_n are called *generating sets* for S_{n+1}. We realize that for any such subset A, it follows that $1 \in A$ and, for $n \geq 2, n \in A$. For $n = 3, \ 4, \ $ and 5, we find the following examples of generating sets:

$$n = 3: \ \{1, 3\}, \ \{1, 2, 3\}$$
$$n = 4: \ \{1, 2, 4\}, \ \{1, 3, 4\}, \ \{1, 2, 3, 4\}$$
$$n = 5: \ \{1, 3, 5\}, \ \{1, 2, 3, 5\},$$
$$\{1, 2, 4, 5\}, \ \{1, 3, 4, 5\}, \ \{1, 2, 3, 4, 5\}.$$

Here we see that the g_5 generating sets for S_6 (where $n = 5$) are obtained from those of S_5 (where $n = 4$) and from those of S_4 (where $n = 3$), by placing 5 in each of the g_4 generating subsets for S_5 and in each of the g_3 generating subsets for S_4. Consequently, $g_5 = g_4 + g_3$ and this particular case generalizes to

$$g_n = g_{n-1} + g_{n-2}, \quad n \geq 3, \quad g_1 = 1 \text{ (for } \{1\}), \quad g_2 = 1 \text{ (for } \{1, 2\}).$$

(Note that we could define $g_0 = 0$, by extending the given recurrence relation to $n \geq 2$ and solving the equation $g_0 = g_2 - g_1$ to obtain $g_0 = 1 - 1 = 0$.) Here we find that

$$g_n = F_n, \quad n \geq 1.$$

(More on generating sets can be found in Reference [26]. A generalization of this idea is found in Reference [54].)

Example 5.3: Next we examine binary strings made up of 0's and 1's. For $n \geq 1$, there are 2^n binary strings of length n—that is, the strings are made up of n symbols, each a 0 or a 1. We wish to count those strings of length n, where there are no consecutive 1's. So we shall let b_n count the number of such strings of length n and learn, for example, that (i) $b_3 = 5$, for the strings 000, 100, 010, 001, 101; and (ii) $b_4 = 8$, for the strings 0000, 1000, 0100, 0010, 0001, 1010, 1001, 0101. In general, when $n \geq 2$, there are two cases to consider for a string s of length n with no consecutive 1's:

(i) s ends in 0: Here the possibilities for the preceding $n - 1$ symbols of s constitute all of the b_{n-1} strings of length $n - 1$ with no consecutive 1's.

(ii) s ends in 1 (actually 01): Now the possibilities for the preceding $n - 2$ symbols of s are counted by the b_{n-2} strings of length $n - 2$ with no consecutive 1's.

Consequently,

$$b_n = b_{n-1} + b_{n-2}, \quad n \geq 2, \quad b_0 = 1, \quad b_1 = 2,$$

and

$$b_n = F_{n+2}, \quad n \geq 0.$$

Is it just a coincidence that the answers for a_n (in Example 5.1) and for b_n (here in Example 5.3) are the same? We can relate these results as follows. When $n = 5$, for instance, we can correspond the subset $\{1, 4\}$ of Example 5.1 with the string 10010 and the subset $\{1, 3, 5\}$ with the string 10101. In general, we line up the integers in S_n as $1, 2, 3, \ldots, n - 1, n$. Then for a string of n 0's and 1's (with no consecutive 1's), we examine the locations for the 1's. If a 1 is in the ith position, for $1 \leq i \leq n$, we then select i from S_n to determine our corresponding subset with no consecutive integers.

Example 5.4:   As in Examples 5.1 and 5.2, we again define $S_n = \{1, 2, 3, \ldots, n - 1, n\}$, for $n \geq 1$. Now we are interested in the functions $f : S_n \to S_n$ that are one-to-one (and consequently, onto) or onto (and consequently, one-to-one). These functions are called the *permutations* of S_n and they number $n!$. For a given $n \geq 1$, we want to determine the number of these permutations f such that $|i - f(i)| \leq 1$, for all $1 \leq i \leq n$— that is, we want to count the permutations f where (i) $f(1) = 1$ or $f(1) = 2$; (ii) $f(n) = n$ or $f(n) = n - 1$; and, (iii) $f(i) = i - 1$ or $f(i) = i$ or $f(i) = i + 1$ for all $2 \leq i \leq n - 1$. When $n = 3$, for instance, out of the six possible permutations for S_3, we find the following three that satisfy the given condition:

(1) $f: S_3 \to S_3$	(2) $f: S_3 \to S_3$	(3) $f: S_3 \to S_3$
$f(1) = 1$	$f(1) = 1$	$f(1) = 2$
$f(2) = 2$	$f(2) = 3$	$f(2) = 1$
$f(3) = 3$	$f(3) = 2$	$f(3) = 3.$

For $n \geq 1$, let p_n count the permutations of S_n that satisfy the stated condition. There are two cases to consider:

(i) $f(n) = n$: Here we can use any of the p_{n-1} permutations $f : S_{n-1} \to S_{n-1}$.

(ii) $f(n) = n - 1$: When this happens it follows that $f(n - 1) = n$, and under these conditions we can then use any of the p_{n-2} permutations $f : S_{n-2} \to S_{n-2}$.

Consequently, we see that

$$p_n = p_{n-1} + p_{n-2}, \quad n \geq 3, \quad p_1 = 1, \quad p_2 = 2,$$

and

$$p_n = F_{n+1}, \quad n \geq 1.$$

Example 5.5: [Olry Terquem (1782-1862)]: Again, we let $S_n = \{1, 2, 3, \ldots, n - 1, n\}$, where $n \geq 1$, but this time we are interested in the subsets of S_n of the form $\{a_1, a_2, a_3, \ldots, a_k\}$, where (i) $a_1 < a_2 < a_3 < \cdots < a_k$ (so $k \leq n$); (ii) a_i is odd for i odd, with $i \leq n$; and, (iii) a_i is even for i even, with $i \leq n$. These sets are called the *alternating subsets* of S_n. When $n = 3$, we find five such subsets— namely, $\varnothing$, $\{1\}$, $\{1, 2\}$, $\{1, 2, 3\}$, and $\{3\}$. For $n = 4$, there are eight such subsets: $\varnothing$, $\{1\}$, $\{1, 2\}$, $\{1, 2, 3\}$, $\{1, 2, 3, 4\}$, $\{1, 4\}$, $\{3\}$, and $\{3, 4\}$. In general, for $n \geq 1$, let t_n count the number of alternating subsets of S_n. Then $t_1 = 2$, $t_2 = 3$, and, for $n \geq 3$, we consider the following:

Suppose that $B = \{b_1, b_2, \ldots, b_k\}$ is an alternating subset of S_n where (i) $b_1 < b_2 < \cdots < b_k$; (ii) b_i is odd for i odd, with $i \leq n$; and, (iii) b_i is even for i even, with $i \leq n$. There are two cases to examine.

(1) If $b_1 = 1$, then $\{b_2 - 1, b_3 - 1, \ldots, b_k - 1\}$ is an alternating subset of S_{n-1}. This implies that there are t_{n-1} alternating subsets of S_n that contain 1.

(2) If $b_1 \neq 1$, then $b_1 \geq 3$ and $\{b_1 - 2, b_2 - 2, \ldots, b_k - 2\}$ is an alternating subset of S_{n-2}. So there are t_{n-2} alternating subsets of S_n that do not contain 1.

Since these two cases cover all the possibilities and have nothing in common, it follows that

$$t_n = t_{n-1} + t_{n-2}, \quad n \geq 3, \quad t_1 = 2, \quad t_2 = 3,$$

and

$$t_n = F_{n+2}, \quad n \geq 1.$$

Example 5.6 (a): This example is due to George Andrews. Let us start with $S_4 = \{1, 2, 3, 4\}$. The set $A = \{2, 4\}$ is a subset of S_4 and is such that 2 (the element 2 from A) $\geq |A|$, the size of A. Likewise $4 \geq |A|$. We call such a subset A a *fat subset* of S_4. The subset $B = \{3, 4\}$ is also a *fat* subset of S_4, since $3 \geq 2 = |B|$ and $4 \geq 2 = |B|$. However, the subset $C = \{1, 2\}$ is not a fat subset of S_4 because $1 \in C$ but $1 \not\geq 2 = |C|$.

In general, for $n \geq 1$, a subset A of S_n is called a *fat* subset of S_n if $x \geq |A|$ for every $x \in A$. How many fat subsets does the set S_n have?

If we let f_n count the number of fat subsets of S_n, then we find that $f_1 = 2$ for the fat subsets $\varnothing$ and $\{1\}$ of $S_1 = \{1\}$, and $f_2 = 3$ for the fat subsets $\varnothing$, $\{1\}$, and $\{2\}$ of $S_2 = \{1, 2\}$. Now for $n \geq 3$, there are two things to consider. If A is a fat subset of S_n and $n \notin A$, then A is a fat subset of S_{n-1}. If, however, $n \in A$, then $1 \notin A$ because with $1, n \in A$, it follows that $|A| \geq 2$ and $1 \ngeq |A|$. Upon removing n and subtracting 1 from each of the remaining integers in A, we obtain a fat subset of S_{n-2}. (Alternatively, we could take any fat subset for S_{n-2}, add 1 to each integer in the set, and then place n into the new resulting set.) Consequently,

$$f_n = f_{n-1} + f_{n-2}, \quad n \geq 3, \quad f_1 = 2, \quad f_2 = 3,$$

and, as in our previous example,

$$f_n = F_{n+2}, \quad n \geq 1.$$

But now we learn a little bit more. Note that for S_4 there are eight fat subsets. This follows because from $S_4 = \{1, 2, 3, 4\}$, there is one way to select the null subset, $\binom{4}{1}$ ways to select a fat subset of size 1, and $\binom{3}{2}$ ways to select a fat subset of size 2 (from $\{2, 3, 4\}$). Consequently, $F_6 = 1 + \binom{4}{1} + \binom{3}{2} = \binom{5}{0} + \binom{4}{1} + \binom{3}{2}$, a sum of binomial coefficients. In like manner, we have $F_7 =$ the number of fat subsets of $S_5 = \binom{6}{0}$ $+ \binom{5}{1} + \binom{4}{2} + \binom{3}{3}$, and for $n \geq 1$, it follows that

$$F_n = \begin{cases} \binom{n-1}{0} + \binom{n-2}{1} + \binom{n-3}{2} + \cdots + \binom{(n-1)/2}{(n-1)/2}, & n \text{ odd} \\ \binom{n-1}{0} + \binom{n-2}{1} + \binom{n-3}{2} + \cdots + \binom{n/2}{(n/2)-1}, & n \text{ even} \end{cases}.$$

Before leaving this example, let us consider the two versions of Pascal's triangle in Fig. 5.1 on p. 18. In Fig. 5.1(a), we add the numbers along the seven diagonal lines indicated:

The resulting sums are

$$1 = 1 = F_1$$
$$1 = 1 = F_2$$
$$1 + 1 = 2 = F_3$$
$$1 + 2 = 3 = F_4$$
$$1 + 3 + 1 = 5 = F_5$$
$$1 + 4 + 3 = 8 = F_6$$
$$1 + 5 + 6 + 1 = 13 = F_7.$$

Now does this pattern continue or are we being misled? Well, consider the version of Pascal's triangle in Fig. 5.1(b). If we compute the same sums along the seven diagonal

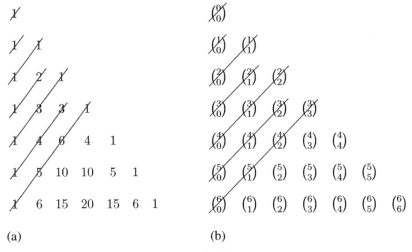

(a)   (b)

FIGURE 5.1

lines indicated in the figure, we find that

$$\binom{0}{0} = 1 = F_1$$

$$\binom{1}{0} = 1 = F_2$$

$$\binom{2}{0} + \binom{1}{1} = 1 + 1 = 2 = F_3$$

$$\binom{3}{0} + \binom{2}{1} = 1 + 2 = 3 = F_4$$

$$\binom{4}{0} + \binom{3}{1} + \binom{2}{2} = 1 + 3 + 1 = 5 = F_5$$

$$\binom{5}{0} + \binom{4}{1} + \binom{3}{2} = 1 + 4 + 3 = 8 = F_6$$

$$\binom{6}{0} + \binom{5}{1} + \binom{4}{2} + \binom{3}{3} = 1 + 5 + 6 + 1 = 13 = F_7.$$

This time the results we found earlier for F_n, when we were counting fat subsets, indicate that this pattern does continue.

This pattern can also be established by the Alternative, or Strong, form of the Principle of Mathematical Induction, and the combinatorial identity

$$\binom{n+1}{r} = \binom{n}{r} + \binom{n}{r-1}, \quad \text{for } n \geq r \geq 1.$$

Note, for instance, how

$$\binom{6}{0} + \binom{5}{1} + \binom{4}{2} + \binom{3}{3} = 1 + 5 + 6 + 1 = 13 = 8 + 5$$

$$= (1 + 4 + 3) + (1 + 3 + 1)$$

$$= \left[\binom{5}{0} + \binom{4}{1} + \binom{3}{2}\right] + \left[\binom{4}{0} + \binom{3}{1} + \binom{2}{2}\right]$$

$$= \binom{5}{0} + \left[\binom{4}{1} + \binom{4}{0}\right] + \left[\binom{3}{2} + \binom{3}{1}\right] + \binom{2}{2}$$

and how

$$\binom{6}{0} = \binom{5}{0}, \quad \binom{5}{1} = \binom{4}{1} + \binom{4}{0}$$

$$\binom{4}{2} = \binom{3}{2} + \binom{3}{1}, \text{ and } \binom{3}{3} = \binom{2}{2}.$$

Example 5.6 (b): Along the same line, let us consider the following three subsets of $S_{10} = \{1, 2, 3, \ldots, 10\}$—namely, $\{2, 6\}$, $\{3, 8, 10\}$, and $\{4, 6, 8, 9\}$. You might wonder what, if anything, these three subsets have in common. Considering the size of each subset, we see that

$|\{2, 6\}| =$ the size of $\{2, 6\} = 2$, the minimal element in $\{2, 6\}$
$|\{3, 8, 10\}| = 3$, the minimal element in $\{3, 8, 10\}$
$|\{4, 6, 8, 9\}| = 4$, the minimal element in $\{4, 6, 8, 9\}$.

So now what we want to determine, for each $n \geq 1$, is the number m_n of subsets A of S_n where the minimal element of A equals $|A|$, the size of A. To motivate the solution, we shall consider the cases for $n = 4$, 5, and 6.

$$n = 4: \quad \{1\}, \ \{2, 3\}, \ \{2, 4\}$$
$$n = 5: \quad \{1\}, \ \{2, 3\}, \ \{2, 4\}, \ \{2, 5\}, \ \{3, 4, 5\}$$
$$n = 6: \quad \{1\}, \ \{2, 3\}, \ \{2, 4\}, \ \{2, 5\}, \ \{3, 4, 5\}$$
$$\{2, 6\}, \ \{3, 4, 6\}, \ \{3, 5, 6\}$$

The first five subsets for $n = 6$ are precisely those for the case where $n = 5$. But what about the last three subsets for $n = 6$? Since 6 is a member of each such subset, the minimal element is at least 2. Turning to the three subsets for $n = 4$, in each case we increase each element in the subset by 1 and then add in the element 6. (Since the largest possible element in any subset of S_4 is 4, there is no danger that 6 will come about as $k + 1$ for some k in a subset of S_4.) Consequently,

$$\{1\} \text{ becomes } \{1 + 1, 6\} = \{2, 6\}$$
$$\{2, 3\} \text{ becomes } \{2 + 1, 3 + 1, 6\} = \{3, 4, 6\}$$
$$\{2, 4\} \text{ becomes } \{2 + 1, 4 + 1, 6\} = \{3, 5, 6\}.$$

So

$$m_6 = m_5 + m_4.$$

The same type of argument can be given for each $n \geq 3$, so we arrive at the recurrence relation

$$m_n = m_{n-1} + m_{n-2}, \quad n \geq 3, \quad m_1 = 1 \text{ (for } \{1\}), \quad m_2 = 1 \text{ (for } \{1\}),$$

and, consequently, $m_n = F_n$, for $n \geq 1$.

We can also obtain the result for $n = 7$, for example, by the following alternative argument:

(1) There is one subset when the minimal element is 1—namely, $\{1\}$.

(2) For the minimal element 2, there are $\binom{5}{1}$ subsets, since we select one of the five elements $3, 4, 5, 6,$ and 7.

(3) When the minimal element is 3, there are $\binom{4}{2}$ subsets—for here two elements are selected from the four elements $4, 5, 6,$ and 7.

(4) There is only one subset when the minimal element is 4—namely, $\{4, 5, 6, 7\}$.

So in total, the number of subsets of S_7, where the minimal element equals the size of the subset, is

$$1 + \binom{5}{1} + \binom{4}{2} + 1 = \binom{6}{0} + \binom{5}{1} + \binom{4}{2} + \binom{3}{3} = 13 = F_7.$$

For the general case, we use what we learned earlier in part (a) of this example and find that for $n \geq 1$,

$$m_n = \begin{cases} \binom{n-1}{0} + \binom{n-2}{1} + \binom{n-3}{2} + \cdots + \binom{(n-1)/2}{(n-1)/2}, & n \text{ odd} \\ \binom{n-1}{0} + \binom{n-2}{1} + \binom{n-3}{2} + \cdots + \binom{n/2}{(n/2)-1}, & n \text{ even} \end{cases} = F_n.$$

EXERCISES FOR CHAPTER 5

1. Tanya and Greta take turns flipping a coin—Tanya first, Greta second, Tanya third, and so on. They continue to do this until heads result, for the first time, on two consecutive flips. How many different sequences of heads and tails could have come up if they stopped after (a) seven flips; (b) the 12th flip, or (c) the nth flip, for $n \geq 2$?

2. Rowyn and Ridge have a collection of blocks. The base of each block is a square with sides of 3 inches. The height of each block is either 1 inch or 2 inches. Suffice it to say that there are abundantly many blocks of each height and that all the blocks of a given height are identical in appearance. Rowyn and Ridge

wish to stack some of these blocks, one on top of another, in order to construct a tower. In how many ways can they stack the blocks so that they construct a tower of height (a) 10 inches; (b) 17 inches; and (c) n inches, for $n \geq 1$?

3. (a) Let $S = \{4, 5, \ldots, 16, 17\}$. How many subsets of S contain no consecutive integers?

   (b) For (fixed, but arbitrary) positive integers m, n, let $T = \{m, m + 1, m + 2, \ldots, m + n - 1, m + n\}$. How many subsets of T contain no consecutive integers?

   (c) Let U be a set of consecutive integers with smallest element 31. What is the largest element in U if the number of subsets of U with no consecutive integers is 55?

   (d) Suppose that W is a set of consecutive integers with largest element 7. If W has 377 subsets containing no consecutive integers, what is the smallest element in W?

4. For Christmas Benjamin received a large box of 40 square blocks, each with sides two inches long. Twenty of the blocks are red and the other 20 are white. For $n \geq 1$, let s denote a linear arrangement of n (≤ 40) blocks. This arrangement can be represented as a sequence of n symbols—each an R (red) or a W (white). A *run* within the arrangement s is a (consecutive) list of maximal length that uses just one of the symbols, R or W. For example, consider the arrangement $RWRRWWWR$, made up of eight blocks. Here we find five runs: namely, R, W, RR, WWW, and R. Benjamin is interested in counting the number of ways he can arrange 10 of his blocks so that the first block is red and all the runs that occur in the arrangement are of odd length. Consequently, Benjamin wants to include in his count the arrangements $RWWWRRRWWW$ and $RRRRWRRRW$, but not arrangements such as $WRRRWWWRRR$ or $RRRWWWWWRRR$. How many such arrangements of 10 blocks are possible?

5. For a fixed positive integer n, let a_n count the number of binary sequences $x_1, x_2, x_3, \ldots, x_n$, where $x_1 \leq x_2$, $x_2 \geq x_3$, $x_3 \leq x_4$, $x_4 \geq x_5, \ldots$, and (i) $x_{n-1} \leq x_n$ for n even, while (ii) $x_{n-1} \geq x_n$ for n odd and greater than 1. For example, when $n = 4$, we want to include the sequence $0, 0, 0, 1$ in our count but not the sequence $1, 0, 0, 1$. Determine a_n.

6. Consider the binary strings of length 2: that is, the strings 00, 01, 10, and 11. If st and uv are two such strings, we have $st \leq uv$ when $s \leq u$ and $t \leq v$. For a fixed positive integer n, let a_n count the number of sequences (of binary strings of length 2) $x_1 y_1$, $x_2 y_2$, $x_3 y_3$, $\ldots$, $x_n y_n$, where $x_1 y_1 \leq x_2 y_2$, $x_2 y_2 \geq x_3 y_3$, $x_3 y_3 \leq x_4 y_4$, $x_4 y_4 \geq x_5 y_5$, $\ldots$, and (i) $x_{n-1} y_{n-1} \leq x_n y_n$ for n even, while (ii) $x_{n-1} y_{n-1} \geq x_n y_n$ for n odd and greater than 1. Determine a_n.

7. (a) Let $A = \{1, 2, 3\}$. For a fixed positive integer n, let a_n count the number of sequences S_1, S_2, S_3, $\ldots$, S_n, where $S_i \subseteq A$, for each $1 \leq i \leq n$, and where $S_1 \subseteq S_2$, $S_2 \supseteq S_3$, $S_3 \subseteq S_4$, $S_4 \supseteq S_5$, $\ldots$, with (i) $S_{n-1} \subseteq S_n$ for n even, while (ii) $S_{n-1} \supseteq S_n$ for n odd and greater than 1. Determine a_n.

   (b) Answer the question in part (a) if $A = \{1, \ldots, m\}$, where $m \geq 2$.

8. To raise money for the campus drive at their university, the sisters of Gamma Kappa Phi sorority are sponsoring a casino night. As a result, two of their pledges—namely, Piret and Columba—have been assigned to stack green poker chips and gold poker chips so that each stack contains ten chips, where no two adjacent chips are allowed to be green. In addition, in order to become sisters of the sorority, they have to come up with 100 such stacks where no two are the same. Being very bright mathematics majors, the two young ladies were relieved, for they knew this task could be accomplished. How did they know?

CHAPTER 6

Compositions and Palindromes

In this chapter, we shall study different ways to write a positive integer as the sum of positive summands, or parts. For instance, for the positive integer 7, the sums $6 + 1$, $3 + 1 + 3$, 7, $1 + 6$, $2 + 1 + 1 + 2 + 1$, and $4 + 1 + 2$ are six such examples, each of which is called a *composition* of 7. First of all, we note that here we consider $6 + 1$ and $1 + 6$ as different compositions of 7. So *order is relevant* when dealing with the compositions of a positive integer. Also, we see that the composition $3 + 1 + 3$ reads the same going from left to right as it does if we read it from right to left. When this happens, we have a special type of composition that is called a *palindrome*.

Example 6.1: There are 16 ways to write 5 as a sum of positive integers, where the order of the summands is relevant. These representations are called the *compositions* of 5 and are listed in Table 6.1. [If the order is not relevant, then compositions $4 + 1$ and $1 + 4$ are considered to be the same *partition* of 5. Likewise, the three compositions $3 + 1 + 1$, $1 + 3 + 1$, and $1 + 1 + 3$ correspond to only one partition of 5. Overall, the 16 compositions of 5 determine seven partitions of 5.]

TABLE 6.1

(1) 5	(5) $2 + 3$	(9) $2 + 2 + 1$	(13) $1 + 2 + 1 + 1$
(2) $4 + 1$	(6) $3 + 1 + 1$	(10) $2 + 1 + 2$	(14) $1 + 1 + 2 + 1$
(3) $1 + 4$	(7) $1 + 3 + 1$	(11) $1 + 2 + 2$	(15) $1 + 1 + 1 + 2$
(4) $3 + 2$	(8) $1 + 1 + 3$	(12) $2 + 1 + 1 + 1$	(16) $1 + 1 + 1 + 1 + 1$

In order to obtain a formula for the number of compositions of an arbitrary positive integer n, let us look at the compositions of 5 once again. In particular, consider the following composition of 5 :

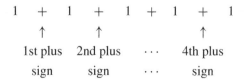

Fibonacci and Catalan Numbers: An Introduction, First Edition. Ralph P. Grimaldi.
© 2012 John Wiley & Sons, Inc. Published 2012 by John Wiley & Sons, Inc.

Here we have five summands, each of which is 1, and four plus signs. For the set $\{1, 2, 3, 4\}$ there are $2^4 = 16$ subsets. But what does this have to do with the compositions of 5?

Consider a subset of $\{1, 2, 3, 4\}$, say $\{1, 3\}$. Now form the following composition of 5:

$$(1 \quad + \quad 1) \quad + \quad (1 \quad + \quad 1) \quad + \quad 1$$
$$\uparrow \qquad\qquad\qquad \uparrow$$
$$\text{1st plus} \qquad\qquad \text{3rd plus}$$
$$\text{sign} \qquad\qquad\qquad \text{sign}$$

Here the subset $\{1, 3\}$ indicates that we should place parentheses around the 1's on either side of the first and third plus signs. This results in the composition

$$2 + 2 + 1.$$

In the same way the subset $\{1, 3, 4\}$ indicates the use of the first, third, and fourth plus signs, giving us

$$(1 \quad + \quad 1) \quad + \quad (1 \quad + \quad 1 \quad + \quad 1)$$
$$\uparrow \qquad\qquad\qquad \uparrow \qquad\quad \uparrow$$
$$\text{1st plus} \qquad\qquad \text{3rd plus} \quad \text{4th plus}$$
$$\text{sign} \qquad\qquad\qquad \text{sign} \qquad \text{sign}$$

or the composition

$$2 + 3.$$

Going in reverse order, we see that the composition $3 + 1 + 1$ comes from

$$(1 + 1 + 1) + 1 + 1$$

and is determined by the subset $\{1, 2\}$ of $\{1, 2, 3, 4\}$. In Table 6.2, we have listed four compositions of 5 along with the corresponding subset of $\{1, 2, 3, 4\}$ that determines each of them.

TABLE 6.2

Composition of 5	Determining Subset of $\{1, 2, 3, 4\}$
(i) $1 + 1 + 1 + 1 + 1$	(i) $\varnothing$
(ii) $1 + 2 + 2$	(ii) $\{2, 4\}$
(iii) $3 + 2$	(iii) $\{1, 2, 4\}$
(iv) 5	(iv) $\{1, 2, 3, 4\}$

Since there is a one-to-one correspondence between the compositions of 5 and the subsets of $\{1, 2, 3, 4\}$, we see once again that there are $16 = 2^4$ compositions of 5. The same type of argument indicates that for an arbitrary positive integer n, there are 2^{n-1} compositions. (This result can also be obtained using the Principle of Mathematical Induction.)

But what does this have to do with the Fibonacci numbers?

Example 6.2: In Reference [1], K. Alladi and V. E. Hoggatt, Jr., examine compositions of a positive integer n, but the only summands they allow are 1's and 2's. Consequently, they find the following compositions for $n = 3, 4,$ and 5.

$$(n = 3): 2 + 1, 1 + 2, 1 + 1 + 1$$
$$(n = 4): 2 + 2, 2 + 1 + 1, 1 + 2 + 1, 1 + 1 + 2, 1 + 1 + 1 + 1$$
$$(n = 5): 2 + 2 + 1, 2 + 1 + 1 + 1, 1 + 2 + 1 + 1, 1 + 1 + 2 + 1,$$
$$1 + 1 + 1 + 1 + 1,$$
$$2 + 1 + 2, 1 + 2 + 2, 1 + 1 + 1 + 2$$

If we let c_n count the number of compositions of n, where the only summands allowed are 1's and 2's, we see that $c_5 = 8 = 5 + 3 = c_4 + c_3$. Note how the first five compositions of 5 can be obtained by appending "$+1$" to each of the five compositions of 4. The last three compositions of 5 are the three compositions of 3 with "$+2$" appended. This idea carries over to the general case, where 5 is replaced by n (≥ 3), and provides us with the recurrence relation

$$c_n = c_{n-1} + c_{n-2}, \quad n \geq 3, \ c_1 = 1, \ c_2 = 2,$$

and so

$$c_n = F_{n+1}, \ n \geq 1.$$

In Example 6.1, we found the number of compositions of 5 to be $2^{5-1} = 2^4$. We did this by determining the number of subsets for the set $\{1, 2, 3, 4\}$ describing the four plus signs—namely, first, second, third, and fourth, in the composition $1 + 1 + 1 + 1 + 1$ of 5. What happens when we restrict the summands of our compositions to be only 1's and 2's? In Table 6.3, we see three of the eight compositions of 5 (where the only summands are 1's and 2's)—each with its own corresponding determining subset of $\{1, 2, 3, 4\}$.

TABLE 6.3

Composition of 5	Determining Subset of $\{1, 2, 3, 4\}$
(i) $1 + 2 + 2$	(i) $\{2, 4\}$
(ii) $1 + 1 + 2 + 1$	(ii) $\{3\}$
(iii) $2 + 1 + 2$	(iii) $\{1, 4\}$

From the third composition in Table 6.2, where the summands were not restricted, we found that the determining subset $\{1, 2, 4\}$ corresponded with the composition $(1 + 1 + 1) + (1 + 1) = 3 + 2$. From this and the three results in Table 6.3, we see that if we cannot have a summand greater than 2, then we must avoid determining subsets that contain consecutive integers. Consequently, the number of compositions of n with only 1's and 2's as summands is the same as the number of subsets of $\{1, 2, 3, 4\}$ with no consecutive integers. From the result in Example 5.1, this was found to be $F_{4+2} = F_6$, confirming our earlier answer that $c_5 = F_{5+1} = F_6$.

Example 6.3: Looking back at the compositions of Example 6.2, let us now focus on those compositions called *palindromes*. They are the compositions that read the same from right to left, as they do from left to right. For example, among the eight compositions of 5 where the only summands are 1's and 2's, there are two that are also palindromes: $2 + 1 + 2$ and $1 + 1 + 1 + 1 + 1$.

Now suppose that we want to determine the number of palindromes among the F_{12} compositions of 11. (Here and throughout this discussion, the only summands are 1's and 2's.) For the integer 11, each palindrome will contain a central summand that is odd. So in this case, 1, is the only possible central summand. To the right of this 1, we write a plus sign followed by any of the F_6 compositions of 5. Then we place the *mirror image* of this same composition (that is, the same composition, but now *in reverse order*), followed by a plus sign, on the left side of this central 1. Consequently, we learn that there are $F_6 = 8$ compositions of 11 (using only 1's and 2's as summands) that are palindromes. In general, for n odd, among the F_{n+1} compositions of n, there are $F_{(n+1)/2}$ compositions that are palindromes.

To determine the number of palindromes among the F_{13} compositions of 12, there are two cases to consider for the central symbol:

(i) If the central symbol is a plus sign, then we place one of the F_7 compositions of 6 on the right of this plus sign and its mirror image on the left.

(ii) If instead of a central plus sign we have a central summand, it must be even, so it must be 2 in this case (as only 1's and 2's may be used as summands). Now we place a plus sign followed by one of the F_6 compositions of 5 on the right of this 2 and the mirror image of the composition, followed by a plus sign, on the left.

So, in total, there are $F_7 + F_6 = F_8$ compositions of 12 that are palindromes.

For the general case, when n is even, among the F_{n+1} compositions of n, there are $F_{n/2} + F_{(n/2)+1} = F_{(n/2)+2} = F_{(n+4)/2}$ compositions that are palindromes.

Example 6.4: In 1972, Leonard Carlitz (1907–1999) of Duke University conceived the following example. Start with two adjacent rows of regular hexagonal cells, as in a beehive. See Fig. 6.1 (a). If a bee starts at the starting cell (S) on the left of the top row, it may move horizontally (one step) to cell 2, as in Fig. 6.1 (b). The same bee may also go to cell 2, from cell S, by first going southeast to cell 1 and then northeast from cell 1 to cell 2, as in Fig. 6.1 (c). In general, all motion is to the right—going

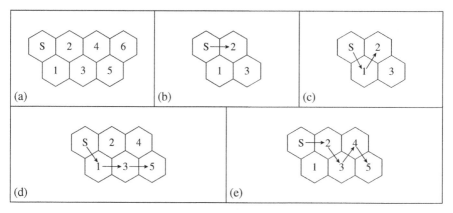

FIGURE 6.1

one step to the east in either row, one step northeast from the bottom row to the top row, or one step southeast from the top row to the bottom row. Using only these types of moves, we want to determine, for $n \geq 1$, the number of ways this bee can travel from the starting cell (S) to the cell with the label n.

For $n = 5$, it turns out that there are eight ways for the bee to travel from the cell S to the cell labeled 5. One way is to travel from cell S to cell 1, then from cell 1 to cell 3, and finally, from cell 3 to cell 5, as shown in Fig. 6.1 (d). This can be recorded as

$$\text{Cell } S \searrow \text{Cell } 1 \to \text{Cell } 3 \to \text{Cell } 5.$$

A second, somewhat longer path, can be recorded as

$$\text{Cell } S \to \text{Cell } 2 \searrow \text{Cell } 3 \nearrow \text{Cell } 4 \searrow \text{Cell } 5.$$

This is shown in Fig. 6.1 (e). Now let us consider the following correspondence. In each of the preceding paths, whenever we see either the symbol $\nearrow$ or the symbol $\searrow$, we shall write the number 1. For the symbol $\to$, we agree to write the number 2. (After all, the one step corresponding to $\to$ in the first row provides the same result as the two steps: $\searrow$ followed by $\nearrow$. In the second row, $\to$ corresponds to the two steps: $\nearrow$ followed by $\searrow$.) Also, we agree to maintain the order in which the 1's and 2's are obtained. Inserting plus signs between these 1's and 2's, we arrive at the following compositions of 5:

$$1 + 2 + 2$$

for our first path and

$$2 + 1 + 1 + 1$$

for the second path. Starting with the compositions of 5 with only 1's and 2's as summands, for the composition

$$1 + 1 + 2 + 1$$

we find the corresponding path

$$\text{Cell } S \searrow \text{Cell } 1 \nearrow \text{Cell } 2 \rightarrow \text{Cell } 4 \searrow \text{Cell } 5.$$

Consequently, the number of paths the bee can take in going from cell S to cell 5 is the same as the number of compositions of 5, where the only summands are 1's and 2's. So from Example 6.2, it follows that the number of these paths is $F_6 = 8$ when $n = 5$, and, in general, is F_{n+1} for $n \geq 1$.

Example 6.5:   In 1901, Eugen Netto (1846–1919) studied compositions of a positive integer n, where any positive integer could be used as a summand—except for 1. If we let e_n count the number of these compositions for $n \geq 1$, we find the following:

TABLE 6.4

n	e_n	Compositions	n	e_n	Compositions
1	0		4	2	$4, 2 + 2$
2	1	2	5	3	$5, 2 + 3, 3 + 2$
3	1	3	6	5	$6, 2 + 4, 3 + 3, 4 + 2, 2 + 2 + 2$

As suggested by the results in Table 6.4, for $n \geq 3$, these e_n compositions of n can be obtained from (i) the e_{n-1} compositions of $n - 1$ by adding 1 to the last summand of each such composition or from (ii) the e_{n-2} compositions of $n - 2$ by appending "+2" to each of these compositions. Consequently, we arrive at

$$e_n = e_{n-1} + e_{n-2}, \quad n \geq 3, \; e_1 = 0, \; e_2 = 1,$$

and

$$e_n = F_{n-1}, \quad n \geq 1.$$

Now let us direct our attention to the palindromes that appear among these F_{n-1} compositions of n. We will consider the case for $n = 15$. For any odd value of n, the central summand of a palindrome must be an odd integer. Since we cannot use the summand 1 in this situation, the smallest central summand we can have is 3. We then place a plus sign followed by one of the F_5 compositions of 6 on the right of this central 3. To the left of this central 3, we place a plus sign and then, to the left of this plus sign, the mirror image of the composition we added on the right. So there are F_5 palindromes of 15 where 3 is the central summand. If the central summand is 5, we carry through in the same way with the F_4 compositions of 5. However, as we continue, we realize that 13 cannot be used as a central summand. For the palindrome $1 + 13 + 1$ contains the summand 1, the only positive integer we cannot use in these compositions. However, we do consider the single-summand composition 15 as a palindrome of itself. Table 6.5 provides all of the central summands we consider, along with the corresponding number of palindromes.

TABLE 6.5

Central Summand	Number of Palindromes
3	F_5
5	F_4
7	F_3
9	F_2
11	F_1
13	F_0
15	1

Consequently, the number of palindromes for 15, under these conditions, is

$$F_5 + F_4 + F_3 + F_2 + F_1 + F_0 + 1$$
$$= \sum_{r=0}^{5} F_r + 1 = (F_7 - 1) + 1 = F_7,$$

by Property 4.5. For any odd integer $n \geq 3$, this result generalizes as

$$F_{(n-5)/2} + F_{((n-5)/2)-1} + F_{((n-5)/2)-2} + \cdots + F_1 + F_0 + 1$$
$$= \sum_{r=0}^{(n-5)/2} F_r + 1 = (F_{((n-5)/2)+2} - 1) + 1 = F_{(n-1)/2}.$$

For the palindromes of 14, the central symbol is a plus sign or one of the seven even integers: 2, 4, ..., 12, 14. Arguing as we did for the case of $n = 15$, we arrive at the results in Table 6.6.

TABLE 6.6

Central Symbol	Number of Palindromes
+	F_6
2	F_5
4	F_4
6	F_3
8	F_2
10	F_1
12	F_0
14	1

Consequently, when summands are restricted to be at least 2, among the F_{13} compositions of 14, the number that are palindromes is

$$\sum_{r=0}^{6} F_r + 1 = (F_8 - 1) + 1 = F_8,$$

by Property 4.5. For the general case, if $n \geq 2$ is an even integer, then we find that among these F_{n-1} compositions of n, the number that are palindromes is

$$F_{(n/2)-1} + F_{(n/2)-2} + \cdots + F_1 + F_0 + 1$$
$$= \left(\sum_{r=0}^{(n/2)-1} F_r \right) + 1 = (F_{((n/2)-1)+2} - 1) + 1$$
$$- F_{((n+2)/2)},$$

again by Property 4.5.

(The reader who is interested in compositions where all summands are allowed, except for a fixed positive integer k, should refer to Reference [14].)

Example 6.6:   Finally, we shall consider compositions of a positive integer n, where the summands are restricted to be *odd* positive integers. We shall let d_n count these compositions of n. Table 6.7 exhibits what happens for $n = 1, 2, 3, 4$, and 5.

TABLE 6.7

n	Compositions of n	d_n
1	1	1
2	$1 + 1$	1
3	$3, 1 + 1 + 1$	2
4	$3 + 1, 1 + 3, 1 + 1 + 1 + 1$	3
5	$3 + 1 + 1, 1 + 3 + 1, 1 + 1 + 1 + 1 + 1$	5
	$5, 1 + 1 + 3$	

Upon examining the results for $n = 3, 4$, and 5, we see that the first three compositions of 5 are obtained from the compositions of 4 by appending "$+1$" to each such composition. The other two compositions of 5 result when we add 2 to the last summand in each of the compositions of 3. Since the same type of argument can be used when we replace 5 by any $n \geq 3$, it follows that

$$d_n = d_{n-1} + d_{n-2}, \quad n \geq 3, d_1 = 1, d_2 = 1,$$

and

$$d_n = F_n, \quad n \geq 1.$$

When we consider the palindromes for this situation, the result is rather straightforward when n is even. We need only consider what happens when the central symbol is a plus sign, for we cannot have an even central summand. Consequently, we find that among these F_n compositions of n, the number of palindromes is $F_{n/2}$, the number of compositions of $n/2$ where all summands are odd.

For n odd, however, let us consider a special case—say, $n = 15$. For the palindromes of any odd integer, the central summand must be odd. Starting with the central summand 1, we place a plus sign on its right and then one of the F_7 compositions of 7, where the summands are all odd. Then, on the left of this central 1, we place a plus sign. Follow this by placing on the left of the plus sign the mirror image of the composition used earlier on the right. Doing the comparable thing for the central summands 3, 5,..., 13, and 15, we arrive at the results in Table 6.8.

TABLE 6.8

Central Summand	Number of Palindromes
1	F_7
3	F_6
5	F_5
7	F_4
9	F_3
11	F_2
13	F_1
15	1

Consequently, when using only odd summands, it follows from Property 4.5 that among the F_{15} compositions of 15, the number that are palindromes is

$$\sum_{r=1}^{7} F_r + 1 = (F_9 - 1) + 1 = F_9.$$

For n odd, where $n \geq 3$, a similar argument shows that the number of palindromes among these F_n compositions of n is

$$\sum_{r=1}^{(n-1)/2} F_r + 1 = \left(F_{((n-1)/2)+2} - 1 \right) + 1 = F_{(n+3)/2},$$

again by Property 4.5.

EXERCISES FOR CHAPTER 6

1. (a) Use the Principle of Mathematical Induction to prove that for $n \geq 1$, there are 2^{n-1} compositions of n.

   (b) For $n \geq 1$, how many of the compositions of n are palindromes?

2. Determine the number of compositions of 24, where (a) each summand is even; (b) each summand is a multiple of 3; (c) there is at least one odd summand; and (d) each summand is odd.

3. Determine the number of compositions of 48, where (a) the only summands are 2's and 4's; (b) the only summands are 3's and 6's.

4. The positive integer n has 1597 compositions where the first summand is 3 and all the other summands are 1's and 2's. Determine n.

5. Each staircase in Elizabeth's apartment house has 10 steps. When Elizabeth's boyfriend Anthony comes to see her, he must climb up eight staircases to reach her apartment on the top floor. At each step of any staircase, other than the last step, Anthony can climb up either one step or two steps. So, for example, Anthony could climb one staircase by taking three double steps, followed by two single steps, followed by one final double step. Another different approach could use two single steps followed by four double steps.

   (a) In how many ways can Anthony climb up one staircase under these conditions?

   (b) If the way Anthony climbs up one staircase is independent of the way he climbs up any other staircase, in how many ways can he ascend the eight staircases to get to see Elizabeth?

6. The positive integer n has 2584 compositions where the last summand is 4 and all the other summands are greater than 1. Determine n.

7. The positive integer n has 610 compositions where the first summand is 6, the last summand is 3, and all the other summands are odd. Determine n.

8. How many compositions of 10 are palindromes with only one even summand?

CHAPTER 7

Tilings: Divisibility Properties of the Fibonacci Numbers

Further properties of the Fibonacci numbers arise as we examine problems that deal with various ways to tile certain types of chessboards.

Example 7.1: Let us start with a $2 \times n$ chessboard, where $n \geq 1$. The case for $n = 3$ is shown in Fig. 7.1 (a). We want to cover such a chessboard using 1×2 (horizontal) dominoes, which we can also use as 2×1 (vertical) dominoes. Such dominoes (or tiles) are shown in Fig. 7.1 (b).

For $n \geq 1$, let q_n count the number of ways we can cover (or tile) a $2 \times n$ chessboard using the 1×2 and 2×1 dominoes. Here $q_1 = 1$, since a 2×1 chessboard requires one 2×1 (vertical) domino. A 2×2 chessboard can be covered in two ways—using two 1×2 (horizontal) dominoes or two 2×1 (vertical) dominoes, as demonstrated

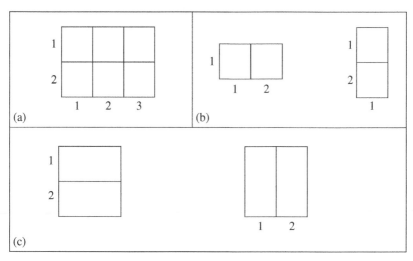

FIGURE 7.1

Fibonacci and Catalan Numbers: An Introduction, First Edition. Ralph P. Grimaldi.
© 2012 John Wiley & Sons, Inc. Published 2012 by John Wiley & Sons, Inc.

in Fig. 7.1 (c). So $q_2 = 2$. For $n \geq 3$, we consider the last (nth) column of a $2 \times n$ chessboard. This column can be covered in two ways;

(i) By one 2×1 (vertical) domino: Then the remaining $2 \times (n-1)$ chessboard can be covered in q_{n-1} ways.

(ii) By the right squares of two 1×2 (horizontal) dominoes placed one on top of the other: Here the remaining $2 \times (n-2)$ chessboard can be covered in q_{n-2} ways.

The two ways mentioned in (i) and (ii) cover all the possibilities and have nothing in common, so we arrive at

$$q_n = q_{n-1} + q_{n-2}, \quad n \geq 3, \quad q_1 = 1, \quad q_2 = 2,$$

and

$$q_n = F_{n+1}.$$

Example 7.2: Now we shall start with a $1 \times n$ chessboard that we wish to tile with 1×1 squares and 1×2 rectangles (or dominoes), as shown in Fig. 7.2 (a). Part (b) of the figure demonstrates the possible square–rectangular tilings of a 1×4 chessboard. So if we let l_n count the number of square–rectangular tilings of a $1 \times n$ chessboard, we see that $l_4 = 5$.

To derive a recurrence relation for l_n, we consider the last (nth) square in a $1 \times n$ chessboard, for $n \geq 3$. There are two cases to examine:

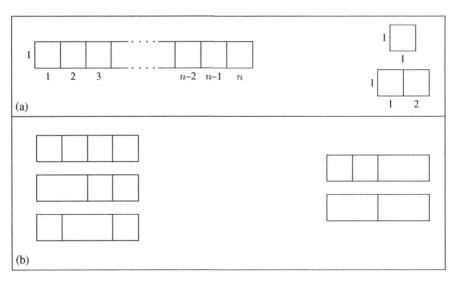

(a)

(b)

FIGURE 7.2

(i) The nth square is covered with a 1×1 square tile: Then the preceding $n - 1$ squares of the remaining $1 \times (n - 1)$ chessboard can be tiled in l_{n-1} ways.

(ii) The nth square and the $(n - 1)$st square are both covered by a 1×2 rectangular tile: This leaves the remaining $n - 2$ squares of the $1 \times (n - 2)$ chessboard, which we can tile in l_{n-2} ways.

Since these two cases cover all the possibilities and have nothing in common, it follows that

$$l_n = l_{n-1} + l_{n-2}, \quad n \geq 3, \ l_1 = 1, \ l_2 = 2,$$

and

$$l_n = F_{n+1}, \quad n \geq 1.$$

as we found for c_n, the number of compositions of n using only 1's and 2's as summands, earlier in Example 6.2.

Using this notion of tilings, we provide a second proof for Property 4.5:

$$\text{For } n \geq 0, \ \sum_{r=0}^{n} F_r = F_{n+2} - 1.$$

This time the proof will be combinatorial—that is, we shall count the same collection of objects in two different ways and then equate the two results. Start with a $1 \times (n + 1)$ chessboard. From the previous argument we know that this chessboard can be tiled in F_{n+2} ways. Now let us inventory these tilings according to where we find the last 1×2 rectangular tile on this chessboard. (From this point on, we shall refer to a 1×2 rectangular tile as simply a rectangle.)

If there are no 1×2 rectangles in the tiling, then we have the one tiling made up of only 1×1 squares—namely, $n + 1$ of them.

For $2 \leq i \leq n + 1$, if the last rectangle occupies squares $i - 1$ and i, then there are 1×1 squares (on the right) at positions $i + 1$ through $n + 1$. (For $i = n + 1$, there are actually no other squares on the right.) The $i - 2$ squares to the left of this last rectangle on a $1 \times (n + 1)$ chessboard can then be tiled in $F_{(i-2)+1} = F_{i-1}$ ways. Consequently, using this approach, we find that the total number of tilings is

$$1 + \sum_{i=2}^{n+1} F_{i-1} = 1 + \sum_{r=1}^{n} F_r = 1 + \sum_{r=0}^{n} F_r.$$

Equating the results obtained by these two different ways of counting all of the tilings of a $1 \times (n + 1)$ chessboard, we learn that

$$\sum_{r=0}^{n} F_r = F_{n+2} - 1,$$

which we knew previously as Property 4.5.

Example 7.3: Now let us establish another property for the Fibonacci numbers, again using a combinatorial argument. Instead of just writing it down and then attempting to prove it, let us list some results that motivate the property. The first result we know quite well.

For $n \geq 0$,

$$\text{(i)} \; F_{n+2} = F_{n+1} + F_n = F_2 F_{n+1} + F_1 F_n$$
$$\text{(ii)} \; F_{n+3} = F_{n+2} + F_{n+1}$$
$$= 2F_{n+1} + 1F_n = F_3 F_{n+1} + F_2 F_n$$
$$\text{(iii)} \; F_{n+4} = F_{n+3} + F_{n+2}$$
$$= (2F_{n+1} + F_n) + (F_{n+1} + F_n)$$
$$= 3F_{n+1} + 2F_n = F_4 F_{n+1} + F_3 F_n$$
$$\text{(iv)} \; F_{n+5} = F_{n+4} + F_{n+3}$$
$$= (F_4 F_{n+1} + F_3 F_n) + (F_3 F_{n+1} + F_2 F_n)$$
$$= (F_4 + F_3) F_{n+1} + (F_3 + F_2) F_n$$
$$= F_5 F_{n+1} + F_4 F_n.$$

These results suggest

Property 7.1: For $m \geq 1$ and $n \geq 0$,

$$F_{n+m} = F_m F_{n+1} + F_{m-1} F_n.$$

Since this is true for any $m \geq 1$ when $n = 0$, and for any $n \geq 1$ when $m = 1$, from this point on we shall consider the result for $m \geq 2$ and $n \geq 1$.

We shall start with the $1 \times (n + m - 1)$ chessboard shown in Fig. 7.3 (a). From our previous result for l_n (in Example 7.2), this chessboard can be tiled in $F_{(n+m-1)+1} = F_{n+m}$ ways. But now we shall count the tilings in a second way, depending on what happens at the nth square on the board. (Recall that we are tiling the chessboards with 1×1 square tiles and 1×2 rectangular tiles.)

(i) Suppose there is a 1×2 rectangle covering squares n and $n + 1$ on the $1 \times (n + m - 1)$ board, as in Fig. 7.3 (b). This situation accounts for $F_{(n-1)+1} = F_n$ tilings of the $n - 1$ squares on the left of the nth square on this chessboard, and $F_{((n+m-1)-(n+1))+1} = F_{m-1}$ tilings of the $(n + m - 1) - (n + 1) = m - 2$ squares on the right of the $(n + 1)$st square. So we have a total of $F_{m-1} F_n$ tilings in this case.

(ii) Any other possible tiling of this $1 \times (n + m - 1)$ chessboard can be *broken* at the vertical edge separating the nth square of the chessboard from the $(n + 1)$st square of the chessboard. This is demonstrated in Fig. 7.3 (c). Under these circumstances, there are F_{n+1} ways to tile the first n squares of the chessboard and $F_{((n+m-1)-n)+1} = F_m$ ways to tile the last $(n + m - 1) - n = $

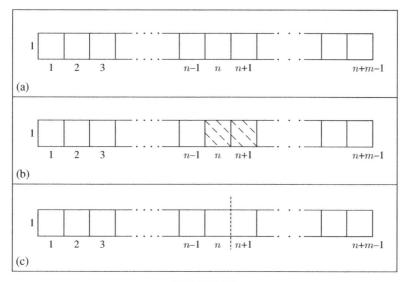

FIGURE 7.3

$m - 1$ squares of the chessboard. Consequently, in this case, there are $F_m F_{n+1}$ possible tilings.

The tilings developed in (i) and (ii) cover all the possible ways to tile a $1 \times (n + m - 1)$ chessboard (and the two cases have nothing in common), so the number of ways to tile such a chessboard is $F_m F_{n+1} + F_{m-1} F_n$.

Equating these two results for the number of ways to tile a $1 \times (n + m - 1)$ chessboard, it now follows that

$$F_{n+m} = F_m F_{n+1} + F_{m-1} F_n.$$

(Once again we have used the combinatorial type of proof. For more on these combinatorial types of proofs involving tilings, the reader should refer to References [3, 6].)

Example 7.4:   Now we shall use Property 7.1 from Example 7.3 to establish another property of the Fibonacci numbers. We know that for $m \geq 1$ and $n \geq 0$,

$$F_{n+m} = F_m F_{n+1} + F_{m-1} F_n.$$

(i) When $n = m \geq 1$,

$$F_{2n} = F_{n+n} = F_n F_{n+1} + F_{n-1} F_n$$
$$= F_n (F_{n+1} + F_{n-1}).$$

So F_n divides F_{2n}, for $n \geq 1$.

(ii) Since

$$F_{3n} = F_{n+2n} = F_{2n}F_{n+1} + F_{2n-1}F_n$$

and F_n divides F_{2n}, it follows that for $n \geq 1$, F_n divides F_{3n}.

(iii) Continuing, this same property tells us that

$$F_{(k+1)n} = F_{kn+n} = F_{kn}F_{n+1} + F_{kn-1}F_n,$$

for $n \geq 1$, $k \geq 1$. From this we see that if F_n divides F_{kn}, then F_n divides $F_{(k+1)n}$.

(iv) Keeping n fixed, where $n \geq 1$, an inductive argument on k, based on the results in (i) and (iii), demonstrates that

$$F_n \text{ divides } F_{kn},$$

for $n \geq 1$, $k \geq 2$. This result is also true for $n \geq 1$ and $k = 1$.

Consequently, we now have the following:

Property 7.2:   For $n \geq 1, k \geq 1$,

$$F_n \text{ divides } F_{kn}.$$

Example 7.5:   Recognizing the interplay between compositions and tilings, the following two observations should not come as much of a surprise!

(i) Example 6.5, which deals with compositions without 1, implies that for $n \geq 1$, the number of ways one can tile a $1 \times n$ chessboard, with rectangular tiles of lengths greater than 1, is F_{n-1}.

(ii) Example 6.6, which deals with compositions using only odd summands, indicates that we can tile a $1 \times n$ chessboard, with rectangular tiles of odd length (and this includes the 1×1 squares) in F_n ways.

EXERCISES FOR CHAPTER 7

1. For $n \geq m \geq 1$, prove that $F_n = F_m F_{n-m+1} + F_{m-1}F_{n-m}$.
2. For $n \geq 1, m \geq 1$, prove that $F_{n+m} = F_{m+1}F_n + F_m F_{n-1}$.
3. For $n \geq 1, m \geq 1$, prove that $F_{n+m} = F_{n+1}F_{m+1} - F_{n-1}F_{m-1}$. (P. Mana, 1969) [39].
4. For $n \geq 0$, prove that $F_{n+1}^2 + F_n^2 = F_{2n+1}$. (E. Lucas, 1876).

5. For $n \geq 2$, prove that $F_n F_{n+1} - F_{n-1} F_{n-2} = F_{2n-1}$.

6. For $n \geq 0$, prove that if 3 divides n, then F_n is even.

7. For $n \geq 0$, prove that if 4 divides n, then F_n is a multiple of 3.

8. For $n \geq 0$, prove that if 5 divides n, then F_n is a multiple of 5.

9. For $n \geq 0$, prove that if 6 divides n, then F_n is a multiple of 4.

10. If p is an odd prime, does it necessarily follow that F_p is a prime?

11. For $n \geq 5$, if F_n is a prime, does it necessarily follow that n is a prime?

12. For $p \geq 1, q \geq 1, r \geq 1$, prove that

$$F_{p+q+r} = F_{p+1} F_{q+1} F_{r+1} + F_p F_q F_r - F_{p-1} F_{q-1} F_{r-1}.$$

13. For $n \geq 1$, prove that $F_{3n} = F_{n+1}^3 + F_n^3 - F_{n-1}^3$. (E. Lucas, 1876).

14. For $n \geq 1$, prove that $\sum_{i=1}^{n} F_i F_{3i} = F_n F_{n+1} F_{2n+1}$. (K. G. Recke, 1969) [46].

15. Jeffery wrote a computer program to calculate and print out the Fibonacci numbers $F_0, F_1, F_2, \ldots, F_{74}, F_{75}$. Upon examining the units digit of each of these numbers, he observed that F_0 and F_{60} both ended in 0, F_1 and F_{61} both ended in 1, F_2 and F_{62} both ended in 1, ..., F_{14} and F_{74} both ended in 7, and F_{15} and F_{75} both ended in 0. When he examined only F_0 through F_{59}, however, no repetitive scheme was apparent. Consequently, Jeffery conjectured: For $n \geq 0$ and $r \geq 0$, with r fixed, F_{60n+r} and F_r have the same units digit. Equivalently, this can be rephrased by saying that $F_{60n+r} - F_r$ is divisible by 10, or that $F_{60n+r} \equiv F_r \pmod{10}$.

   Prove Jeffery's conjecture. (Due to this result, we say that the sequence of units digits of the Fibonacci numbers has *period* 60.)

16. For $n \geq 1$, prove that $F_{2n} = F_n^2 + 2F_{n-1} F_n$.

CHAPTER 8

Chess Pieces on Chessboards

This time we want to use our chessboards to study how to place certain types of chess pieces in prescribed ways.

Example 8.1: In the game of chess, a king can move (when positioned on a standard 8×8 chessboard) one space to the left or right, or one space north or south, or one space in one of the four diagonal directions: northeast, southeast, northwest, or southwest. If an opponent has a chess piece in any of these (possibly) eight neighboring positions, then the king can *take* that piece.

Here we shall restrict ourselves to a $1 \times n$ chessboard where n is a positive integer. Our objective is to count the number of ways one can place *nontaking* kings on such a chessboard. For example, in the case of $n = 5$, we can label the squares of the chessboard from 1 to 5, as shown in Fig. 8.1(a). In Fig. 8.1(b), we find four of the ways we can place nontaking kings on this chessboard. By simply listing the

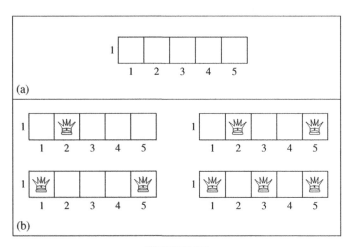

FIGURE 8.1

Fibonacci and Catalan Numbers: An Introduction, First Edition. Ralph P. Grimaldi.
© 2012 John Wiley & Sons, Inc. Published 2012 by John Wiley & Sons, Inc.

labels of the squares where we can place the nontaking kings, we find the following possibilities:

(1) 1	(7) 1, 4
(2) 2	(8) 1, 5
(3) 3	(9) 2, 4
(4) 4	(10) 2, 5
(5) 5	(11) 3, 5
(6) 1, 3	(12) 1, 3, 5

(13) $\varnothing$ (Leave the 1×5 chessboard clear.)

So there are $13 (= F_7)$ ways in which we can place nontaking kings on a 1×5 chessboard. If we let k_n count the number of ways we can do this for a $1 \times n$ chessboard, we could derive the recurrence relation for k_n as we did earlier in many other cases. However, here we may have recognized that this is not really a "new" example where the Fibonacci numbers arise. A slight change in vocabulary shows us that we are simply asking for the number of subsets of $\{1, 2, 3, 4, 5\}$ where there are no consecutive integers. This issue was settled back in Example 5.1. So with $k_1 = 2$ and $k_2 = 3$, it follows that $k_n = F_{n+2}$. (This is the answer we found in Example 5.1.)

Example 8.2: Now let us examine the problem of placing nontaking bishops on a $2 \times n$ chessboard, where $n \geq 1$.

When a bishop is on a chessboard, that piece can only move diagonally, forward (in one turn) or backward (in one turn). Consequently, the situation of placing nontaking bishops on a $1 \times n$ chessboard is not very interesting, for they can never attack one another.

For a $2 \times n$ chessboard, we shall start by considering the case for $n = 8$—as shown in Fig. 8.2. Suppose that a bishop occupies the white square in the first column of this 2×8 chessboard. Since there are only two rows for this chessboard, that bishop can only move one square (at a time) along a diagonal. To avoid the chance of this bishop taking another, we cannot place a second bishop in the white square in column 2 of this board. Should there be a bishop in the white square of column 5 of this board, then no other bishop can be placed on the white square for either column 4 or column 6. We see that, in general, when a bishop starts on a white square, he can only move onto another white square. Consequently, bishops on white squares have no interaction with those on the black squares, and vice versa.

Replace 8 by n. Now envision straightening out the zigzag path a bishop would travel along. At this point we see that the number of ways of placing nontaking bishops along the zigzag path through the n white squares (or the n black squares) on a $2 \times n$ chessboard is the same as the number of ways of placing nontaking kings on a $1 \times n$ chessboard—namely, F_{n+2}. Therefore, the number of ways to place nontaking bishops on a $2 \times n$ chessboard is F_{n+2}^2.

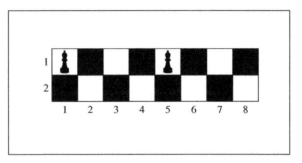

FIGURE 8.2

Using this idea of placing nontaking bishops on a $2 \times n$ chessboard, in Reference [51] we find a combinatorial proof of

$$F_{n+2}^2 = 2(F_{n+1}^2 + F_n^2) - F_{n-1}^2, \quad n \geq 1,$$

or

$$F_n^2 = 2(F_{n-1}^2 + F_{n-2}^2) - F_{n-3}^2, \quad n \geq 3.$$

(This result is equivalent to that given in Exercise 18 of Chapter 4.)

Example 8.3: From kings in Example 8.1 and bishops in Example 8.2, we now turn our attention to rooks (or castles). In chess, one is allowed, on one turn, to move a rook horizontally or vertically (but not diagonally) over as many unoccupied spaces as one wishes. So for the chessboard in Fig. 8.3(a), a rook located at square 1 can only move vertically down to square 2 (and could then take a rook that happens to be at square 2). For this same chessboard, a rook at square 2 can move vertically up to square 1 or horizontally to square 3 or square 4 (if square 3 is not occupied). When a rook at square 2 does move, it takes any rook that happens to be in the square to which it moves.

To determine the number of ways for placing nontaking rooks on a chessboard, we shall introduce the idea of the *rook polynomial*. For a given chessboard C, this polynomial will be denoted by $r(C, x)$ and, for any nonnegative integer k, the coefficient of x^k in $r(C, x)$ will be the number of ways one can place k nontaking rooks on the chessboard C. This coefficient will be denoted by $r_k(C, x)$. For any chessboard C, we have $r_0(C, x) = 1$, the number of ways to place no nontaking rooks on C. Also, $r_1(C, x)$, the number of ways to place one nontaking rook on C, is simply the number of squares on the chessboard C.

The rook polynomial for the chessboard in Fig. 8.3(a) is found to be

$$r(C, x) = 1 + 5x + 5x^2 + x^3,$$

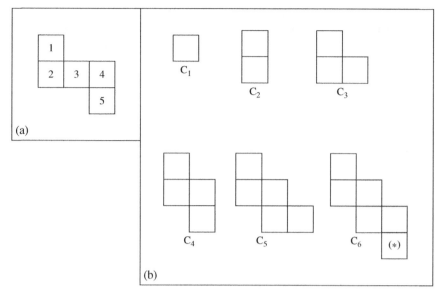

FIGURE 8.3

for there are (i) Five ways to place one nontaking rook on C—simply use any one of the five available squares; (ii) Five ways to place two nontaking rooks—use squares 1 and 3, squares 1 and 4, squares 1 and 5, squares 2 and 5, or squares 3 and 5; and (iii) One way to place three nontaking rooks on C—use squares 1, 3, and 5. Unfortunately, as our chessboards get larger and more complex, we cannot hope to write down their rook polynomials so readily. Yet there is hope. For we shall find that we can often decompose a large chessboard into smaller subboards that are easier to deal with. Before we observe this, however, let us examine the rook polynomials for the chessboards C_1, C_2, C_3, C_4, and C_5, as shown in Fig. 8.3(b). We find the following:

Chessboard	Rook Polynomial
C_1	$1 + x$
C_2	$1 + 2x$
C_3	$1 + 3x + x^2$
C_4	$1 + 4x + 3x^2$
C_5	$1 + 5x + 6x^2 + x^3$

and observe that

$$r(C_3, x) = 1 + 3x + x^2 = (1 + 2x) + x(1 + x) = r(C_2, x) + x\, r(C_1, x)$$
$$r(C_4, x) = 1 + 4x + 3x^2 = (1 + 3x + x^2) + x(1 + 2x)$$
$$= r(C_3, x) + x\, r(C_2, x)$$
$$r(C_5, x) = 1 + 5x + 6x^2 + x^3 = (1 + 4x + 3x^2) + x(1 + 3x + x^2)$$
$$= r(C_4, x) + x\, r(C_3, x).$$

To understand why this happens, we first observe that both chessboards C_5 and C_6, in Fig. 8.3(b), have only three columns. Then we consider the chessboard C_6 in Fig. 8.3(b)—in particular, the square where we find $(*)$. When we try to place k nontaking rooks on this board (for $0 \le k \le 3$), exactly one of the following occurs:

(i) We do not use the square marked with $(*)$: Let C_6' be the subboard obtained from C_6 by deleting the square marked with $(*)$. Then $C_6' = C_5$ and $r_k(C_6', x) = r_k(C_5, x)$.

(ii) We use the square marked with $(*)$—so we place a rook on this square: Now we cannot use any other square of C_6 in the same row or column as the square marked $(*)$. Upon removing the square marked $(*)$ and all other squares in the same row or column as this square, we arrive at the subboard C_6''. Now we see that $C_6'' = C_4$ and, for the remaining $k - 1$ rooks, we have $r_{k-1}(C_6'', x) = r_{k-1}(C_4, x)$.

Consequently, for $0 \le k \le 3$,

$$r_k(C_6, x) = r_k(C_5, x) + r_{k-1}(C_4, x),$$

where we agree that $r_{-1}(C, x) = 0$ for any chessboard C. Then it follows that

$$r_k(C_6, x)x^k = r_k(C_5, x)x^k + r_{k-1}(C_4, x)x^k, \quad \text{for } 0 \le k \le 3,$$

so

$$
\begin{aligned}
r(C_6, x) &= \sum_{k=0}^{3} r_k(C_6, x)x^k \\
&= \sum_{k=0}^{3} r_k(C_5, x)x^k + \sum_{k=0}^{3} r_{k-1}(C_4, x)x^k \\
&= \sum_{k=0}^{3} r_k(C_5, x)x^k + x \sum_{k=0}^{3} r_{k-1}(C_4, x)x^{k-1} \\
&= \sum_{k=0}^{3} r_k(C_5, x)x^k + x \sum_{k=0}^{2} r_k(C_4, x)x^k \\
&= r(C_5, x) + x\, r(C_4, x).
\end{aligned}
$$

For the general situation, we have another recursive situation—namely,

$$r(C_1, x) = 1 + x, \quad r(C_2, x) = 1 + 2x$$

and

$$r(C_n, x) = r(C_{n-1}, x) + x\, r(C_{n-2}, x),$$

for $n \ge 3$.

But what does this have to do with the Fibonacci numbers? If we let r_n count the total number of ways we can place nontaking rooks on the chessboard C_n, then after setting $x = 1$ in $r(C_n, x)$, we find that $r_n = r(C_n, 1)$. Therefore,

$$r_1 = r(C_1, 1) = 1 + 1 = 2,$$
$$r_2 = r(C_2, 1) = 1 + 2 = 3, \text{ and}$$
$$r_n = r(C_n, 1) = r(C_{n-1}, 1) + 1 \cdot r(C_{n-2}, 1) = r_{n-1} + r_{n-2},$$

for $n \geq 3$.

Consequently, for $n \geq 1$,

$$r_n = F_{n+2}.$$

(The technique used here to decompose a given chessboard can be applied to more general chessboards. The interested reader can find this discussed in Section 8.4 of Reference [25].)

CHAPTER 9

Optics, Botany, and the Fibonacci Numbers

As we shall soon see, the Fibonacci numbers even make their way into situations in the natural sciences.

Example 9.1: The following application was introduced in Reference [43] and then solved in Reference [42]. In the science of optics, the branch of physics where one investigates the propagation of light, another instance of the Fibonacci numbers arises. Start by considering a glass plate with two reflective faces, as shown in Fig. 9.1(a). Here Face 1 is the face on the left side of the glass plate, while Face 2 is the face on the right side. In Fig. 9.1(b), we see a single reflection (or change in direction) that occurs at Face 1, when we view this glass plate from the side.

If we now place two such plates back-to-back, as in Fig. 9.1(c), then we have four reflecting faces. Faces 1 and 2 belong to the plate on the left; Faces 3 and 4 belong to the plate on the right. When a light ray falls on this stack of two glass plates, let s_n count the number of different paths the ray can take when it is reflected n times, where $n \geq 0$. For example, in Fig. 9.1(d), we see the case where there are no reflections, so there is only one possible path and $s_0 = 1$. Part (e) of Fig. 9.1 shows us the two different paths that can occur with one reflection. So $s_1 = 2$. Parts (f), (g), and (h) of Fig. 9.1 demonstrate what happens when the ray is reflected two, three and four times, respectively. Here we see that $s_2 = 3$, $s_3 = 5$, and $s_4 = 8$. Note how in these three parts of Fig. 9.1, we have indicated below each stack (of two glass plates) (i) the face where the last reflection occurs, and (ii) the face where the previous reflection occurs.

Examine the first five paths when $n = 4$. In each case, the last reflection occurs at Face 4. Consequently, the previous reflection takes place at an odd-numbered face— that is, Face 1 or Face 3. Note that the last reflection for all the paths for $n = 3$ occurs at Faces 1 and 3. So we simply take each path for $n = 3$ and, instead of letting the ray emerge through Face 4, we add a new reflection at Face 4 and have the ray emerge through Face 1. In this way, the first five paths for $n = 4$ are readily obtained from the five paths for $n = 3$.

Fibonacci and Catalan Numbers: An Introduction, First Edition. Ralph P. Grimaldi.
© 2012 John Wiley & Sons, Inc. Published 2012 by John Wiley & Sons, Inc.

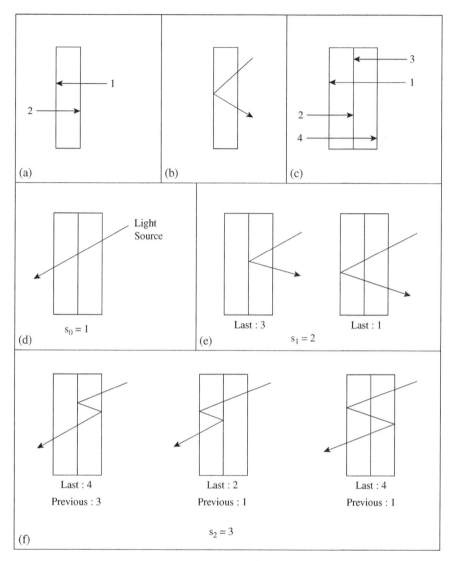

FIGURE 9.1

How do we account for the remaining three paths when $n = 4$? In these three cases, the last reflection takes place at Face 2. Therefore, the previous reflection should occur at Face 1 or Face 3. Looking back at Fig. 9.1(c) we see that this could not occur at Face 3, so the previous reflection occurs at Face 1. (Note, however, that a reflection at Face 1 can be preceded by a reflection at either of the even-numbered faces—that is, either Face 2 or Face 4.) So, to obtain the last three paths for $n = 4$, we start with a path for $n = 2$. Now, however, instead of allowing the ray to emerge through Face 1,

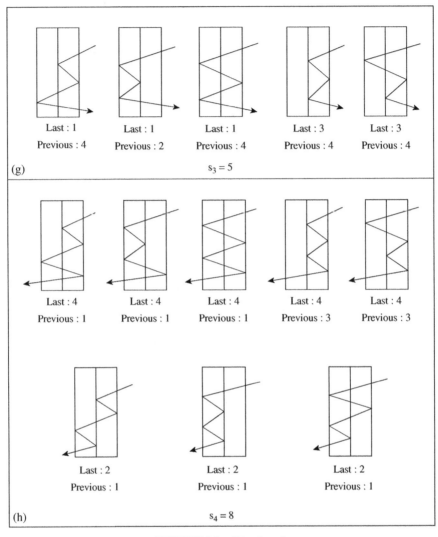

FIGURE 9.1   (*Continued*)

we add a reflection at Face 1 followed by a reflection at Face 2. At this point, the ray emerges through Face 1, having been reflected two additional times.

Consequently, we now see why

$$s_4 = s_3 + s_2.$$

To generalize this situation, consider the case where n is even, $n \geq 2$. Then each of the s_n paths emerges through Face 1 and the last reflection occurs at Face 2 or Face 4.

(i) If the last reflection occurs at Face 4, then the previous reflection occurs at Face 1 or Face 3. Now since $n - 1$ is odd, each of the s_{n-1} paths (for the $n - 1$ case) emerges through Face 4 and the last reflection occurs at Face 1 or Face 3. So here we can take each of these s_{n-1} paths and add a reflection at Face 4 before allowing the ray to then emerge through Face 1.

(ii) Otherwise, the last reflection occurs at Face 2. Since the previous reflection cannot occur at Face 3, it occurs at Face 1. (A reflection at Face 1 can be preceded by a reflection at either Face 2 or Face 4.) So now we start with any of the s_{n-2} paths (for the $n - 2$ case). Since $n - 2$ is even, these paths emerge through Face 1. Now, however, we do not allow the ray to emerge until we have added a reflection at Face 1 followed by one at Face 2. Consequently, for n even, $n \geq 2$, we have $s_n = s_{n-1} + s_{n-2}$.

(A comparable argument, which we assign as an exercise for this chapter, provides the same recurrence relation when n is odd, $n \geq 3$.)

Therefore, it follows that

$$s_n = s_{n-1} + s_{n-2}, \quad n \geq 2, \ s_0 = 1, \ s_1 = 2,$$

and

$$s_n = F_{n+2}.$$

Example 9.2: Although the famous "Problem of the Rabbits" is not very true to reality, we shall now find an instance where the Fibonacci numbers do arise at times in nature. Turning to the study of botany, the branch of biology that deals with plant life, we find the Fibonacci numbers sometimes occurring when we count the number of petals for certain flowers. For example, the number of petals that occur for the enchanter's nightshade is 2 ($= F_3$), while for the iris and the trilium, we find 3 ($= F_4$) petals. In Table 9.1 we list a small assortment of flowers with the corresponding number of petals that naturally occur for each of them.

TABLE 9.1

Flower	Number of Petals
Enchanter's Nightshade	2 ($= F_3$)
Iris, Trilium	3 ($= F_4$)
Buttercup, Columbine	5 ($= F_5$)
Celandine	8 ($= F_6$)
Chamomile, Corn Marigold	13 ($= F_7$)
Aster, Black-eyed Susan	21 ($= F_8$)

There are cases where a certain type of flower may not always have the same number of petals. For instance, the delphinium has 5 ($= F_5$) petals in some cases and 8 ($= F_6$) petals in other cases.

For more on the presence of the Fibonacci numbers in flowers and other plants, we direct the reader to the study of *phyllotaxis*, the botanical name for leaf arrangement.

Within this branch of botany, one finds that in certain botanical structures, such as artichokes, pine cones, pineapples, and sunflowers, the number of rows of scales that wind in one direction is a Fibonacci number and the number of rows of scales that wind in the other direction is the Fibonacci number that precedes or follows the first one. Chapter 4 of Reference [15] and Chapter 3 of Reference [38] provide examples for this type of natural occurrence of the Fibonacci numbers. An introductory article on this topic is provided in Reference [40]. More advanced work in this area can be found in References [34, 57].

EXERCISE FOR CHAPTER 9

1. Verify the recurrence relation in Example 9.1 for the case where n is odd.

CHAPTER 10

Solving Linear Recurrence Relations: The Binet Form for F_n

Up to this point, if we wanted to know a specific Fibonacci number—for instance, F_{37}—we had to start with $F_0 = 0$, $F_1 = 1$, and use the recurrence relation $F_n = F_{n-1} + F_{n-2}$, $n \geq 2$, to compute

$$(1) \quad F_2 = F_1 + F_0 = 1 + 0 = 1$$
$$(2) \quad F_3 = F_2 + F_1 = 1 + 1 = 2$$
$$(3) \quad F_4 = F_3 + F_2 = 2 + 1 = 3$$
$$(4) \quad F_5 = F_4 + F_3 = 3 + 2 = 5$$

$$\vdots \quad \vdots \quad \vdots$$

$$(35) \quad F_{36} = F_{35} + F_{34} = 9,227,465 + 5,702,887$$
$$= 14,930,352$$
$$(36) \quad F_{37} = F_{36} + F_{35} = 14,930,352 + 9,227,465$$
$$= 24,157,817.$$

Now our objective is to somehow determine F_{37}, but without performing any of these 36 calculations. To do so, we want to derive an explicit formula for the general term F_n in terms of n—not in terms of previous values in the sequence of Fibonacci numbers. We shall accomplish this by introducing the following idea:

Definition 10.1: For real constants $C_0, C_1, C_2, \ldots, C_k$, with $C_0 \neq 0$ and $C_k \neq 0$, an expression of the form

$$C_0 a_n + C_1 a_{n-1} + C_2 a_{n-2} + \cdots + C_k a_{n-k} = 0,$$

where $n \geq k$, is called a *kth-order linear homogeneous recurrence relation with constant coefficients*. When the right side of this recurrence relation is not 0, then the relation is referred to as *nonhomogeneous*.

Fibonacci and Catalan Numbers: An Introduction, First Edition. Ralph P. Grimaldi.
© 2012 John Wiley & Sons, Inc. Published 2012 by John Wiley & Sons, Inc.

We shall focus our concern on the case where $k = 2$, $C_0 = 1$, and $C_2 \neq 0$. For example,

$$a_n - 5a_{n-1} + 6a_{n-2} = 0,$$

or

$$a_n = 5a_{n-1} - 6a_{n-2},$$

is a second-order linear homogeneous recurrence relation with constant coefficients.

To solve such a recurrence relation, set $a_n = Ar^n$, where A and r are nonzero constants. Upon substituting this expression into the given recurrence relation, we find that

$$Ar^n = 5Ar^{n-1} - 6Ar^{n-2}.$$

Dividing by A and by r^{n-2}, we have

$$r^2 = 5r - 6 \quad \text{or} \quad r^2 - 5r + 6 = 0,$$

a quadratic equation in r. This quadratic equation in r is called the *characteristic equation*. Since $r^2 - 5r + 6 = (r - 2)(r - 3) = 0$, the roots of the characteristic equation are $r = 2$ and $r = 3$, and they are called the *characteristic roots*. Consequently, the general solution for

$$a_n = 5a_{n-1} - 6a_{n-2} \quad \text{or} \quad a_n - 5a_{n-1} + 6a_{n-2} = 0$$

has the form

$$a_n = c_1 2^n + c_2 3^n,$$

where c_1 and c_2 are arbitrary constants. We verify that this is a (actually, *the*) general solution as follows: Consider the recurrence relation

$$a_n - 5a_{n-1} + 6a_{n-2} = 0.$$

Substituting $a_n = c_1 2^n + c_2 3^n$ into the left-hand side of this recurrence relation, we find that

$$(c_1 2^n + c_2 3^n) - 5(c_1 2^{n-1} + c_2 3^{n-1}) + 6(c_1 2^{n-2} + c_2 3^{n-2})$$
$$= c_1 2^{n-2}(2^2 - 5(2) + 6) + c_2 3^{n-2}(3^2 - 5(3) + 6)$$
$$= c_1 2^{n-2}(4 - 10 + 6) + c_2 3^{n-2}(9 - 15 + 6)$$
$$= c_1 2^{n-2}(0) + c_2 3^{n-2}(0) = 0,$$

verifying directly that $a_n = c_1 2^n + c_2 3^n$ is indeed the general solution.

Should we also know the value of a_n for two specific values of n—generally the *initial values* (or *initial conditions*) a_0 and a_1—then we can use these values to determine c_1 and c_2. For example, if $a_0 = 1$ and $a_1 = 4$, then it follows that

$$1 = a_0 = c_1 \cdot 2^0 + c_2 \cdot 3^0 = c_1 + c_2$$
$$4 = a_1 = c_1 \cdot 2^1 + c_2 \cdot 3^1 = 2c_1 + 3c_2,$$

and we find that $c_1 = -1$ and $c_2 = 2$. So

$$a_n = (-1)2^n + 2 \cdot 3^n, \quad n \geq 0,$$

is the (unique) solution of the *initial value problem*

$$a_n - 5a_{n-1} + 6a_{n-2} = 0, \quad n \geq 2, \ a_0 = 1, \ a_1 = 4.$$

It is also the (unique) solution of the equivalent initial value problem

$$a_{n+2} - 5a_{n+1} + 6a_n = 0, \quad n \geq 0, \ a_0 = 1, \ a_1 = 4.$$

Before we continue, we must inform the reader that what we have done here to solve a second-order linear homogeneous recurrence relation with constant coefficients requires that the resulting quadratic equation in r has *distinct* real roots. This is the only case we shall deal with in this text. The other two cases—complex conjugate roots and repeated real roots—are covered in References [11, 25, 47].

Now let us turn to the second-order recurrence relation we are most interested in at this time—namely,

$$F_n = F_{n-1} + F_{n-2}, \quad n \geq 2, \ F_0 = 0, \ F_1 = 1.$$

Again we start with a substitution: $F_n = Ar^n$, $A \neq 0$, $r \neq 0$. Upon substituting this into the recurrence relation, this time we find that

$$Ar^n = Ar^{n-1} + Ar^{n-2}.$$

Dividing through by A and r^{n-2}, we are then led to the characteristic equation

$$r^2 - r - 1 = 0.$$

The quadratic formula tells us that the characteristic roots are

$$r = \frac{-(-1) \pm \sqrt{(-1)^2 - 4(1)(-1)}}{2(1)} = \frac{1 \pm \sqrt{5}}{2}.$$

It has become standard to assign

$$\alpha = \frac{1 + \sqrt{5}}{2} \quad \text{and} \quad \beta = \frac{1 - \sqrt{5}}{2}.$$

Consequently,

$$F_n = c_1 \alpha^n + c_2 \beta^n, \quad n \geq 0.$$

With

$$0 = F_0 = c_1 + c_2$$

and

$$1 = F_1 = c_1 \alpha + c_2 \beta = c_1 \left(\frac{1 + \sqrt{5}}{2} \right) + c_2 \left(\frac{1 - \sqrt{5}}{2} \right),$$

it follows that $c_1 = 1/\sqrt{5}$ and $c_2 = -1/\sqrt{5}$. Consequently, we can express F_n explicitly by

$$F_n = \frac{1}{\sqrt{5}} \alpha^n - \frac{1}{\sqrt{5}} \beta^n, \quad n \geq 0.$$

Before proceeding, let us reveal some of the fascinating properties exhibited by α and β.

First and foremost,

$$\alpha = \frac{1 + \sqrt{5}}{2} \approx 1.61803398\ldots$$

is commonly known as the *golden ratio*, divine proportion, divine section (by Johannes Kepler), golden mean, sacred chapter (by Leonardo da Vinci), or sacred ratio. This number was known to the Greeks more than 1600 years before the time of Leonardo of Pisa (also known as Fibonacci). Even before the Greeks, this amazing constant was used by the ancient Egyptians in the construction of the Great Pyramid of Giza (c. 3070 B. C.).

In about 1909, the American mathematician Mark Barr designated this constant by ϕ, the Greek letter *phi*—this in honor of the great Greek sculptor Phideas (c. 490 B. C. – 420 B. C.).

Among the many properties satisfied by α and β, we have

$$\alpha^2 = \alpha + 1 \qquad\qquad\qquad \beta^2 = \beta + 1$$
$$\alpha\beta = -1 \quad \alpha^{-1} = -\beta \qquad \beta^{-1} = -\alpha$$
$$\alpha - \beta = \sqrt{5} \qquad\qquad\qquad \alpha + \beta = 1$$
$$\alpha^2 + \beta^2 = 3 \qquad\qquad\qquad \alpha^2 - \beta^2 = \sqrt{5}$$

Other properties of α and β are given in the exercises for this chapter.

Since $\alpha - \beta = \sqrt{5}$, we may now rewrite our explicit formula for F_n as

$$F_n = \frac{\alpha^n - \beta^n}{\alpha - \beta}, \quad n \geq 0.$$

This representation of F_n is called the *Binet form* for the Fibonacci numbers. It was discovered in 1843 by the French mathematician Jacques Phillipe Marie Binet (1786–1856). However, in reality, the formula had been discovered earlier—in 1718—by the French mathematician Abraham DeMoivre (1667–1754). Shortly after Jacques Binet's discovery—in 1844, the French mathematician and engineer Gabriel Lamé also found this result, independent of the work of the other two French mathematicians.

Using the Binet form for F_n, one can derive many more identities involving the Fibonacci numbers. We shall provide three of them here and list others, including Cassini's identity in Property 4.9, in the exercises for this chapter.

Property 10.1:

$$\lim_{n\to\infty} \frac{F_{n+1}}{F_n} = \alpha.$$

Proof: Since $\alpha = (1 + \sqrt{5})/2$ and $\beta = (1 - \sqrt{5})/2$, it follows that $|\beta/\alpha| = \left|(1 - \sqrt{5})/(1 + \sqrt{5})\right| < 1$. Consequently, as $n \to \infty$, we find that $|\beta/\alpha|^n \to 0$ and $(\beta/\alpha)^n \to 0$. Therefore,

$$\lim_{n\to\infty} \frac{F_{n+1}}{F_n} = \lim_{n\to\infty} \frac{(\alpha^{n+1} - \beta^{n+1})/(\alpha - \beta)}{(\alpha^n - \beta^n)/(\alpha - \beta)} = \lim_{n\to\infty} \frac{\alpha^{n+1} - \beta^{n+1}}{\alpha^n - \beta^n}$$

$$= \lim_{n\to\infty} \frac{\alpha - \beta(\beta/\alpha)^n}{1 - (\beta/\alpha)^n} = \frac{\alpha - \beta(0)}{1 - 0} = \alpha.$$

$\square$

For our other two properties, it will help to recall the Binomial theorem:

For real variables x, y, and n, a nonnegative integer,

$$(x+y)^n = \binom{n}{0}x^0 y^n + \binom{n}{1}x^1 y^{n-1} + \cdots + \binom{n}{n-1}x^{n-1}y^1 + \binom{n}{n}x^n y^0$$

$$= \sum_{k=0}^{n} \binom{n}{k}x^k y^{n-k},$$

where

$$\binom{n}{k} = \frac{n!}{k!(n-k)!} \text{ and } 0! = 1.$$

Property 10.2:

$$\sum_{k=0}^{n} \binom{n}{k}2^k F_k = F_{3n}.$$

Proof: Along with the Binet form for F_n, the following will prove useful here: Since $\alpha^2 = \alpha + 1$, it follows that

$$2\alpha + 1 = (\alpha + 1) + \alpha = \alpha^2 + \alpha = \alpha(\alpha + 1) = \alpha(\alpha^2) = \alpha^3.$$

Likewise we have $2\beta + 1 = \beta^3$. Now let us see where these results come into play.

$$\sum_{k=0}^{n} \binom{n}{k}2^k F_k = \sum_{k=0}^{n} \binom{n}{k}2^k \left(\frac{\alpha^k - \beta^k}{\alpha - \beta}\right)$$

$$= \frac{1}{\alpha - \beta}\sum_{k=0}^{n} \binom{n}{k}2^k \alpha^k - \frac{1}{\alpha - \beta}\sum_{k=0}^{n} \binom{n}{k}2^k \beta^k$$

$$= \frac{1}{\alpha - \beta}\sum_{k=0}^{n} \binom{n}{k}(2\alpha)^k 1^{n-k} - \frac{1}{\alpha - \beta}\sum_{k=0}^{n} \binom{n}{k}(2\beta)^k 1^{n-k}$$

$$= \frac{1}{\alpha - \beta}(2\alpha + 1)^n - \frac{1}{\alpha - \beta}(2\beta + 1)^n$$

$$= \frac{1}{\alpha - \beta}(\alpha^3)^n - \frac{1}{\alpha - \beta}(\beta^3)^n = \frac{\alpha^{3n} - \beta^{3n}}{\alpha - \beta} = F_{3n}.$$

□

Property 10.3:

$$\sum_{k=0}^{2n+1} \binom{2n+1}{k}F_k^2 = 5^n F_{2n+1}.$$

Proof: Once again we shall start with the Binet form for F_n. This time it is useful to recall that $\alpha\beta = -1$.

$$\sum_{k=0}^{2n+1} \binom{2n+1}{k} F_k^2$$

$$= \left[\frac{1}{(\alpha - \beta)}\right]^2 \sum_{k=0}^{2n+1} \binom{2n+1}{k} \left(\alpha^k - \beta^k\right)^2$$

$$= \left[\frac{1}{(\alpha - \beta)}\right]^2 \left[\sum_{k=0}^{2n+1} \binom{2n+1}{k} (\alpha^2)^k - 2\sum_{k=0}^{2n+1} \binom{2n+1}{k} (\alpha\beta)^k\right.$$

$$\left. + \sum_{k=0}^{2n+1} \binom{2n+1}{k} (\beta^2)^k\right]$$

$$= \left[\frac{1}{(\alpha - \beta)}\right]^2 \left[\sum_{k=0}^{2n+1} \binom{2n+1}{k} (\alpha^2)^k\right.$$

$$\left. -2\sum_{k=0}^{2n+1} \binom{2n+1}{k} (-1)^k 1^{(2n+1)-k} + \sum_{k=0}^{2n+1} \binom{2n+1}{k} (\beta^2)^k\right]$$

$$= \left[\frac{1}{(\alpha - \beta)}\right]^2 \left[\sum_{k=0}^{2n+1} \binom{2n+1}{k} (\alpha^2)^k\right.$$

$$\left. -2[(-1) + 1]^{2n+1} + \sum_{k=0}^{2n+1} \binom{2n+1}{k} (\beta^2)^k\right]$$

$$= \left[\frac{1}{(\alpha - \beta)}\right]^2 \left[\sum_{k=0}^{2n+1} \binom{2n+1}{k} (\alpha^2)^k + \sum_{k=0}^{2n+1} \binom{2n+1}{k} (\beta^2)^k\right]$$

$$= \left[\frac{1}{(\alpha - \beta)}\right]^2 \left[\sum_{k=0}^{2n+1} \binom{2n+1}{k} (\alpha^2)^k 1^{(2n+1)-k}\right.$$

$$\left. + \sum_{k=0}^{2n+1} \binom{2n+1}{k} (\beta^2)^k 1^{(2n+1)-k}\right]$$

$$= \left[\frac{1}{(\alpha - \beta)}\right]^2 \left[(1 + \alpha^2)^{2n+1} + (1 + \beta^2)^{2n+1}\right]$$

$$= \left(\frac{1}{5}\right)\left[(2 + \alpha)^{2n+1} + (2 + \beta)^{2n+1}\right].$$

Since $\alpha^2 = \alpha + 1$, it follows that $(2 + \alpha)^2 = 4 + 4\alpha + \alpha^2 = 4(1 + \alpha) + \alpha^2 = 5\alpha^2$. Likewise, $(2 + \beta)^2 = 5\beta^2$. So

$$\left(\frac{1}{5}\right)\left[(2+\alpha)^{2n+1} + (2+\beta)^{2n+1}\right] = \left(\frac{1}{5}\right)\left[\frac{(2+\alpha)^{2n+2}}{(2+\alpha)} + \frac{(2+\beta)^{2n+2}}{(2+\beta)}\right]$$

$$= \left(\frac{1}{5}\right)\left[\frac{((2+\alpha)^2)^{n+1}}{(2+\alpha)} + \frac{((2+\beta)^2)^{n+1}}{(2+\beta)}\right]$$

$$= \left(\frac{1}{5}\right)\left[\frac{(5\alpha^2)^{n+1}}{(2+\alpha)}\right] + \left(\frac{1}{5}\right)\left[\frac{(5\beta^2)^{n+1}}{(2+\beta)}\right]$$

$$= 5^n\left[\frac{\alpha^{2n+2}}{(2+\alpha)}\right] + 5^n\left[\frac{\beta^{2n+2}}{(2+\beta)}\right]$$

$$= (5^n)(\alpha^{2n+1})\left(\frac{\alpha}{2+\alpha}\right) + (5^n)(\beta^{2n+1})\left(\frac{\beta}{2+\beta}\right).$$

Now $\alpha^2 = \alpha + 1 \Rightarrow \alpha^2 + 1 = \alpha + 2 \Rightarrow \alpha^2 - \alpha\beta = 2 + \alpha \Rightarrow \alpha(\alpha - \beta) = 2 + \alpha \Rightarrow \alpha/(2 + \alpha) = 1/(\alpha - \beta)$. Similarly, $\beta/(2 + \beta) = -1/(\alpha - \beta)$. So

$$(5^n)(\alpha^{2n+1})\left(\frac{\alpha}{2+\alpha}\right) + (5^n)(\beta^{2n+1})\left(\frac{\beta}{2+\beta}\right)$$

$$= (5^n)(\alpha^{2n+1})\left(\frac{1}{\alpha-\beta}\right) - (5^n)(\beta^{2n+1})\left(\frac{1}{\alpha-\beta}\right)$$

$$= (5^n)\left[\frac{\alpha^{2n+1} - \beta^{2n+1}}{\alpha - \beta}\right] = 5^n F_{2n+1}.$$

$\square$

EXERCISES FOR CHAPTER 10

1. Solve the initial value problem: $a_n = 10a_{n-1} - 21a_{n-2}$, $n \geq 2, a_0 = 1$, $a_1 = 1$.

2. Solve the initial value problem: $a_n = -3a_{n-1} + 10a_{n-2}$, $n \geq 2, a_0 = 2$, $a_1 = 3$.

3. For $n \geq 1$, let a_n count the number of ways in which Jennifer and Dustin can tile a $3 \times n$ chessboard using square 1×1 tiles and square 2×2 tiles. Find and solve a recurrence relation (with initial conditions) for a_n.

4. Verify that (i) $\alpha\beta = -1$; (ii) $\alpha^{-1} = -\beta$; and, (iii) $\beta^{-1} = -\alpha$.

5. Verify the following:
   (a) (i) $\alpha + \beta = 1$ (ii) $\alpha - \beta = \sqrt{5}$

(b) (i) $\alpha^2 + \beta^2 = 3$ (ii) $\alpha^2 - \beta^2 = \sqrt{5}$

(c) (i) $\alpha^3 + \beta^3 = 4$ (ii) $\alpha^3 - \beta^3 = 2\sqrt{5}$

6. Verify that (i) $\alpha = 1/(\alpha - 1)$ and (ii) $\beta = 1/(\beta - 1)$.

7. Verify that (i) $1/\alpha^n = 1/\alpha^{n+1} + 1/\alpha^{n+2}$ and (ii) $1/\beta^n = 1/\beta^{n+1} + 1/\beta^{n+2}$.

8. Verify that (i) $\sum_{n=1}^{\infty} 1/\alpha^n = \alpha$ and (ii) $\sum_{n=0}^{\infty} 1/\alpha^n = \alpha^2$.

9. Verify that $\sum_{n=1}^{\infty} 1/\alpha^{2n} = 1/\alpha$.

10. Verify that $\alpha\sqrt{3 - \alpha} = \sqrt{\alpha + 2}$.

11. Verify that $\sqrt{3 - \beta} = (1/2)\sqrt{10 + 2\sqrt{5}}$.

12. Determine $\lim_{n \to \infty} F_n/F_{n+1}$.

13. Determine $\lim_{n \to \infty} 2n/(n + 1 + \sqrt{5n^2 - 2n + 1})$. (V. E. Hoggatt, Jr., and D. A. Lind, 1967) [31].

14. Verify that $(2 + \beta)^2 = 5\beta^2$.

15. Verify that $\beta/(2 + \beta) = -1/(\alpha - \beta)$.

16. Determine the points of intersection of the parabola $y = x^2 - 1$ and the line $y = x$.

17. Find the points of intersection of the hyperbola $y = 1 + 1/x$ and the line $y = x$.

18. (a) For $n \geq 1$, prove that $\alpha^n = \alpha F_n + F_{n-1}$.

   (b) State and prove a comparable result for β.

19. For a fixed integer $n \geq 0$, determine the value of $\sum_{k=0}^{n} \binom{n}{k}\alpha^{3k-2n}$. (H. Freitag, 1975) [21].

20. Find $\sum_{k=0}^{\infty} |\beta|^k$.

21. For positive real numbers a, b, c, if $a/b = c/a$, then a is called the *mean proportion* for b and c.

   (a) Determine the mean proportion for (i) 3 and 12; (ii) 4 and 25; and, (iii) 5 and 10.

   (b) Let D, E, and F be the points on the line segment shown in Fig. 10.1. If r is the length of the segment DE, s is the length of the segment EF, and r is the mean proportion for $r + s$ and s, determine r/s.

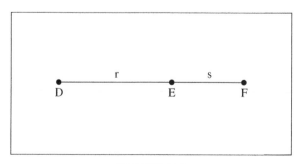

FIGURE 10.1

22. Determine the value of c if $c > 0$ and $c = \int_0^1 x^c dx$.

23. Suppose that the lengths of the sides of a right triangle form the geometric progression s, rs, $r^2 s$. Determine the value of r (a) if s is the length of the hypotenuse; (b) if $r^2 s$ is the length of the hypotenuse.

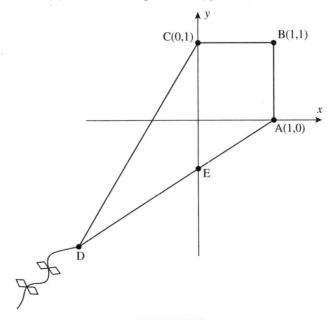

FIGURE 10.2

24. Matt and his brother Sean are constructing a kite—as shown in Fig. 10.2— where the distances AB and CB are each one yard. If they want the area of trapezoid $ABCE$ to equal the area of triangle CDE, where D is on the line $y = x$, determine (i) the coordinates of D; (ii) the coordinates of E; and, (iii) the length of CE.

25. Al inscribes a triangle in a rectangle, as shown in Fig. 10.3. If the three resulting right triangles (within the rectangle and outside the inscribed triangle) are equal in area, determine b/a and c/d.

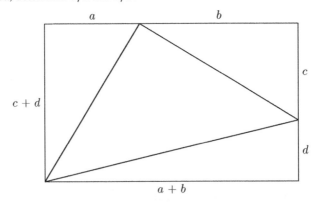

FIGURE 10.3

26. Find the value of the constant c satisfying the equation

$$n^n + (n + c)^n = (n + 2c)^n$$

in the limit as n goes to infinity. (A. Brousseau, 1964) [8].

27. For $n \geq 2$, use the Binet form for the Fibonacci numbers to prove that $F_n F_{n+1} - F_{n-1} F_{n-2} = F_{2n-1}$. (E. Lucas, 1876).

28. Use the Binet form for F_n to establish Cassini's identity in Property 4.9—that is, for $n \geq 1$, $F_{n-1} F_{n+1} - F_n^2 = (-1)^n$.

29. For $n \geq 0$, use the Binet form for F_n to prove that $F_{n+1}^2 + F_n^2 = F_{2n+1}$. (E. Lucas, 1876).

30. (a) For $n \geq 1$, prove that $F_{n+1}^2 - F_{n-1}^2 = F_{2n}$. (E. Lucas, 1876).

    (b) For $n \geq 1$, let T_n be an isosceles trapezoid with bases of length F_{n-1} and F_{n+1}, and sides of length F_n. Prove that the area of T_n is $(\sqrt{3}/4)F_{2n}$. (Note that when $n = 1$, the trapezoid degenerates into a triangle. However, the formula is still correct.)

31. For $n \geq 1$, prove that

$$\sum_{i=1}^{n} F_{4i-2} = F_{2n}^2.$$

32. For $m \geq n \geq 0$, prove that $F_{2m+1} F_{2n+1} = F_{m+n+1}^2 + F_{m-n}^2$. (S. B. Tadlock, 1965) [53].

33. For $n \geq 0$, $k \geq 0$, prove that $F_n^2 + F_{n+2k+1}^2 = F_{2k+1} F_{2n+2k+1}$. (B. Sharpe, 1965) [49].

34. For $n \geq 0, k \geq 0$, prove that $F_{n+2k}^2 - F_n^2 = F_{2k} F_{2n+2k}$. (B. Sharpe, 1965) [49].

35. Prove that

$$\sum_{i=0}^{\infty} \frac{F_i}{2^i} = 2.$$

36. Prove that

$$\sum_{i=0}^{\infty} \frac{F_{i+1}}{2^i} = 4. \quad \text{(J. H. Butchart, 1968) [12].}$$

37. For $n \geq 0$, prove that $\sum_{i=0}^{n}(-1)^{n-i} \binom{n}{i} F_{2i} = F_n$.

38. For $n \geq 0$, prove that $\sum_{i=0}^{n}(-1)^i \binom{n}{i} F_{2i} = (-1)^n F_n$. (H. W. Gould, 1963) [24].

39. For $n \geq 0, j \geq 0, j$ fixed, prove that $\sum_{i=0}^{n} \binom{n}{i} F_{i+j} = F_{2n+j}$.

40. For $n \geq 0$, prove that $\sum_{i=0}^{2n}(-1)^i \binom{2n}{i} 2^{i-1} F_i = 0$. (J. L. Brown, Jr., 1967) [10].

41. For $n \geq m \geq 0$, prove that $\alpha^m F_{n-m+1} + \alpha^{m-1} F_{n-m} = \alpha^n$.

42. (a) For $n \geq 1$, prove that

$$\frac{1}{F_n F_{n+2}} = \frac{1}{F_n F_{n+1}} - \frac{1}{F_{n+1} F_{n+2}}.$$

  (b) Determine

$$\sum_{n=1}^{\infty} \frac{1}{F_n F_{n+2}}. \quad \text{(A. Brousseau, 1969) [9].}$$

43. In Fig. 10.4, $\triangle BAC$ is an isosceles triangle with $AB = AC$ and $AB/BC = AC/BC = \alpha$, the golden ratio. (Such a triangle is called a *golden triangle*.) Suppose that BD is drawn as shown with $AD/DC = \alpha$.

  (a) Prove that $BC/DC = \alpha$.

  (b) Prove that $\triangle ABC$ is similar to $\triangle CBD$.

  (c) Prove that $\angle DBA = \angle DBC$, thus establishing that BD is the bisector of $\angle ABC$.

44. Let $\triangle BAC$ be a golden triangle, as in Fig. 10.4, and suppose that BD is drawn so that $AD/DC = \alpha$. Prove that $\triangle CBD$ is a golden triangle.

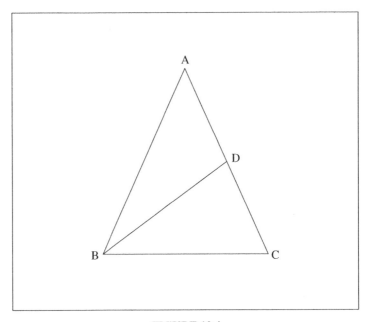

FIGURE 10.4

45. If $\triangle BAC$ is a golden triangle with $\angle B = \angle C$, as in Fig. 10.4, what is the measure of $\angle A$?

46. For the golden triangle in Fig. 10.4, what is the ratio of the area of $\triangle BAC$ to the area of $\triangle ADB$?

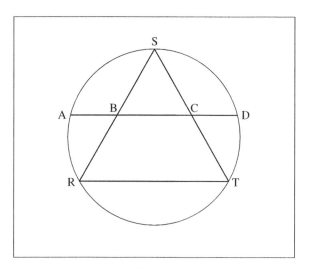

FIGURE 10.5

47. In order to join the geometry club at the St. Ann Academy, Aileen must solve a challenge problem proposed by members Maggie and Eileen. Maggie first draws a circle and then inscribes the equilateral triangle RST in it, as shown in Fig. 10.5. Then Eileen draws the chord AD, so that AD intersects RS at B, the midpoint of RS, and AD intersects ST at C, the midpoint of ST. Aileen is then challenged to prove that $BC/AB = \alpha$. Provide a proof to help Aileen meet this challenge.

48. For $n \geq 0$, prove that

$$\alpha^{-n} = (-1)^{n+1}(\alpha F_n - F_{n+1}).$$

49. Prove that

$$\sum_{i=0}^{\infty} \frac{F_i}{3^{i+1}} = \frac{1}{5}.$$

50. (a) For $n \geq 0$, prove that $\left| F_n - (\alpha^n/\sqrt{5}) \right| < (1/2)$. [From this it follows that $-(1/2) < F_n - (\alpha^n/\sqrt{5}) < (1/2)$, or $(\alpha^n/\sqrt{5}) - (1/2) < F_n < (\alpha^n/\sqrt{5}) + (1/2)$. This indicates that F_n is the integer closest to $(\alpha^n/\sqrt{5})$.]

   (b) For which values of n do we find $F_n > (\alpha^n/\sqrt{5})$?

(c) For which values of n do we find $F_n < (\alpha^n/\sqrt{5})$?

(d) For which values of n do we find $F_n = \left\lfloor \alpha^n/\sqrt{5} \right\rfloor$, where $\lfloor x \rfloor$ denotes the *floor* of x? (*Note*: For each real number x, (i) $\lfloor x \rfloor = x$, for x an integer; and, (ii) $\lfloor x \rfloor =$ the integer to the left of x on the real number line, when x is not an integer. The floor of x is sometimes referred to as the *greatest integer* in x.)

(e) For which values of n do we find $F_n = \left\lceil \alpha^n/\sqrt{5} \right\rceil$, where $\lceil x \rceil$ denotes the *ceiling* of x? (*Note*: For each real number x, (i) $\lceil x \rceil = x$, for x an integer; and, (ii) $\lceil x \rceil =$ the integer to the right of x on the real number line, when x is not an integer.)

51. (a) We can use the Binet form for F_n to extend the Fibonacci numbers so that the subscripts are negative integers. In doing so, for $n > 0$, we define

$$F_{-n} = \frac{\alpha^{-n} - \beta^{-n}}{\alpha - \beta}.$$

Use this definition to prove that for $n > 0$, $F_{-n} = (-1)^{n+1} F_n$.

(b) Following the idea set forth in part (a), for $n > 0$, we define $L_{-n} = \alpha^{-n} + \beta^{-n}$. Use this definition to prove that for $n > 0$, $L_{-n} = (-1)^n L_n$, where $L_n = \alpha^n + \beta^n$.

(c) For $n > 0$, prove that $\alpha F_{-n} + F_{-(n+1)} = \alpha^{-n}$.

52. Nettie needs a square of paper for an origami construction. So she cuts a rectangular piece of paper with width w and length l, where $w < l < 2w$, into two pieces, a $w \times w$ square and a $w \times (l - w)$ rectangle. While folding her origami animal, she happens to notice that the $w \times (l - w)$ rectangle is similar (in the geometric sense) to the original piece of paper. Under these circumstances, what is the ratio of l to w? (These rectangles are referred to as *golden rectangles*.)

More on α and β: Applications in Trigonometry, Physics, Continued Fractions, Probability, the Associative Law, and Computer Science

Before we provide some instances where the constants α and β come into play, in Reference [22] we find mention of a rather interesting observation: the ratio of one's height to the height of one's navel is approximately 1.618—according to extensive empirical observations made by European scholars. Yes, the remarkable and mysterious number α continuously arises in all kinds of unexpected places!

Example 11.1: Let us start by recalling some trigonometric identities:

$$(1)\ \sin 2\theta = 2\sin\theta\cos\theta$$

$$(2)\ \cos 2\theta = \cos^2\theta - \sin^2\theta = 2\cos^2\theta - 1$$

$$\begin{aligned}(3)\ \cos 3\theta &= \cos(2\theta + \theta) = \cos 2\theta\cos\theta - \sin 2\theta\sin\theta \\ &= \cos\theta(2\cos^2\theta - 1) - (2\sin\theta\cos\theta)\sin\theta \\ &= 2\cos^3\theta - \cos\theta - 2\cos\theta\sin^2\theta \\ &= 2\cos^3\theta - \cos\theta - 2\cos\theta(1 - \cos^2\theta) \\ &= 2\cos^3\theta - \cos\theta + 2\cos^3\theta - 2\cos\theta \\ &= 4\cos^3\theta - 3\cos\theta\end{aligned}$$

Now let $\theta = \pi/10$. Then

$$\frac{\pi}{2} = 5\theta = 2\theta + 3\theta.$$

Fibonacci and Catalan Numbers: An Introduction, First Edition. Ralph P. Grimaldi.
© 2012 John Wiley & Sons, Inc. Published 2012 by John Wiley & Sons, Inc.

So the angles 2θ and 3θ are complementary angles, and since the sine and cosine are cofunctions, it follows that

$$\sin 2\theta = \cos 3\theta.$$

Therefore, from the above trigonometric identities, we find that

$$2 \sin \theta \cos \theta = \sin 2\theta = \cos 3\theta = 4 \cos^3 \theta - 3 \cos \theta.$$

Upon dividing by $\cos \theta$ (as $\cos \theta \neq 0$), it then follows that

$$2 \sin \theta = 4 \cos^2 \theta - 3 = 4 (1 - \sin^2 \theta) - 3 = -4 \sin^2 \theta + 1$$

and

$$4 \sin^2 \theta + 2 \sin \theta - 1 = 0, \quad \text{a quadratic equation in } \sin \theta.$$

Consequently,

$$\sin \theta = \frac{-2 \pm \sqrt{2^2 - 4(4)(-1)}}{2(4)} = \frac{-2 \pm \sqrt{20}}{8} = \frac{-1 \pm \sqrt{5}}{4}.$$

Since $\theta = \pi/10$ is in the first quadrant, $\sin \theta > 0$, so $\sin(\pi/10) = \sin \theta = (-1 + \sqrt{5})/4 = -(1/2) \cdot ((1 - \sqrt{5})/2) = -(1/2)\beta = (-(1/2))(-(1/\alpha)) = 1/(2\alpha)$, since $\alpha\beta = -1$. Meanwhile,

$$\cos \frac{\pi}{5} = \cos 2\theta = \cos^2 \theta - \sin^2 \theta = 1 - 2 \sin^2 \theta = 1 - 2 \left(\frac{1}{2\alpha} \right)^2 = 1 - \frac{1}{2\alpha^2}$$

$$= \frac{2\alpha^2 - 1}{2\alpha^2} = \frac{(2\alpha^2 - 1)\beta^2}{2\alpha^2\beta^2} = \frac{2\alpha^2\beta^2 - \beta^2}{2\alpha^2\beta^2}$$

$$= \frac{2 - (\beta + 1)}{2} \quad \text{(Since } \alpha\beta = -1 \text{ and } \beta^2 = \beta + 1.)$$

$$= \frac{1 - \beta}{2} = \frac{\alpha}{2} \quad \text{(Since } \alpha + \beta = 1.)$$

With $\cos(\pi/5) = \alpha/2$, we now find that

$$\cos\frac{3\pi}{5} = \cos 3\left(\frac{\pi}{5}\right) = 4\cos^3\frac{\pi}{5} - 3\cos\frac{\pi}{5}$$

$$= 4\left(\frac{\alpha}{2}\right)^3 - 3\left(\frac{\alpha}{2}\right) = \frac{1}{2}\alpha^3 - \frac{3}{2}\alpha$$

$$= \frac{1}{2}\alpha(\alpha^2 - 3) = \frac{1}{2}\alpha(-\beta^2) \quad \text{(Since } \alpha^2 + \beta^2 = 3.)$$

$$= \frac{1}{2}(-\beta)(\alpha\beta) = \frac{1}{2}\beta \quad \text{(Since } \alpha\beta = -1.)$$

So $\alpha = 2\cos(\pi/5)$ and $\beta = 2\cos(3\pi/5)$.

Now from the Binet form for F_n, we obtain the trigonometric form for the Fibonacci numbers—namely,

$$F_n = \frac{\alpha^n - \beta^n}{\alpha - \beta} = \frac{\alpha^n - \beta^n}{\sqrt{5}}$$

$$= \frac{(2\cos(\pi/5))^n - (2\cos(3\pi/5))^n}{\sqrt{5}}$$

$$= \frac{1}{\sqrt{5}}(2^n)\left[\cos^n\left(\frac{\pi}{5}\right) - \cos^n\left(\frac{3\pi}{5}\right)\right], \quad n \geq 0.$$

This form for F_n was established by W. Hope-Jones in 1921.

Example 11.2:   Electrostatics is the branch of physics that deals with phenomena due to the attractions and repulsions of electric charges according to the distances between them, but with no dependence upon their motion. In 1972, while a student at the Indian Statistical Institute, Basil Davis studied the following problem dealing with the potential energy of a system comprising two negative charges and one positive charge.

Suppose that two negative charges of $-q$ are located at the points Y and Z, while one positive charge of $+q$ is located at the point X—as on the line shown in Fig. 11.1 on p. 68. If we assign a to the distance between X and Y, and b to the distance between Y and Z, then we can write the following:

  (i) For the charges at X and Y, the potential energy is $k((+q)(-q))/a$;

 (ii) For the charges at Y and Z, the potential energy is $k((-q)(-q))/b$; and

(iii) For the charges at X and Z, the potential energy is $k((+q)(-q))/(a+b)$, where k is a constant. [In the mks system, $k \doteq 8.99 \times 10^9 \ (Nm^2)/C^2$ (where N: newtons, m: meters, and C: coulombs) and the potential energy is given in joules.]

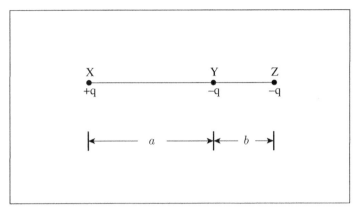

FIGURE 11.1

We want to determine the ratio a/b so that the total potential energy of the system is 0. Consequently, we examine the equation

$$k\frac{(+q)(-q)}{a} + k\frac{(-q)(-q)}{b} + k\frac{(+q)(-q)}{a+b} = 0.$$

First we divide through by the constant k to obtain

$$\frac{(+q)(-q)}{a} + \frac{(-q)(-q)}{b} + \frac{(+q)(-q)}{a+b} = 0.$$

Dividing through by q^2, we then find that

$$-\frac{1}{a} + \frac{1}{b} - \frac{1}{a+b} = 0.$$

Multiplying by $ab(a+b)$, we arrive at

$$-b(a+b) + a(a+b) - ab = 0$$
$$-ab - b^2 + a^2 + ab - ab = 0$$
$$a^2 - ab - b^2 = 0$$
$$\left(\frac{a}{b}\right)^2 - \left(\frac{a}{b}\right) - 1 = 0, \text{ a quadratic in } \frac{a}{b}.$$

Therefore, $(a/b) = (-(-1) \pm \sqrt{(-1)^2 - 4(1)(-1)})/(2(1)) = (1 \pm \sqrt{5})/2$ and as $a > 0$ and $b > 0$, it follows that for the total potential energy of the system to be 0,

$$\frac{a}{b} = \frac{1+\sqrt{5}}{2} = \alpha, \quad \text{the golden ratio.}$$

Example 11.3: (a) Here we want to evaluate $\sqrt{1 + \sqrt{1 + \sqrt{1 + \sqrt{1 + \cdots}}}}$. So let us start with

$$c = \sqrt{1 + \sqrt{1 + \sqrt{1 + \sqrt{1 + \cdots}}}}$$

Then upon squaring both sides of this equation, we see that

$$c^2 = 1 + \sqrt{1 + \sqrt{1 + \sqrt{1 + \sqrt{1 + \cdots}}}}$$

or

$$c^2 = 1 + c.$$

Consequently,

$$c^2 - c - 1 = 0$$

and

$$c = \frac{-(-1) \pm \sqrt{(-1)^2 - 4(1)(-1)}}{2(1)} = \frac{1 \pm \sqrt{5}}{2}.$$

Since $c > 0$, it follows that

$$c = \frac{1 + \sqrt{5}}{2}, \quad \text{the golden ratio.}$$

(b) This time let us start with the following idea. The expression

$$2 + \cfrac{1}{3 + \frac{1}{7}}$$

is an example of a *finite continued fraction*. It can be simplified as

$$2 + \cfrac{1}{3 + \frac{1}{7}} = 2 + \frac{1}{\frac{22}{7}} = 2 + \frac{7}{22} = \frac{44 + 7}{22} = \frac{51}{22}.$$

Since the notation can quickly become quite cumbersome, the above expression is often written as [2; 3, 7]. In general, if a_1 is a nonnegative integer and a_2, a_3, ..., a_n

are positive integers, then the finite continued fraction

$$a_1 + \cfrac{1}{a_2 + \cfrac{1}{a_3 + \cfrac{1}{\cdots + \cfrac{1}{a_{n-1} + \frac{1}{a_n}}}}}$$

can also be represented by

$$[a_1;\ a_2,\ a_3,\ \ldots,\ a_{n-1},\ a_n].$$

So now consider the following four finite continued fractions:

$$(1)\ 1 \qquad (2)\ 1 + \frac{1}{1} \qquad (3)\ 1 + \cfrac{1}{1 + \frac{1}{1}} \qquad (4)\ 1 + \cfrac{1}{1 + \cfrac{1}{1 + \frac{1}{1}}}.$$

Simplifying these finite continued fractions, we find the following:

$$(1)\ 1 = 1 = \frac{1}{1} = \frac{F_2}{F_1}$$

$$(2)\ 1 + \frac{1}{1} = 1 + 1 = 2 = \frac{2}{1} = \frac{F_3}{F_2}$$

$$(3)\ 1 + \cfrac{1}{1 + \frac{1}{1}} = 1 + \frac{1}{1 + 1} = 1 + \frac{1}{2} = \frac{3}{2} = \frac{F_4}{F_3}$$

$$(4)\ 1 + \cfrac{1}{1 + \cfrac{1}{1 + \frac{1}{1}}} = 1 + \cfrac{1}{1 + \frac{1}{1+1}} = 1 + \cfrac{1}{1 + \frac{1}{2}} = 1 + \frac{1}{\frac{3}{2}} = 1 + \frac{2}{3} = \frac{5}{3} = \frac{F_5}{F_4}.$$

At this point we turn our attention to the *infinite continued fraction*

$$c = 1 + \cfrac{1}{1 + \cfrac{1}{1 + \cfrac{1}{1 + \frac{1}{1 + \cdots}}}} = [1;\ 1,\ 1,\ 1, \ldots].$$

Note here that

$$c = 1 + \frac{1}{c},$$

so $c^2 = c + 1$ or $c^2 - c - 1 = 0$. Consequently,

$$c = \frac{-(-1) \pm \sqrt{(-1)^2 - 4(1)(-1)}}{2(1)} = \frac{1 \pm \sqrt{5}}{2}.$$

But since $c > 0$, we have

$$c = \frac{1 + \sqrt{5}}{2},$$

another instance of the golden ratio.

(Alternatively, from the continued fractions in (1)–(4), the pattern implies that $c = \lim_{n \to \infty} \frac{F_{n+1}}{F_n} = \alpha$.)

Our next two examples involve some probability theory.

Example 11.4: Suppose that S is the sample space for an experiment $\mathcal{E}$. If A, B are events from S with $A \cup B = S$, $A \cap B = \varnothing$, $P(A) = p$, the probability that event A occurs, and $P(B) = p^2$, the probability that event B occurs, what is the value of p? Since $A \cap B = \varnothing$, we have

$$1 = P(S) = P(A \cup B) = P(A) + P(B).$$

Consequently,

$$1 = p + p^2 \quad \text{or} \quad p^2 + p - 1 = 0.$$

By the quadratic formula, the roots of this equation are $p = (-1 \pm \sqrt{1^2 - 4(1)(-1)})/(2(1)) = (-1 \pm \sqrt{5})/2$. Since $(-1 - \sqrt{5})/2 < 0$, it follows that $p = (-1 + \sqrt{5})/2 = -\beta$. (Let us not be misguided by that negative sign in front of β. After all, $\beta < 0$, so $-\beta \approx 0.61803398... > 0$.)

Example 11.5: Derek and Ryne toss a loaded coin, where p (> 0) is the probability the coin comes up heads when it is tossed. The first of these two budding gamblers to obtain a head is the winner. Ryne goes first but if he tosses a tail, then Derek gets two chances. If he tosses two tails, then Ryne again tosses the coin, and if his toss is a tail, then Derek again goes twice (if his first toss is a tail). This continues until someone tosses a head. What value of p makes this a fair game (that is, a game where both Ryne and Derek have probability $1/2$ of winning)?

Now the probability that Ryne wins this game is

$$p + (1 - p)(1 - p)^2 p + (1 - p)(1 - p)^2(1 - p)(1 - p)^2 p + \cdots,$$

where the first summand—namely, p—accounts for when Ryne tosses a head on the first toss. The second summand accounts for when (i) Ryne tosses a tail on the first toss—hence the factor $1 - p$; (ii) Derek tosses a tail on both of his tosses—hence the factor $(1 - p)^2$; and, finally, Ryne wins by tossing a head on his next toss—as indicated by the factor p. In like manner, the third summand accounts for when Ryne wins—after he tossed a tail on both his first and second turns, while Derek tossed two tails on both of his first and second turns.

The above expression for when Ryne wins this game can also be expressed as

$$p[1 + (1-p)^3 + (1-p)^6 + (1-p)^9 + \cdots]$$
$$= p \left[\frac{1}{[1-(1-p)^3]} \right].$$

For this game to be fair, we need to have

$$\frac{1}{2} = \frac{p}{[1-(1-p)^3]},$$

so we see that

$$p = \left(\frac{1}{2}\right)[1 - (1-p)^3]$$
$$2p = [1 - (1-p)^3] = 1 - (1 - 3p + 3p^2 - p^3)$$
$$2p = 3p - 3p^2 + p^3, \text{ and}$$
$$0 = p^3 - 3p^2 + p = p(p^2 - 3p + 1).$$

With $p > 0$, it follows that $p^2 - 3p + 1 = 0$ or

$$p = \frac{-(-3) \pm \sqrt{(-3)^2 - 4(1)(1)}}{2(1)} = \frac{3 \pm \sqrt{5}}{2}.$$

Since $p < 1$, we find that

$$p = \frac{3 - \sqrt{5}}{2} = \left[\frac{1-\sqrt{5}}{2}\right]^2 = \beta^2.$$

(More on the powers of α and β can be found in Exercise 33 of Chapter 13.)

Our next example for this chapter deals with the Associative Law for a certain binary operation. This result was established in 1936 by the Scottish-American mathematician and science fiction author Eric Temple Bell (1883–1960).

Example 11.6: Let a, b, and c be fixed real numbers with $ab = 1$. If **R** denotes the set of real numbers, then the function

$$f : \mathbf{R} \times \mathbf{R} \to \mathbf{R}$$

where

$$f(x, y) = a + bxy + c(x + y)$$

is an example of a binary operation on **R**. For this binary operation to be associative, it needs to satisfy the condition that

$$f(f(x, y), z) = f(x, f(y, z)), \quad \text{for all real numbers } x, y, z.$$

We would like to know if there are any values of c for which this happens. Computing $f(f(x, y), z)$, we find that

$$f(f(x, y), z) = f(a + bxy + c(x + y), z)$$
$$= a + b[(a + bxy + c(x + y))z] + c[(a + bxy + c(x + y)) + z]$$
$$= a + ac + c^2x + bcxy + b^2xyz + bcxz + c^2y + bcyz + abz + cz.$$

Meanwhile,

$$f(x, f(y, z)) = f(x, a + byz + c(y + z))$$
$$= a + b[x(a + byz + c(y + z))] + c[x + (a + byz + c(y + z))]$$
$$= a + ac + abx + cx + c^2y + c^2z + b^2xyz + bcxy + bcxz + bcyz.$$

For the binary operation f to be associative, we need to observe that

$$f(f(x, y), z) = f(x, f(y, z))$$
$$\Rightarrow c^2x + (ab + c)z = abx + cx + c^2z.$$

Now with $ab = 1$, it follows that

$$c^2x + z + cz = x + c^2z + cx$$

or

$$(c^2 - c - 1)x = (c^2 - c - 1)z.$$

Since x, z are arbitrary, this requires that $c^2 - c - 1 = 0$. Consequently, $c = \alpha$ or $c = \beta$—that is, c is the golden ratio or its negative reciprocal.

Our final example is truly amazing in that it has to do with estimating the number of divisions performed in the Euclidean algorithm to find gcd(a, b)—that is, the greatest common divisor of the positive integers a and b. Needless to say, no one was concerned about this type of computational efficiency at the time when Leonardo of Pisa published his *Liber Abaci*.

Example 11.7: The following application will give us one more example where the golden ratio arises—this time in a computer science setting. This application deals with Gabriel Lamé's work in estimating the number of divisions performed in the

Euclidean algorithm to find gcd(a, b), where a and b are integers with $a \geq b \geq 2$. To find this estimate, we need the following property of the Fibonacci numbers, which can be established by the Alternative, or Strong, form of the Principle of Mathematical Induction. This is left as an exercise.

Property 11.1:   For $n \geq 3$, $F_n > \alpha^{n-2}$.

To address this situation, we need to recall the following version of the division algorithm: Given the positive integers s, t with $s \geq t$, there exist unique positive integers q, r so that

$$s = qt + r, \quad 0 \leq r < t,$$

where q is called the *quotient* and r the *remainder.*

Now let us start with two integers a and b, where $a \geq b \geq 2$. For the sake of the notation, we shall let $r_0 = a$ and $r_1 = b$. In the Euclidean algorithm, we apply the division algorithm, successively as follows, until we arrive at a remainder of 0:

$$
\begin{aligned}
r_0 &= q_1 r_1 + r_2, & 0 < r_2 < r_1 \\
r_1 &= q_2 r_2 + r_3, & 0 < r_3 < r_2 \\
r_2 &= q_3 r_3 + r_4, & 0 < r_4 < r_3 \\
&\ \ \vdots \quad\quad \vdots & \vdots \\
r_{n-2} &= q_{n-1} r_{n-1} + r_n, & 0 < r_n < r_{n-1} \\
r_{n-1} &= q_n\, r_n.
\end{aligned}
$$

Then, at this point, we learn that gcd(a, b) = gcd(r_0, r_1) = r_n, the last nonzero remainder.

From the subscript on r we see that n divisions have been carried out to determine r_n, the gcd(a, b). Furthermore, $q_i \geq 1$, for all $1 \leq i \leq n-1$, and $q_n \geq 2$ because $r_n < r_{n-1}$. Examining the $n - 1$ nonzero remainders $r_n, r_{n-1}, r_{n-2}, ..., r_2$, along with $r_1 (= b)$, we find that

$$
\begin{aligned}
r_n &> 0, \quad \text{so} \quad r_n \geq 1 = F_2 \\
[q_n &\geq 2, r_n \geq 1] \Rightarrow r_{n-1} = q_n r_n \geq 2 \cdot 1 = 2 = F_3 \\
r_{n-2} &= q_{n-1} r_{n-1} + r_n \geq 1 \cdot r_{n-1} + r_n \geq F_3 + F_2 = F_4 \\
&\ \vdots \quad\quad \vdots \quad\quad \vdots \quad\quad\quad\quad\quad\quad\quad \vdots \quad \vdots \\
r_2 &= q_3 r_3 + r_4 \geq 1 \cdot r_3 + r_4 \geq F_{n-1} + F_{n-2} = F_n \\
b = r_1 &= q_2 r_2 + r_3 \geq 1 \cdot r_2 + r_3 \geq F_n + F_{n-1} = F_{n+1}.
\end{aligned}
$$

Consequently, if n divisions are performed in the Euclidean algorithm in determining gcd(a, b), where $a \geq b \geq 2$, then $b \geq F_{n+1}$. Therefore, by virtue of Property 11.1, we may now write

$$b > \alpha^{(n+1)-2} = \alpha^{n-1} = \left(\frac{1 + \sqrt{5}}{2}\right)^{n-1}.$$

So, as a result, we find that

$$b > \alpha^{n-1} \Rightarrow \log_{10} b > \log_{10}(\alpha^{n-1}) = (n-1)\log_{10}\alpha > \frac{n-1}{5},$$

since $\log_{10}\alpha = \log_{10}[(1 + \sqrt{5})/2] \doteq 0.208988 > 0.2 = 1/5$.

At this point, suppose that $10^{k-1} \le b < 10^k$, so that the base 10 (decimal) representation of b has k digits. Then

$$k = \log_{10} 10^k > \log_{10} b > \frac{n-1}{5} \quad \text{and} \quad n < 5k + 1.$$

Since n and k are positive integers and $n < 5k + 1$, it follows that $n \le 5k$, and this final inequality now completes a proof for the following:

Lamé's Theorem: Let a and b be positive integers with $a \ge b \ge 2$. Then the number of divisions needed in the Euclidean algorithm to determine gcd(a, b) is at most 5 times the number of decimal digits in b.

Furthermore, as $b \ge 2$ in Lamé's theorem, it follows that $\log_{10} b \ge \log_{10} 2$, so $5\log_{10} b \ge 5\log_{10} 2 = \log_{10} 2^5 = \log_{10} 32 > 1$. From our previous work, we know that $n - 1 < 5\log_{10} b$, so

$$n < 1 + 5\log_{10} b < 5\log_{10} b + 5\log_{10} b = 10\log_{10} b,$$

and we may write $n \in O(\log_{10} b)$[1]—that is, for n large, there is a constant C so that $n \doteq C \log_{10} b$. [Hence, the number of divisions needed in the Euclidean algorithm to determine gcd(a, b) for positive integers a and b, where $a \ge b \ge 2$, is $O(\log_{10} b)$—that is, on the order of the number of decimal digits in b.]

Before proceeding with further new material, let us mention for the interested reader that more on the golden ratio—in such areas as geometry, probability, and fractals—can be found in Reference [56].

---

[1] Let $\mathbf{Z}^+$ denote the set of positive integers and $\mathbf{R}$ the set of real numbers. For functions f, $g : \mathbf{Z}^+ \to \mathbf{R}$, we say that g *dominates* f (or f is *dominated* by g) if there exist constants $M \in \mathbf{R}^+$ (the set of positive real numbers) and $k \in \mathbf{Z}^+$ such that $|f(n)| \le M |g(n)|$ for all $n \in \mathbf{Z}^+$, where $n \ge k$. When f is dominated by g, we say that f is of *order* (*at most*) g and we use what is referred to as "big-Oh" notation to denote this. We write $f \in O(g)$, where $O(g)$ is read "order g" or "big-Oh of g." (More on this topic can be found in Chapter 5 of Reference [25] and Chapter 2 of Reference [47].)

EXERCISES FOR CHAPTER 11

1. In Example 11.1 we found that $\cos(\pi/5) = \alpha/2$. Express each of the following in terms of α: (i) $\sin(\pi/5)$; (ii) $\cos(\pi/10)$, and (iii) $\sin(\pi/10)$.

2. Determine the value of $\sqrt{1 - \sqrt{1 - \sqrt{1 - \sqrt{1 - \cdots}}}}$.

3. Convert each of the following finite continued fractions to a rational number: (a) $[2; 1, 4, 9]$ and (b) $[3; 3, 7, 2, 4]$.

4. For $n \geq 3$, prove that (a) $F_n > \alpha^{n-2}$ and (b) $F_n < \alpha^{n-1}$.

5. Consider the right triangle ACB, where the measure of angle B is $36°$ and the length of AB is s, as shown in Fig. 11.2. If this triangle is revolved about the vertical axis (containing side AC), determine the following for the resulting right circular cone: (a) The circumference of the base; (b) The area of the base; (c) The volume of the cone; and, (d) The lateral surface area of the cone.

6. Consider the isosceles triangle BAC, as shown in Fig. 11.3. Here the line segment BD bisects angle B. If the measure of angle A is $36°$, prove that triangle CBD is a golden triangle.

7. Consider the regular pentagon $ABCDE$ inscribed in the circle shown in Fig. 11.4.

   (a) Use the law of sines and the double angle formula for the sine to show that $AC/AM = 2\cos 36°$.

   (b) Explain why $\cos 18° = \sin 72° = 4\sin 18° \cos 18°(1 - 2\sin^2 18°)$.

   (c) Show that $\sin 18°$ is a root of the cubic polynomial equation $8x^3 - 4x + 1 = 0$, and show that $\sin 18° = (\sqrt{5} - 1)/4$.

   (d) Verify that $AC/AM = \alpha$.

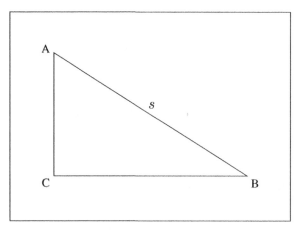

FIGURE 11.2

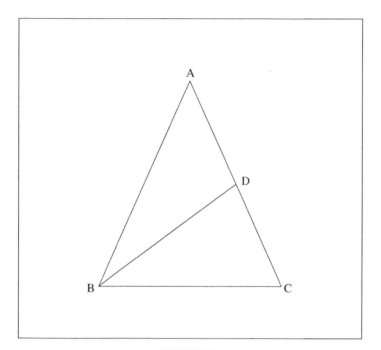

FIGURE 11.3

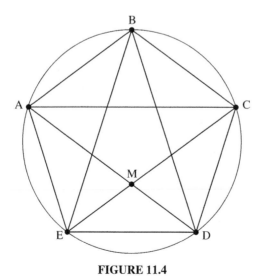

FIGURE 11.4

8. Using DeMoivre's theorem, one finds that the five complex fifth roots of unity are given by

$$\cos\frac{2\pi n}{5} + i \sin\frac{2\pi n}{5}, \quad n = 0, 1, 2, 3, 4.$$

When $n = 0$, one obtains the root $\cos 0° + i \sin 0° = 1$. Determine the other four roots and express them in terms of α. [DeMoivre's theorem is named after the French-born mathematician Abraham de Moivre (1667–1754). He was a good friend of Sir Isaac Newton (1642–1727).]

9. Let $\triangle QRS$ be an isosceles triangle with $QR = QS$. Mark the point T on QR so that $QT = ST = RS$. Prove that $2\cos Q = QR/RS = \alpha$.

10. Let AB be one side of a regular decagon inscribed in a circle with center O and radius 1. Show that the length of AB is $\sqrt{2 - \alpha}$ $(= 1/\alpha)$.

11. A regular pentagon is inscribed in a circle with center O and radius 1. What is the radius of the circle (with center O) that is inscribed in the pentagon?

12. Determine the area of a regular pentagon of side 1.

13. (a) Olivia and her uncle Michael plan to carve a race car out of a piece of wood in the shape of a rectangular solid. The dimensions (in feet) of this piece of wood are length $= l$, width $= w$, and height $= 1$, with $l > w$. Furthermore, the volume of this piece of wood is 1 cubic foot and the diagonal is 2 feet long. Determine the values of l and w.

    (b) What is the total surface area of the six faces of the block in part (a)?

14. Sandra draws a circle with center O and then she draws a secant from the outside point P. The secant starts at P, intersecting the circle first at point Q and then at point R. Finally, Sandra draws a tangent from P to the point T on the circle, so that $PT = QR$. What is PR/QR?

15. (a) Show that $\sin 18° = 1/(2\alpha)$.

    (b) A regular decagon is inscribed in a circle with center at O and radius r. If s is the length of a side of the decagon, verify that $r = s\alpha$. [This occurrence of α was observed, among others, in the manuscript *De Divina Proportione*, which was written in Milan between 1496 and 1498 by Fra Luca Bartolomeo de Pacioli (1445–1517) and published in Venice in 1509. Fra Luca Pacioli was a Franciscan friar and Italian mathematician who collaborated with Leonardo da Vinci (1452–1519) and is often regarded as the "Father of Accounting."]

    (c) Determine the area of the decagon in part (b).

CHAPTER 12

Examples from Graph Theory: An Introduction to the Lucas Numbers

Let us start with some basic ideas from graph theory.

An *undirected graph* G is made up of a nonempty set V of *vertices* and a set E (possibly empty) of *edges* between the vertices in V. For example, in Fig. 12.1(a), we have $V = \{u, v, w\}$ and $E = \{uv, vw, vv\}$, where it is understood, for example, that $uv = vu$. Here we say that v, w are *neighbors* and u, v are also neighbors—but u, w are not neighbors. Also, the edge vv shown in the figure is an example of a *loop* (at v).

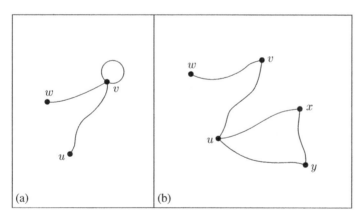

FIGURE 12.1

In Fig. 12.1(b), we have a second undirected graph—this time $V = \{u, v, w, x, y\}$ and $E = \{uv, vw, ux, xy, yu\}$. Here the edges ux, xy, and yu constitute a *cycle*. Since this cycle contains three edges, we say that it has *length* 3.

Now let us turn our attention to the vertices w and x. Notice that we can get from w to x by going from w to v, then from v to u, and then from u to x. We can also travel from w to x by going from w to v, then from v to u, then from u to y, then from y (back) to u, and finally from u to x. These two ways of traveling from w to x are examples of

Fibonacci and Catalan Numbers: An Introduction, First Edition. Ralph P. Grimaldi.
© 2012 John Wiley & Sons, Inc. Published 2012 by John Wiley & Sons, Inc.

w–x walks. In general, a *w–x* walk in *G* is a sequence of vertices in *V*, beginning with *w* and ending with *x*, such that consecutive vertices in the sequence are neighbors. When no edge is repeated in a *w–x* walk, then it is called a *w–x trail.* However, in a trail a vertex may be repeated. When no vertex is repeated (and, consequently, no edge), the result is called a *w–x path.* Consequently, our first *w–x* walk—namely,

$$w, v, u, x$$

—is a path (of length 3). Our second *w–x* walk—that is,

$$w, v, u, y, u, x$$

—is neither a trail, nor a path, from *w* to *x*. This walk has length 5.

The following example deals with the notion of walks in an undirected graph.

Example 12.1: After returning home from a hard day at work, Ernest and Brenda like to jog to maintain their healthy lifestyles (not to mention the nice deduction they receive on their health insurance premiums). They have a lovely home at vertex *u*, as shown in Fig. 12.2, where they can jog around the one-mile circular drive. If they wish to jog two miles, they can either (i) jog around the one-mile circular drive twice or (ii) jog from *u* to *v* (a distance of one mile), where they wave to their friend Flo who owns her home at that vertex, and then jog back home (to *u*) from *v*. Should they wish to jog three miles on these roads, it turns out that they have three possible routes. For four miles, there are five possibilities.

At this point, if we want to know how many different possibilities Ernest and Brenda have, if they want to jog *n* miles on these roads, then we need to determine the number of walks from *u* to *u* of length *n*, for the graph shown in Fig. 12.2. (Since these walks start and end at the same vertex, they are called *closed* walks.) So let w_n count the number of closed *u–u* walks of length *n* in the graph. We know that $w_1 = 1$ and $w_2 = 2$. To determine w_n for $n \geq 3$, consider the last edge in the walk.

(i) If the last edge is the loop *uu*, then the $n - 1$ edges in any of the w_{n-1} walks could have been used before this final one-mile jog around the circular drive.

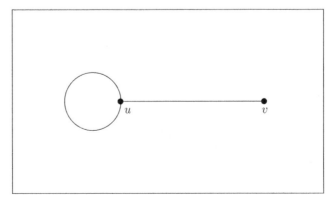

FIGURE 12.2

(ii) If the last edge is not the loop uu, then it is the edge from v to u. This means that the preceding edge is the one from u to v. But prior to these two edges, any of the w_{n-2} walks could have been traveled before this final two-mile jog to and from where Flo's home is situated (at v).

Since these two cases have nothing in common and cover all the possibilities we have

$$w_n = w_{n-1} + w_{n-2}, \quad n \geq 3, \ w_1 = 1, \ w_2 = 2,$$

so

$$w_n = F_{n+1}, \quad n \geq 1.$$

[If we change the vocabulary a bit, we may recognize that what we are also counting is the number of compositions of n, where the only permissible summands are 1's (for the one-mile jogs around the loop uu) and 2's (for the two-mile jogs from u to v, followed by the return from v to u). This is exactly what we did in Example 6.2, where we learned that c_n, the number of compositions of n, where the only summands are 1's and 2's, is likewise F_{n+1}, for $n \geq 1$.]

Our next example introduces us to the idea of a perfect matching in an undirected graph.

Example 12.2: Let $G = (V, E)$ be an undirected graph, where $|V|$, the number of vertices in V, is an even number—say $2n$. A subset M of E is called a *perfect matching* for the graph if $|M|$, the number of edges in M, is n and the edges in M cover all the vertices in V—that is, for every vertex $v \in V$, there is one edge e in M such that v is a vertex on edge e.

For example, in the graph of Fig. 12.3(i), we find two perfect matchings: (1) $\{ab, cd\}$; and (2) $\{ad, bc\}$. The graph in Fig. 12.3 (ii) has six edges and four vertices—

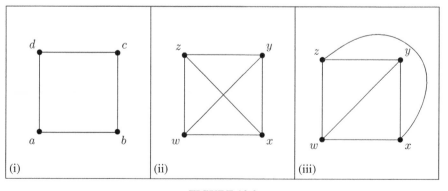

FIGURE 12.3

not five. A vertex does not occur just because two edges happen to cross in a particular depiction of a graph. Note that one could redraw the graph in part (ii) so that it would appear as in part (iii). No matter, both graphs have vertex set $V = \{w, x, y, z\}$ and edge set $E = \{wx, wy, wz, xy, xz, yz\}$. Furthermore, both graphs have three perfect matchings: (1) $\{wx, yz\}$; (2) $\{wz, xy\}$; and, (3) $\{wy, xz\}$.

Now consider the graphs in Fig. 12.4.

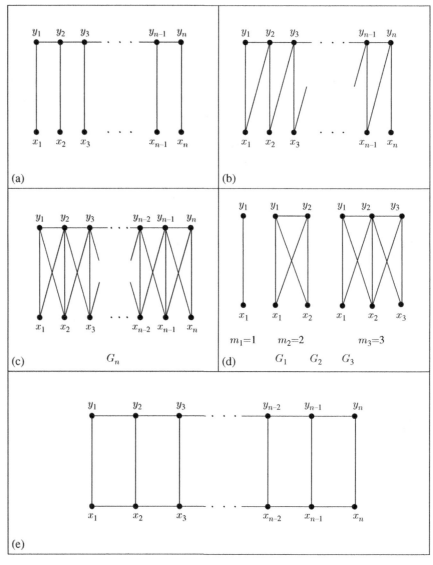

FIGURE 12.4

The graph in Fig. 12.4(a) has only one perfect matching: $\{x_1 y_1, x_2 y_2, \ldots, x_n y_n\}$. Likewise, there is only one perfect matching for the graph in Fig. 12.4(b). Not so, for the graph in Fig. 12.4(c). In Fig. 12.4(d) we have the cases for when $n = 1, 2, 3$ for the graph in Fig. 12.4(c). We see that when $n = 1$, there is one perfect matching: $\{x_1 y_1\}$. For $n = 2$, we find two perfect matchings: $\{x_1 y_1, x_2 y_2\}$ and $\{x_1 y_2, x_2 y_1\}$. When $n = 3$, three perfect matchings are present—namely, (1) $\{x_1 y_1, x_2 y_2, x_3 y_3\}$; (2) $\{x_1 y_2, x_2 y_1, x_3 y_3\}$; and, (3) $\{x_1 y_1, x_2 y_3, x_3 y_2\}$. Should we consider the graph G_4, we would find five, not four, perfect matchings. Can we determine the number of perfect matchings for the general case?

If we let m_n count the number of perfect matchings for the graph G_n in Fig. 12.4(c), there are two considerations:

1. One way we can account for vertex x_n is to use the edge $x_n y_n$. But then we cannot use the edges $x_n y_{n-1}, x_{n-1} y_n$, and $y_{n-1} y_n$. So to have a perfect matching for the other $2n - 2$ vertices, we examine the resulting graph after we delete the vertices x_n, y_n and the edges $x_{n-1} y_n, x_n y_{n-1}, y_{n-1} y_n$, and $x_n y_n$. For the resulting graph—namely, G_{n-1}—there are m_{n-1} possible perfect matchings.

2. The only other way to account for vertex x_n is to use the edge $x_n y_{n-1}$. Then, to continue developing our perfect matching, we must use the edge $x_{n-1} y_n$ in order to account for the vertex y_n. At this point, we have accounted for the four vertices x_{n-1}, x_n, y_{n-1}, and y_n, and have the two edges $x_n y_{n-1}$ and $x_{n-1} y_n$ as part of a perfect matching. If we remove these four vertices and two edges from the graph, along with the edges $x_n y_n, y_{n-1} y_n, x_{n-1} y_{n-1}, x_{n-2} y_{n-1}, x_{n-1} y_{n-2}$, and $y_{n-2} y_{n-1}$, which we can no longer use, then the resulting graph—in this case, G_{n-2}—provides m_{n-2} possible perfect matchings.

Consequently, since the considerations in (1) and (2) cover all possibilities and have nothing in common, it follows that

$$m_n = m_{n-1} + m_{n-2}, \quad n \geq 3, \; m_1 = 1, \; m_2 = 2,$$

so

$$m_n = F_{n+1}, \quad n \geq 1.$$

What about the graph in Fig. 12.4(e)? This graph is called the *ladder graph on n rungs*. If we let l_n count the number of perfect matchings for this graph (on n rungs), again we have two considerations:

1. If we use edge $x_n y_n$ to account for vertex x_n (and vertex y_n), then we cannot use edges $x_{n-1} x_n$ and $y_{n-1} y_n$. So to account for the other $2n - 2$ vertices, we need a perfect matching for the resulting ladder graph on $n - 1$ rungs. There are l_{n-1} perfect matchings in this case.

2. If we do not use the edge $x_n y_n$, then the only way left to account for the vertex x_n and the vertex y_n is to use the edges $x_{n-1}x_n$ and $y_{n-1}y_n$. Then we cannot use any of the edges $x_{n-2}x_{n-1}$, $x_{n-1}y_{n-1}$, $y_{n-2}y_{n-1}$, and $x_n y_n$. Upon removing all four vertices and all six edges mentioned above, we arrive at the ladder graph on $n-2$ rungs, which has l_{n-2} perfect matchings. With $l_1 = 1$ and $l_2 = 2$, we once again arrive at a familiar sight—namely,

$$l_n = l_{n-1} + l_{n-2}, \quad n \geq 3, \ l_1 = 1, \ l_2 = 2,$$

and

$$l_n = F_{n+1}, \quad n \geq 1.$$

The reader interested in graph theory should, hopefully, find these examples dealing with perfect matchings worthwhile. But suppose we want to see if there is some place, beyond graph theory, where this idea of a perfect matching arises. To this end, we present the following example developed in Reference [23]. Further material on this idea is given in References [29, 55].

Example 12.3: Start by considering the graph H_n consisting of a zigzag arrangement made up of n regular hexagons—as shown in Fig. 12.5, for the cases of $n = 1, 2$, and 3. We see that the graph H_1 has 6 (= 4 + 2) vertices and 6 (= 5 + 1) edges. The graph H_2 can be obtained from that of H_1 by adding on five new edges and four new vertices, as shown in Fig. 12.5(b). Thus H_2 has 6 + 4 (= 10 = 4 · 2 + 2) vertices and 6 + 5 (= 11 = 5 · 2 + 1) edges. Following suit, the graph H_3 has 14 (= 4 · 3 + 2) vertices and 16 (= 5 · 3 + 1) edges. In general, the graph H_n has $4n + 2$ vertices and $5n + 1$ edges, and it represents the molecular graph of a *catafusene* (more formally, a cata-condensed benzenoid polycyclic hydrocarbon whose molecular formula is given by $C_{4n+2}H_{2n+4}$, where **C** is for carbon and **H** is for hydrogen). When $n = 1$, the catafusene C_6H_6 is called *benzene* and appears as in Fig. 12.6 (a). Note how in Fig. 12.6 (b) we have a second drawing of benzene—different according to where we have located the double bonds for the carbon atoms. If we are not permitted to rotate one figure so that it looks like the other, then we would consider these configurations

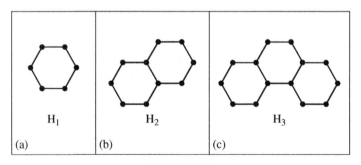

FIGURE 12.5

H
|
H˗C=C˗H
C   C
| ||
C   C
H˗C˗C˗H
|
H
(a)

H
|
H˗C˗C=C˗H
C
|| |
C   C
H˗C˗C=C˗H
|
H
(b)

The Two Kekulé Structures for Benzene

FIGURE 12.6

as distinct. From Reference [44, Pp. 187–188], we learn that these configurations (the two *Kekulé* structures for benzene) are the most stable valence-bond structures that can be written for this known hexagonal planar configuration. (Modern quantum mechanical theory suggests that the two *distinct* structures of benzene are *in resonance* and the actual structure of benzene is a *hybrid* of the proposed Kekulé structures.)

Now we shall extend what we have learned about benzene and look into the Kekulé structures of the larger catafusenes. We let k_n count the number of Kekulé structures for the catafusene $C_{4n+2}H_{2n+4}$. And now it is time to recognize that k_n is also the number of perfect matchings for the graph H_n—each such perfect matching made up of $(1/2)(4n + 2) = 2n + 1$ edges. We know that $k_1 = 2$. For $n = 2$, the catafusene $C_{10}H_8$ is called *naphthalene*. It has three Kekulé structures, for there are three perfect matchings for the graph H_2, as shown in Fig. 12.7. Consequently, $k_2 = 3$.

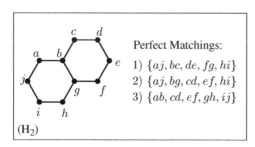

Perfect Matchings:
1) $\{aj, bc, de, fg, hi\}$
2) $\{aj, bg, cd, ef, hi\}$
3) $\{ab, cd, ef, gh, ij\}$

FIGURE 12.7

For the general formula, consider the zigzag arrangement of n regular hexagons as shown in Fig. 12.8 on p.86. (Here the arrangement is for the case where n is odd. However, the same argument follows for when n is even.) As we have seen throughout many of our other examples, there are two cases to consider:

Case 1: If we do not use the edge vw, then we are forced into using the edge uv (to account for the vertex v) and the edge xw (to account for the vertex w). Upon removing the five edges tu, uv, vw, wx, and xy, along with the four vertices u, v, w, and x, we no longer have the hexagon marked with $(*)$—but we still have the edge ty, along with the vertices t and y. So at this point we

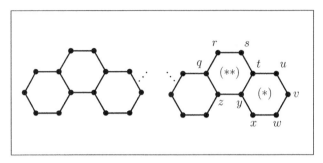

FIGURE 12.8

have reduced the problem to finding a perfect matching in the graph H_{n-1}. Consequently, there are k_{n-1} perfect matchings in this case.

Case 2: This time we will use the edge vw. Now we are faced with having to use the edge tu (to account for the vertex u) and the edge yx (to account for the vertex x). Furthermore, at this point, the edge ty cannot be used. So the only way we can account for the vertex s is to use the edge rs. Starting with the graph H_n, after we remove the 10 edges $qr, rs, st, tu, uv, vw, wx, xy, yz$, and yt, along with the eight vertices r, s, t, u, v, w, x, and y, we still have the edge qz, along with the vertices q and z. At this point the two hexagons marked with $(*)$ and $(**)$ have been removed. To complete the perfect matching we have developed up to this point, we must now determine a perfect matching for H_{n-2}. There are k_{n-2} perfect matchings available in this case.

Since the situations in these two cases cover all possibilities and have nothing in common, it follows that

$$k_n = k_{n-1} + k_{n-2}, \quad n \geq 3, \ k_1 = 2, \ k_2 = 3,$$

so

$$k_n = F_{n+2}, \quad n \geq 1.$$

[In particular, the catafusene *phenanthrene* $C_{14}H_{10}$ (where $n = 3$) has $F_5 = 5$ Kekulé structures, the catafusene *chrysene* $C_{18}H_{12}$ (where $n = 4$) has $F_6 = 8$ Kekulé structures, and the catafusene *picene* $C_{22}H_{14}$ (where $n = 5$) has $F_7 = 13$ Kekulé structures.]

Our next example deals with a special type of undirected graph called a tree.

Example 12.4:   Up to this point, all of the undirected graphs we have studied, have been made up of one piece. In Fig. 12.9 (i) we have an example of an undirected graph that is made up of two pieces, which are called *components*. The undirected graph in Fig. 12.9 (ii) has three components. Thus we say that the undirected graphs in Fig. 12.9

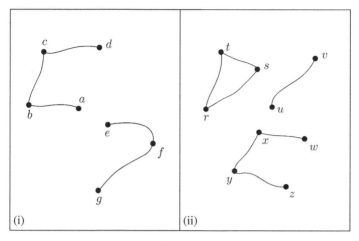

FIGURE 12.9

are *disconnected.* On the other hand, when an undirected graph G is *connected*, then for any vertices v and w in G, there is a path from v to w (as well as a path from w to v).

Each of the graphs in Fig. 12.10 is connected. The graph in Fig. 12.10 (i) has a loop at vertex f and a cycle of length 4, made up from the edges be, ed, dc, and cb. Meanwhile, the graphs in parts (ii) and (iii) of Fig. 12.10 are connected and have no loops or cycles. Under these circumstances, we call these two graphs *trees*. The tree in Fig. 12.10 (iii) is also a path. Note that the tree in Fig. 12.10 (ii) has six vertices and five edges, while the tree in Fig. 12.10 (iii) has four vertices and three edges. In general, if the undirected graph G is a tree with vertex set V and edge set E, then $|V| = |E| + 1$.

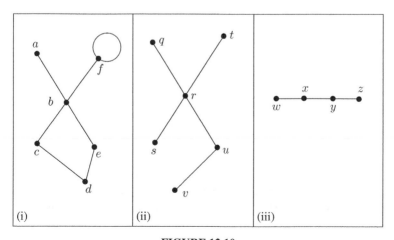

FIGURE 12.10

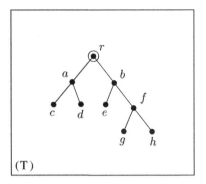

FIGURE 12.11

At times we want to have some extra structure in a tree. The tree T shown in Fig. 12.11 is called a *rooted binary tree* with the circled vertex r as the *root*. We call vertex a the *left child* of r and vertex b the *right child*. Likewise, vertex c is the left child of vertex a, while vertex h is the right child of vertex f. The graph made up of the vertices a, c, and d and the edges ac and ad form a *subgraph* of T that is a tree. This subgraph is called the *left subtree of T* (*rooted at a*). The vertices f, g, and h together with the edges fg and fh constitute the *right subtree of b*. Also, the vertex e by itself constitutes the left subtree of b. So a tree can consist of just one vertex (and no edges). Each of the vertices c, d, e, g, and h is incident with just one edge and is called a *leaf* of T. The other vertices—namely, r, a, b, and f—are called the *internal vertices* of T. The adjective *binary* tells us that each internal vertex has at most two children.

Now we shall examine a special collection of rooted binary trees.
The rooted *Fibonacci trees* T_n, $n \geq 1$, are defined recursively as follows:

(1) T_1 is the rooted tree consisting of only the root.
(2) T_2 is the same as T_1—it too is a rooted tree that consists of only one vertex.
(3) For $n \geq 3$, T_n is the rooted binary tree with T_{n-1} as its left subtree and T_{n-2} as its right subtree.

The first six rooted Fibonacci trees are shown in Fig. 12.12. For $n \geq 1$, we want to determine the following:
(1) l_n, the number of leaves in T_n; (2) v_n, the (total) number of vertices in T_n; (3) i_n, the number of internal vertices in T_n; and, (4) e_n, the number of edges in T_n.

(1) To determine l_n, we see from part (3) of the recursive definition of T_n that the number of leaves in T_n is the sum of the number of leaves in its left subtree T_{n-1} and the number of leaves in its right subtree T_{n-2}. Consequently, we find that

$$l_n = l_{n-1} + l_{n-2}, \quad n \geq 3, \ l_1 = 1, \ l_2 = 1,$$

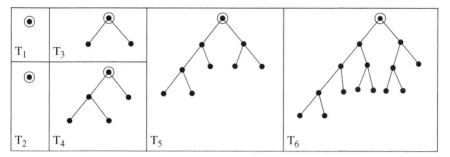

FIGURE 12.12

and

$$l_n = F_n.$$

(2) The number of vertices in T_n, however, is not quite the sum of the numbers of vertices in its left subtree T_{n-1} and its right subtree T_{n-2}. For we must also account for the root of T_n. Consequently, this leads to

$$v_n = v_{n-1} + v_{n-2} + 1, \quad n \geq 3, \ v_1 = 1, \ v_2 = 1.$$

This is an example of a linear *nonhomogeneous* recurrence relation (with initial conditions). Methods for solving such recurrence relations are given in References [11, 25, 47]. However, here we shall take a different approach. Examining the six Fibonacci trees in Fig. 12.12, we find the following:

$$v_1 = 1 = 2 \cdot 1 - 1 = 2F_1 - 1$$
$$v_2 = 1 = 2 \cdot 1 - 1 = 2F_2 - 1$$
$$v_3 = 3 = 2 \cdot 2 - 1 = 2F_3 - 1$$
$$v_4 = 5 = 2 \cdot 3 - 1 = 2F_4 - 1$$
$$v_5 = 9 = 2 \cdot 5 - 1 = 2F_5 - 1$$
$$v_6 = 15 = 2 \cdot 8 - 1 = 2F_6 - 1.$$

These results lead us to conjecture that for all $n \geq 1$, $v_n = 2F_n - 1$. To establish this conjecture, we shall use the Alternative, or Strong, form of the Principle of Mathematical Induction.

Proof: From the results we listed, we can see that the conjecture is certainly true for $n = 1$ and $n = 2$ (as well as $n = 3, 4, 5,$ and 6). This will establish the basis step for our inductive proof. Next we assume the result true for all positive integers $n = 1, 2,$

..., $k - 1, k \ (\geq 2)$. Now for $n = k + 1$, we have

$$
\begin{aligned}
v_n &= v_{k+1} \\
&= v_k + v_{k-1} + 1 \\
&= (2F_k - 1) + (2F_{k-1} - 1) + 1 \ \text{(By the induction hypothesis.)} \\
&= 2(F_k + F_{k-1}) - 1 \\
&= 2F_{k+1} - 1.
\end{aligned}
$$

Consequently, by the Alternative, or Strong, form of the Principle of Mathematical Induction, it follows that

$$
v_n = 2F_n - 1, \quad n \geq 1. \qquad \square
$$

(3)   Since $v_n = l_n + i_n$, for $n \geq 1$, we find that

$$
i_n = v_n - l_n = (2F_n - 1) - F_n = F_n - 1, \quad n \geq 1.
$$

(4)   Finally, as we mentioned earlier, the number of vertices in any tree is one more than the number of edges, so

$$
e_n = v_n - 1 = (2F_n - 1) - 1 = 2F_n - 2, \quad n \geq 1.
$$

More on rooted Fibonacci trees can be found in References [7, 27]. Reference [36] includes a proof that the edges of a rooted Fibonacci tree can be partitioned into two subsets which result in isomorphic graphs.

Example 12.5: In this example we introduce one of the many properties that arise in the study of graph theory. Let us start with the graph G_1 in Fig. 12.13. Here the vertex set $V = \{a, b, c, d, e\}$. If we now restrict our attention to only the vertices in $W = \{a, b, d\}$, we see that there is no edge in G_1 connecting any two of the vertices in W—that is, none of the possible edges ab, ad, or bd appear in the graph. This observation leads us to the following:

Definition 12.1: Let G be an undirected graph with vertex set V. A subset W of V is called *independent* if there is no edge in G with its vertices in W.

Consequently, when dealing with an independent set of vertices, no two of the vertices in the set can be neighbors.

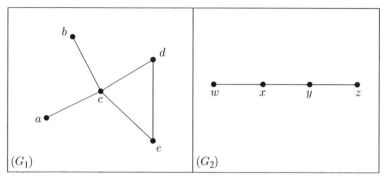

FIGURE 12.13

For the path graph G_2 of Fig. 12.13, we find the following independent sets of vertices.

Size of Independent Set	Independent Set(s)
0	$\varnothing$
1	$\{w\}, \{x\}, \{y\}, \{z\}$
2	$\{w, y\}, \{w, z\}, \{x, z\}$.

So G_2 has eight independent subsets of vertices—including $\varnothing$. Finally, we shall consider the path on n vertices shown in Fig. 12.14 and let i_n count the number of independent sets of vertices that exist for this graph. We find that $i_1 = 2$, for $\varnothing$ and $\{v_1\}$, while $i_2 = 3$, for $\varnothing$, $\{v_1\}$, and $\{v_2\}$. For $n \geq 3$, if we try to select an independent subset of $\{v_1, v_2, ..., v_{n-2}, v_{n-1}, v_n\}$, there are two cases for us to examine:

(1) If we do not use the vertex v_n, then we can use any of the i_{n-1} independent subsets of $\{v_1, v_2, ..., v_{n-2}, v_{n-1}\}$.

(2) If we do use the vertex v_n, then we cannot use the vertex v_{n-1}. But, in addition, we can then use any of the i_{n-2} independent subsets of $\{v_1, v_2, ..., v_{n-2}\}$.

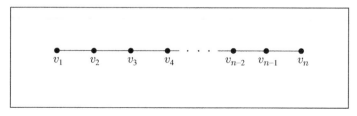

FIGURE 12.14

Since the subsets in these two cases cover all the possibilities and have nothing in common, it follows that

$$i_n = i_{n-1} + i_{n-2}, \quad n \geq 3, \ i_1 = 2, \ i_2 = 3,$$

so

$$i_n = F_{n+2}, \quad n \geq 1.$$

This result should not be very surprising! After all, except for a change in vocabulary, we have already come upon this result in two earlier situations:

(1) In Example 5.1, where we counted the number of subsets of $\{1, 2, 3, ..., n\}$ where there are no consecutive integers.
(2) In Example 8.1, where we determined the number of ways to place nontaking kings on a $1 \times n$ chessboard.

Should the reader feel cheated by these three examples that are really just one, let us consider something definitely new. Start by taking the path in Fig. 12.14 and adding the edge $v_n v_1$. The resulting undirected graph is the cycle on n vertices shown in Fig. 12.15 (a). Now we shall let c_n count the number of independent sets of vertices for $V = \{v_1, v_2, v_3, ..., v_{n-2}, v_{n-1}, v_n\}$ in this cycle. Again we consider two cases:

(i) Suppose we do not use the vertex v_n. Then we remove from the cycle the vertex v_n along with the edges $v_{n-1} v_n$ and $v_1 v_n$. Left behind is a path on the $n - 1$ vertices: $v_1, v_2, v_3, ..., v_{n-2}, v_{n-1}$. From our prior work with paths, we know that this undirected graph has $i_{n-1} = F_{(n-1)+2} = F_{n+1}$ independent sets of vertices.
(ii) On the other hand, if we do use the vertex v_n, then we cannot use either of the vertices v_1 or v_{n-1}. Upon removing the three vertices v_1, v_{n-1}, and v_n, along with the four edges $v_1 v_2$, $v_1 v_n$, $v_{n-2} v_{n-1}$, and $v_{n-1} v_n$, from the given cycle, this time we arrive at the path on the $n - 3$ vertices: $v_2, v_3, ..., v_{n-2}$. This path determines $i_{n-3} = F_{(n-3)+2} = F_{n-1}$ independent sets of vertices for the cycle.

Cases (i) and (ii) cover all the possibilities and have nothing in common, so

$$c_n = i_{n-1} + i_{n-3} = F_{n+1} + F_{n-1}, \quad n \geq 4.$$

Consequently, we can write

$$c_{n-1} = F_n + F_{n-2}$$
$$c_{n-2} = F_{n-1} + F_{n-3}.$$

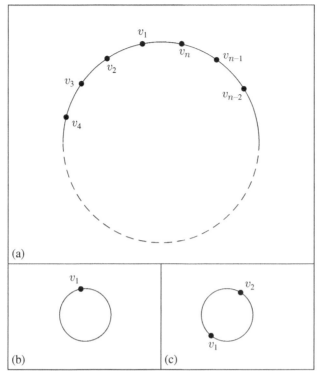

FIGURE 12.15

From this it follows that

$$c_{n-1} + c_{n-2} = (F_n + F_{n-2}) + (F_{n-1} + F_{n-3})$$
$$= (F_n + F_{n-1}) + (F_{n-2} + F_{n-3})$$
$$= F_{n+1} + F_{n-1}$$
$$= c_n.$$

So we now have the recurrence relation $c_n = c_{n-1} + c_{n-2}$, which is exactly like the recurrence relation for the Fibonacci numbers. Consequently, we get the same characteristic equation—namely, $r^2 - r - 1 = 0$—and the same characteristic roots: $(1 + \sqrt{5})/2 = \alpha$ and $(1 - \sqrt{5})/2 = \beta$. Therefore, $c_n = k_1\alpha^n + k_2\beta^n$, where k_1 and k_2 are constants. But what about the initial conditions? It should not be surprising that the initial conditions here will be different from those for the Fibonacci numbers.

Although cycles are generally defined when there are at least three edges, here we shall allow the cases for when there is only one edge (for a loop) or two edges (for a *circuit* in a *multigraph*). When $n = 1$, we see from Fig. 12.15 (b) that for this loop there is only one independent set—namely, $\varnothing$—since v_1 is a neighbor of itself. So $c_1 = 1$. For $n = 2$, consider the circuit on the vertices v_1, v_2 in Fig. 12.15 (c).

Here there are three independent sets: $\varnothing$, $\{v_1\}$, and $\{v_2\}$. So $c_2 = 3$. Consequently, we arrive at

$$c_n = c_{n-1} + c_{n-2}, \quad n \geq 3, \ c_1 = 1, \ c_2 = 3.$$

From

$$1 = c_1 = k_1\alpha + k_2\beta \quad \text{and} \quad 3 = c_2 = k_1\alpha^2 + k_2\beta^2,$$

it follows that

$$\alpha = k_1\alpha^2 + k_2\alpha\beta = k_1\alpha^2 - k_2, \quad \text{since } \alpha\beta = -1.$$

Subtracting $\alpha = k_1\alpha^2 - k_2$ from $3 = k_1\alpha^2 + k_2\beta^2$, we arrive at

$$3 - \alpha = k_2\beta^2 + k_2 = (\beta^2 + 1)k_2 = (\beta + 2)k_2, \quad \text{because } \beta^2 = \beta + 1.$$

So

$$3 - \alpha = (3 - \alpha)k_2, \quad \text{as } \alpha + \beta = 1,$$

and

$$k_2 = 1, \quad \text{since } 3 - \alpha \neq 0.$$

And now

$$1 = k_1\alpha + k_2\beta \Rightarrow 1 = k_1\alpha + \beta \Rightarrow k_1\alpha = 1 - \beta = \alpha \Rightarrow k_1 = 1, \quad \text{as } \alpha \neq 0.$$

Hence,

$$c_n = \alpha^n + \beta^n, \quad n \geq 1.$$

What we have here is the number sequence known as the Lucas (pronounced, Lucah) numbers—sort of the first cousin of the Fibonacci numbers. Here "Lucas" refers to the same renowned French number theorist François Edouard Anatole Lucas, who was mentioned in Chapter 2. We denote these numbers by L_n, for $n \geq 0$, and define them recursively by

$$L_n = L_{n-1} + L_{n-2}, \quad n \geq 2, \ L_0 = 2, \ L_1 = 1.$$

(Here we have extended this number sequence to include the case for $n = 0$.)
    The formula

$$L_n = \alpha^n + \beta^n, \quad n \geq 0,$$

is called the *Binet form for L_n*. These numbers arise in other examples and possess a great number of interesting properties. In the next chapter, we shall look into some of these examples and properties. In the meantime, we provide a list of the first 25 Lucas numbers in Table 12.1.

TABLE 12.1

$L_0 = 2$	$L_5 = 11$	$L_{10} = 123$	$L_{15} = 1364$	$L_{20} = 15,127$
$L_1 = 1$	$L_6 = 18$	$L_{11} = 199$	$L_{16} = 2207$	$L_{21} = 24,476$
$L_2 = 3$	$L_7 = 29$	$L_{12} = 322$	$L_{17} = 3571$	$L_{22} = 39,603$
$L_3 = 4$	$L_8 = 47$	$L_{13} = 521$	$L_{18} = 5778$	$L_{23} = 64,079$
$L_4 = 7$	$L_9 = 76$	$L_{14} = 843$	$L_{19} = 9349$	$L_{24} = 103,682$

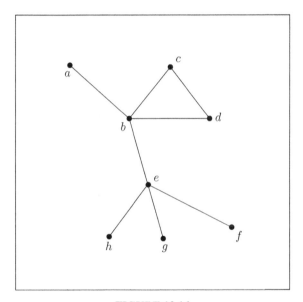

FIGURE 12.16

Example 12.6: Consider the undirected graph in Fig. 12.16, where $V = \{a, b, c, d, e, f, g, h\}$ and $E = \{ab, bc, bd, be, cd, ef, eg, eh\}$. In this graph the subsets $V_1 = \{a, d, f\}$ and $V_2 = \{b, f, g, h\}$ are both independent sets of vertices. However, there is an important difference between these two independent sets. Notice that we could enlarge V_1 to the subset $\{a, d, f, g\}$ which is still independent. For V_2, however, any attempt to add another vertex to V_2 results in a subset of V that is no longer independent. This example leads us now to the following idea.

Definition 12.2: Let G be an undirected graph with vertex set V and edge set E and no loops. An independent subset W of V is called *maximal independent* if for each $v \in V$ where $v \notin W$, the vertex set $W \cup \{v\}$ is not independent.

When W is a maximal independent subset of V, then for each vertex $u \in V$ we see that (i) if $u \in W$, then u has no neighbor in W; and, (ii) if $u \notin W$, then u has a neighbor in W.

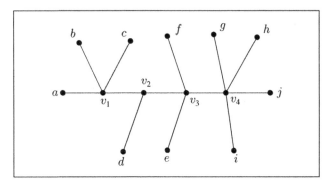

FIGURE 12.17

Now we shall consider a special type of tree called a *caterpillar*. The tree in Fig. 12.17 is an example of a caterpillar. The vertices v_1, v_2, v_3, and v_4 together with the edges v_1v_2, v_2v_3, and v_3v_4 constitute a path that forms the *spine* of the caterpillar. Every other vertex of the caterpillar is a neighbor of one (and only one) of the vertices on the spine. Note that the vertices a and j, and the edges av_1 and v_4j, are *not* part of the spine. Each of the edges of the caterpillar that is not part of the spine is called a *hair* of the caterpillar. In Table 12.2, we have listed some of the maximal independent sets of vertices for the caterpillar in Fig. 12.17, along with the vertices (if any) on the spine that occur in the maximal independent set.

TABLE 12.2

Maximal Independent Subsets W	Spine Vertices in W
$\{v_1, d, e, f, v_4\}$	v_1, v_4
$\{a, b, c, v_2, e, f, v_4\}$	v_2, v_4
$\{v_1, d, e, f, g, h, i, j\}$	v_1
$\{a, b, c, d, e, f, v_4\}$	v_4
$\{a, b, c, d, e, f, g, h, i, j\}$	There are none.

We realize from the results in Table 12.2 that we cannot select vertices on the spine that are neighbors. Furthermore, if we do not select a vertex v on the spine, then we must select all the vertices, not on the spine, that are neighbors of v. Consequently, a maximal independent set of vertices for this caterpillar is determined by a subset of vertices on the spine with no consecutive subscripts. As we learned in Example 5.1, this is $8 = F_6$.

In general, for $n \geq 1$, let us consider a caterpillar whose spine is a path made up of n vertices (and $n - 1$ edges). We label the vertices along the spine as v_1, v_2, v_3, ..., v_{n-1}, v_n, so that the edges of the spine are v_1v_2, v_2v_3, ..., $v_{n-1}v_n$. Furthermore, we assume that there is at least one hair for each vertex on the spine. (So there are no *bald* spots.) Then the number of maximal independent sets of vertices for this caterpillar is the number of subsets of $\{1, 2, 3, ..., n-1, n\}$ with no consecutive integers. From Example 5.1 this number is F_{n+2}.

EXERCISES FOR CHAPTER 12

1. Consider the ladder graph in Fig. 12.4 (e). For $n = 10$, the ladder has 10 rungs.
   (a) How many perfect matchings are there for the ladder graph with 10 rungs?
   (b) How many of the perfect matchings in part (a) contain the edge $x_4 y_4$?
2. Let n be an even integer with $n \geq 4$.
   (a) How many perfect matchings are there for the path with vertex set $\{v_1, v_2, v_3, ..., v_{n-1}, v_n\}$ and edge set $\{v_1 v_2, v_2 v_3, ..., v_{n-1} v_n\}$?
   (b) How many perfect matchings are there for the cycle with vertex set $\{v_1, v_2, v_3, ..., v_{n-1}, v_n\}$ and edge set $\{v_1 v_2, v_2 v_3, ..., v_{n-1} v_n, v_n v_1\}$?
3. The graph in Fig. 12.4 (a) is often referred to as the *comb graph* with n *teeth*. Set up and solve a recurrence relation with initial condition(s) to count the number of independent sets of vertices for this graph.
4. Consider the graph in Fig. 12.18. Set up and solve a recurrence relation with initial condition(s) to count the number of perfect matchings for this graph.
5. Set up and solve a recurrence relation with initial condition(s) to count the number of independent sets of vertices for the graph in Fig. 12.18.

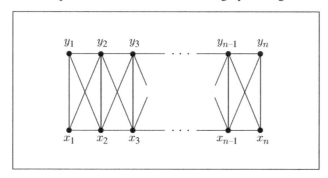

FIGURE 12.18

6. For the caterpillar in Fig. 12.19 on p. 98, why is $\{a, b, c, v_2, v_6, g, h, i\}$ not a maximal independent set of vertices? Does this contradict the result in Example 12.6?
7. For $n \geq 0$, prove that $\gcd(L_n, L_{n+1}) = 1$. (This is comparable to Property 4.1.)
8. For $n \geq 0$, prove that $\gcd(L_n, L_{n+2}) = 1$. (This is comparable to Property 4.2.)
9. Provide an example to show that $\gcd(L_n, L_{n+3}) \neq 1$ for some $n \geq 0$.
10. For $n \geq 0$, prove that $\sum_{r=0}^{n} L_r = L_{n+2} - 1$. What is $\sum_{r=1}^{n} L_r$? (This is comparable to Property 4.5.)

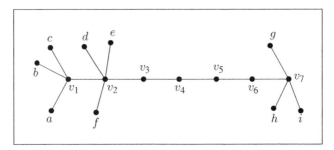

FIGURE 12.19

11. For $n \geq 0$, prove that $\sum_{r=0}^{n} L_r^2 = L_n L_{n+1} + 2$. What is $\sum_{r=1}^{n} L_r^2$? (This is comparable to Property 4.6.)

12. For $n \geq 1$, prove that $\sum_{r=1}^{n} L_{2r-1} = L_{2n} - 2$. (This is comparable to Property 4.7.)

13. For $n \geq 0$, prove that $\sum_{r=0}^{n} L_{2r} = L_{2n+1} + 1$. What is $\sum_{r=1}^{n} L_{2r}$? (This is comparable to Property 4.8.)

14. For $n \geq 1$, prove that $L_{n-1} L_{n+1} - L_n^2 = 5(-1)^{n-1}$. (This is comparable to Cassini's identity in Property 4.9.)

15. Use the Principle of Mathematical Induction to prove that for $n \geq 1$,

$$\sum_{r=1}^{n} r L_r = L_1 + 2L_2 + 3L_3 + \cdots + nL_n = nL_{n+2} - L_{n+3} + 4.$$

(This formula is an example of a *weighted sum* involving the Lucas numbers.)

16. For $n \geq 0$, prove that $\sum_{k=0}^{n} \binom{n}{k} L_k = L_{2n}$.

17. For $n \geq 0$, prove that

$$\sum_{k=0}^{n} \binom{n}{k} L_k L_{n-k} = 2^n L_n + 2.$$

18. Determine $\lim_{n \to \infty} (F_n / L_n)$.

19. Determine $\lim_{n \to \infty} (L_{n+1} / L_n)$.

20. For $k \, (\geq 1)$ fixed, determine $\lim_{n \to \infty} (F_{n+k} / L_n)$. (T. P. Dence, 1968) [18].

21. For $k \, (\geq 1)$ fixed, determine $\lim_{n \to \infty} (L_{n+k} / F_n)$. (T. Koshy, 1998) [37].

22. For $n \geq 0$, prove that $\sum_{k=0}^{n} (-1)^{n-k} \binom{n}{k} L_{2k} = L_n$.

23. For $n \geq 0$, prove that $L_n^2 + L_{n+1}^2 = L_{2n} + L_{2n+2}$. (T. Koshy, 1999)

24. For $n \geq 0$, prove that

$$L_{4n} = L_{2n}^2 - 2.$$

25. For $n \geq 0$, prove that

$$\left(L_n^2 + L_{n+1}^2 + L_{n+2}^2\right)^2 = 2(L_n^4 + L_{n+1}^4 + L_{n+2}^4).$$

26. For $n \geq 0$, prove that

$$L_{3n} = L_n[L_{2n} - (-1)^n].$$

27. For $n \geq 0$, prove that L_{3n} is even.
28. For $n \geq 1$, prove that $L_n - 3L_{n-1}$ is divisible by 5.
29. For $n \geq 0$, $m\ (\geq 0)$ fixed, prove that

$$\sum_{i=0}^{n} \binom{n}{i} L_{mi} L_{mn-mi} = 2^n L_{mn} + 2L_m^n.$$

30. For $n \geq 0$, prove that

$$\sum_{i=0}^{n} (-1)^{n-i} \binom{n}{i} L_i = (-1)^n L^n.$$

CHAPTER 13

The Lucas Numbers: Further Properties and Examples

At this point we have seen many examples where the Fibonacci numbers arise—and there are still more to come. In addition, in the text and, especially, in the exercises, we found that there are numerous properties that this number sequence exhibits. The Lucas numbers likewise exhibit many interesting properties. In fact, now that we are aware of both of these number sequences, we find that there are some properties inter-relating the two sequences. Our next example will provide some of these properties. Many more appear in the exercises for this chapter.

Example 13.1: Previously, in Example 12.5, we let i_n count the number of independent sets of vertices for a path on n vertices and c_n the number of independent sets of vertices for a cycle on n vertices. In that example we learned that

$$c_n = i_{n-1} + i_{n-3}, \quad n \geq 4.$$

Since c_n turned out to be the Lucas number L_n and $i_n = F_{n+2}$, this result translates into the following:

Property 13.1:
$$L_n = F_{n+1} + F_{n-1}, \quad n \geq 1.$$

And from this the next property follows.

Property 13.2:
$$\begin{aligned} L_n &= F_{n+1} + F_{n-1} \\ &= (F_{n+2} - F_n) + (F_n - F_{n-2}) \\ &= F_{n+2} - F_{n-2}, \quad n \geq 2. \end{aligned}$$

This second property can be used to provide additional examples where the Lucas numbers arise:

Fibonacci and Catalan Numbers: An Introduction, First Edition. Ralph P. Grimaldi.
© 2012 John Wiley & Sons, Inc. Published 2012 by John Wiley & Sons, Inc.

(i) In Example 6.2 we learned that the number of compositions of n, where the only summands allowed are 1's and 2's, is F_{n+1}. Of these F_{n+1} compositions of n, the number that start and end with the summand 2 equals the number of compositions of $n - 4$, using only 1's and 2's as summands. That number is F_{n-3}. Consequently, the number of compositions of n (using only 1's and 2's as summands) that do not start and end with the summand 2 is the same as the number of compositions of n (using only 1's and 2's as summands) where 1 is the first or last summand. This number is

$$F_{n+1} - F_{n-3} = F_{(n-1)+2} - F_{(n-1)-2} = L_{n-1}.$$

Likewise, the number of these F_{n+1} compositions that do not start with "1 + 1" and do not end with "1 + 1" is L_{n-1}. In addition, those compositions of n that do not start and end with the summand 1 number $F_{n+1} - F_{n-1} = F_n$.

(ii) In Example 6.4, the number of paths from the starting cell S to cell n was found to be F_{n+1}. Here we find that

   (1) L_{n-1} of these paths do not start with the horizontal move Cell $S \rightarrow$ Cell 2 and do not end with the horizontal move Cell $(n - 2) \rightarrow$ Cell n.

   (2) For n even, there are L_{n-1} paths that do not start with Cell $S \searrow$ Cell 1 $\nearrow$ Cell 2 and do not end with Cell $(n - 2) \searrow$ Cell $(n - 1) \nearrow$ Cell n.

   (3) For n odd, there are L_{n-1} paths that do not start with Cell $S \searrow$ Cell 1 $\nearrow$ Cell 2 and do not end with Cell $(n - 2) \nearrow$ Cell $(n - 1) \searrow$ Cell n.

   (4) For n even, there are F_n paths that do not start with Cell $S \searrow$ Cell 1 and end with Cell $(n - 1) \nearrow$ Cell n.

   (5) For n odd, there are F_n paths that do not start with Cell $S \searrow$ Cell 1 and end with Cell $(n - 1) \searrow$ Cell n.

(iii) For the tilings of the $2 \times n$ chessboard in Example 7.1, we find the following:

   (1) The number of tilings that do not have two (horizontal) dominoes covering the four squares in columns 1 and 2 and two (horizontal) dominoes covering the four squares in columns $n - 1, n$ is $F_{n+1} - F_{n-3} = L_{n-1}$.

   (2) The number of tilings that do not have two (vertical) dominoes covering the four squares in columns 1 and 2 and two (vertical) dominoes covering the four squares in columns $n - 1, n$ is L_{n-1}.

   (3) The number of tilings that do not have a (vertical) domino covering the two squares in each of columns 1 and n is F_n.

(iv) Comparable results take place for the tilings of the $1 \times n$ chessboard in Example 7.2:

   (1) The number of tilings of the $1 \times n$ chessboard that do not lead off and end with a 1×2 rectangular tile is L_{n-1}.

   (2) The number of tilings that do not lead off and end with two 1×1 square tiles is L_{n-1}.

(3) The number of tilings that do not lead off with and end with a 1×1 square tile is F_n.

(v) Finally, let us examine the perfect matchings for the ladder graph in Fig. 12.4 (e) of Example 12.2. Here we find that

(1) The number of perfect matchings that exclude the pair of edges $x_1 y_1$, $x_2 y_2$ and exclude the pair of edges $x_{n-1} y_{n-1}$, $x_n y_n$ is L_{n-1}.

(2) The number of perfect matchings for the ladder graph with n rungs that exclude the pair of edges $x_1 x_2$, $y_1 y_2$ and exclude the pair of edges $x_{n-1} x_n$, $y_{n-1} y_n$ is L_{n-1}.

(3) The number of perfect matchings that exclude both of the edges $x_1 y_1$ and $x_n y_n$ is F_n.

In Examples 13.3 and 13.4 we shall examine two special types of partially ordered sets (also called posets)—namely, the fence and the closed fence, or corral. Before doing so, however, we review the following background material.

Definition 13.1: For a set A, preferably not empty, any subset of $A \times A$ is called a (*binary*) *relation* on A. If $\mathcal{R}$ is a relation on a set A—that is, $\mathcal{R} \subseteq A \times A$—then if $(x, y) \in \mathcal{R}$, we may also write $x\mathcal{R}y$, and in either case say that "x is related to y" (under $\mathcal{R}$). Furthermore, we say that

(i) $\mathcal{R}$ is *reflexive* if $(a, a) \in \mathcal{R}$ (or $a\mathcal{R}a$) for all $a \in A$.

(ii) $\mathcal{R}$ is *antisymmetric* if for all $a, b \in A$, $[(a, b) \in \mathcal{R}$ and $(b, a) \in \mathcal{R}] \Rightarrow a = b$.

(iii) $\mathcal{R}$ is *transitive* if for all $a, b, c \in A$, $[(a, b) \in \mathcal{R}$ and $(b, c) \in \mathcal{R}] \Rightarrow (a, c) \in \mathcal{R}$.

Definition 13.2: If $\mathcal{R}$ is a relation on A, then $\mathcal{R}$ is called a *partial order* if $\mathcal{R}$ is reflexive, antisymmetric, and transitive. In this case, we say that $(A, \mathcal{R})$—the set A with the partial order $\mathcal{R}$—is a *partially ordered set* or *poset*.

Example 13.2: Consider the set A of all positive (integer) divisors of 12. This is often denoted by $D(12)$, where $D(12) = \{1, 2, 3, 4, 6, 12\}$. We define the relation $\mathcal{R}$ on $D(12)$ by $(x, y) \in \mathcal{R}$, or $x\mathcal{R}y$, when x (exactly) divides y. So $\mathcal{R} = \{(1, 1), (1, 2), (1, 3), (1, 4), (1, 6), (1, 12), (2, 2), (2, 4), (2, 6), (2, 12), (3, 3), (3, 6), (3, 12), (4, 4), (4, 12), (6, 6), (6, 12), (12, 12)\}$. We see that $\mathcal{R}$ is a poset and realize that the antisymmetric property follows because we have restricted ourselves to *positive* integer divisors.

We can represent a relation $\mathcal{R}$ on a set A by means of a *directed graph*. The elements of A provide the vertices of the graph. For $x, y \in A$, if $(x, y) \in \mathcal{R}$—that is, $x\mathcal{R}y$—then we draw the directed edge $\overrightarrow{xy}$ from the vertex x to the vertex y. For the poset $D(12)$ with its partial order $\mathcal{R}$, as given above, when $c, d \in D(12)$ and c (exactly) divides d, then $c\mathcal{R}d$ and we draw the directed edge $\overrightarrow{cd}$. The resulting directed graph for this partial order is shown in Fig. 13.1(a). Since a partial order is reflexive, there is a loop at each vertex. Next, due to the antisymmetric property, for (vertices) x,

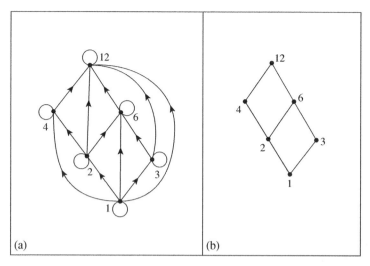

FIGURE 13.1

$y \in D(12)$, if $x \neq y$, then the directed graph cannot contain both the directed edge $\overrightarrow{xy}$ and the directed edge $\overrightarrow{yx}$. Finally, this directed graph demonstrates the transitive property: for (vertices) $x, y \in D(12)$, if there is a directed *path* from x to y, then the directed edge $\overrightarrow{xy}$ also occurs in the graph.

Unfortunately, this directed graph is somewhat cluttered. To simplify the situation, we introduce the concept of the *Hasse diagram* for a (finite) partial order. In Fig. 13.1(b) we find the Hasse diagram for the partial order in part (a) of the figure. How did we arrive at this result? Since we know that a partial order is reflexive, we do not bother to place a loop at each of the vertices in $D(12)$. Also, whenever we have a directed path from x to y in a Hasse diagram, the transitive property for a partial order tells us that we must also have the directed edge $\overrightarrow{xy}$ in our directed graph. For the Hasse diagram, however, we simply acknowledge that the edge exists but do not bother to include it in the diagram. Finally, we adopt the convention that the edges are all directed upward, but do not bother to direct the edges in the diagram. As a result, the Hasse diagram in Fig. 13.1(b) is certainly a simpler and less cluttered counterpart to the directed graph in Fig. 13.1(a).

At this point we need to introduce one final idea.

Definition 13.3: If $(A, \mathcal{R})$ is a partial order, a subset I of A is called an *order ideal* (or *semiorder*) of $(A, \mathcal{R})$, if for all $x, y \in A$, when $x \in I$ and $y\mathcal{R}x$, then $y \in I$.

Now for a new instance where the Lucas numbers arise!

Example 13.3: We shall start with the Hasse diagrams for the three partial orders in Fig. 13.2 on p. 104. These partial orders are examples of *fences* or *zigzags*. We shall denote them by $(Z_n, \mathcal{R})$, where n is the number of vertices in the Hasse diagram.

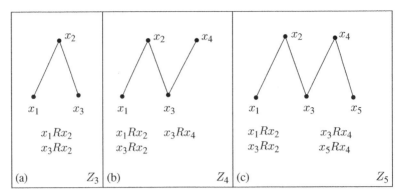

FIGURE 13.2

The relation $\mathcal{R}$ is determined as follows: If x_k and x_{k+1} are in Z_n, then (1) $x_k \mathcal{R} x_{k+1}$ when k is odd; and (2) $x_{k+1} \mathcal{R} x_k$ when k is even. For $(Z_3, \mathcal{R})$ and $(Z_4, \mathcal{R})$, we find the following order ideals:

Partial Order	Order Ideals
$(Z_3, \mathcal{R})$	$\varnothing$, $\{x_1\}$, $\{x_3\}$, $\{x_1, x_3\}$, $\{x_1, x_2, x_3\}$
$(Z_4, \mathcal{R})$	$\varnothing$, $\{x_1\}$, $\{x_3\}$, $\{x_1, x_3\}$, $\{x_1, x_2, x_3\}$
	$\{x_3, x_4\}$, $\{x_1, x_3, x_4\}$, $\{x_1, x_2, x_3, x_4\}$

To determine the number of order ideals for $(Z_5, \mathcal{R})$, we have two options:

(1) We use x_5: Here we simply take any order ideal for $(Z_4, \mathcal{R})$ and add in x_5.
(2) We do not use x_5: Now we cannot use x_4, because for any order ideal I of $(Z_5, \mathcal{R})$,

$$[x_4 \in I \text{ and } x_5 \mathcal{R} x_4] \Rightarrow x_5 \in I.$$

So in this case, we are left with any of the order ideals for $(Z_3, \mathcal{R})$.

Consequently, the partial order $(Z_5, \mathcal{R})$ has 8 ($= F_6$) + 5 ($= F_5$) = 13 ($= F_7$) order ideals.

Before we try to immediately generalize this result, a little more caution is needed. So let us consider the partial order $(Z_6, \mathcal{R})$, as shown in Fig. 13.3. As we try to count the number of order ideals for this partial order, once again there are two cases to consider. There is, however, a difference.

(1) Suppose that I is an order ideal of $(Z_6, \mathcal{R})$ and that $x_6 \in I$. Then $x_5 \in I$ and we can add these two elements to any of the order ideals of $(Z_4, \mathcal{R})$.

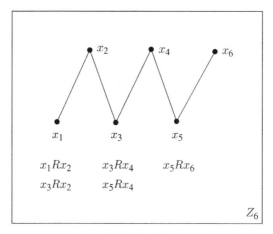

FIGURE 13.3

(2) On the other hand, suppose that I is an order ideal of $(Z_6, \mathcal{R})$ and that $x_6 \notin I$. Then I can be any of the order ideals of $(Z_5, \mathcal{R})$.

Therefore, the partial order $(Z_6, \mathcal{R})$ has $8 \, (= F_6) + 13 \, (= F_7) = 21 \, (= F_8)$ order ideals.

Note how the results in (1) and (2) for $(Z_5, \mathcal{R})$ are the opposite of those in (1) and (2) for $(Z_6, \mathcal{R})$. However, we now see that the following is true whether n is even or odd. If we let o_n count the number of order ideals for the partial order $(Z_n, \mathcal{R})$, then

$$o_n = o_{n-1} + o_{n-2}, \quad n \geq 3, \ o_1 = 2, \ o_2 = 3,$$

and

$$o_n = F_{n+2}, \quad n \geq 1.$$

Turning our attention to the partially ordered fences in Fig. 13.4 on p. 106, we find the following:

	Order Ideals
Figure 13.4 (a)	$\varnothing, \{y_1\}, \{y_0, y_1\}$
Figure 13.4 (b)	$\varnothing, \{y_1\}, \{y_0, y_1\}, \{y_1, y_2\}, \{y_0, y_1, y_2\}$
Figure 13.4 (c)	$\varnothing, \{y_1\}, \{y_0, y_1\},$
	$\{y_3\}, \{y_1, y_3\}, \{y_0, y_1, y_3\}, \{y_1, y_2, y_3\}, \{y_0, y_1, y_2, y_3\}$

Here we see that the number of order ideals for the partial order in Fig. 13.4 (c) is $8 \, (= F_6)$, the sum of the numbers of order ideals for the partial orders in Fig. 13.4 (a) and Fig. 13.4 (b). Comparable to our preceding result for $\{x_1, x_2, \ldots, x_n\}$, here one

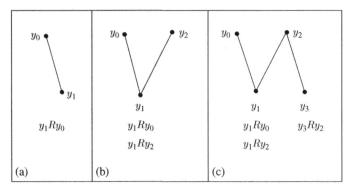

FIGURE 13.4

finds that the number of order ideals for $\{y_0, y_1, y_2, \ldots, y_n\}$, with the relation $\mathcal{R}$ as shown in Fig. 13.4, is F_{n+3}, for $n \geq 1$.

Our new examples have provided more places where the Fibonacci numbers arise. But where are the Lucas numbers? To answer this we introduce a new type of partial order called the *closed fence* or *corral*.

Example 13.4: In Fig. 13.5 we find the closed fences for 2 and 4 elements, respectively. For this type of partial order, the underlying set has an even number of elements and, for $n \geq 4$, we see that the difference between the closed fence and the fence on n vertices is that for the closed fence we also have $x_1 \mathcal{R} x_n$. The two closed fences shown in Fig. 13.5 have the following order ideals.

	Order Ideals
Fig. 13.5 (a)	$\varnothing$, $\{x_1\}$, $\{x_1, x_2\}$
Fig. 13.5 (b)	$\varnothing$, $\{x_1\}$, $\{x_3\}$, $\{x_1, x_3\}$,
	$\{x_1, x_2, x_3\}$, $\{x_1, x_3, x_4\}$, $\{x_1, x_2, x_3, x_4\}$

So the first closed fence has 3 ($= L_2$) order ideals, while the second closed fence has 7 ($= L_4$) order ideals. To determine the number of order ideals for the closed fence when $n = 6$, we use the results in Fig. 13.6 to lead the way. For the closed fence in Fig. 13.6 (a) there are two considerations:

(1) If we place x_6 in the order ideal, then we must include x_1 and x_5, since $x_1 \mathcal{R} x_6$ and $x_5 \mathcal{R} x_6$. Upon removing x_1, x_5, and x_6 from the Hasse diagram in Fig. 13.6 (a), we arrive at the Hasse diagram in Fig. 13.6 (b). So here we will add x_1, x_5, and x_6 to any order ideal for the fence in Fig. 13.6 (b). From our earlier result for Fig. 13.4 (b) we know that there are 5 ($= F_5$) such order ideals.

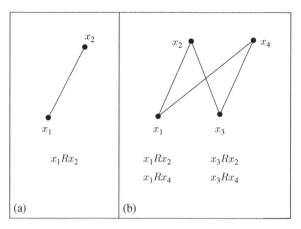

FIGURE 13.5

(2) If x_6 is not placed in the order ideal, upon removing x_6 from the Hasse diagram in Fig. 13.6 (a), we obtain the Hasse diagram in Fig. 13.6 (c). Our previous work with Fig. 13.2 (c) indicates that this case provides 13 $(= F_7)$ order ideals.

So, in total, this closed fence (on six vertices) has $5 + 13 = F_5 + F_7 = 18 = L_6$ order ideals.

The preceding argument can be generalized to show that the closed fence on n vertices, where n is even, has

$$F_{n-1} + F_{n+1} = L_n$$

order ideals, for $n \geq 2$. This follows from Property 13.1.

(More on partial orders, fences, and closed fences can be found in Reference [48].)

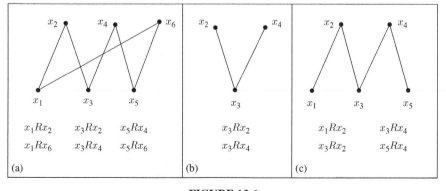

FIGURE 13.6

In Theorem 13.1 we shall come upon another recursive formula to determine the Fibonacci numbers. This time, however, we shall determine F_{n+1} by using only its immediate predecessor F_n. Consequently, we cannot expect this result to be linear.

This formula, along with a comparable one for the Lucas numbers, was discovered in 1963 by S. L. Basin, while at the Sylvania Electronic Systems at Mountain View, California. To derive these formulas, we need the following new properties, which the reader will be asked to verify in the exercises for this chapter.

Property 13.3:
$$F_n + L_n = 2F_{n+1}, \quad \text{for } n \geq 0.$$

Property 13.4:
$$2L_{n+1} - L_n = 5F_n, \quad \text{for } n \geq 0.$$

Property 13.5:
$$L_n^2 - 5F_n^2 = 4(-1)^n, \quad \text{for } n \geq 0.$$

Theorem 13.1:   For $n \geq 0$,

$$\text{(a) } F_{n+1} = \frac{F_n + \sqrt{5\,F_n^2 + 4(-1)^n}}{2}, \text{ and}$$

$$\text{(b) } L_{n+1} = \frac{L_n + \sqrt{5[L_n^2 - 4(-1)^n]}}{2}.$$

We shall prove part (a) and leave the proof of part (b) for the Exercises.

Proof: From Property 13.3 we know that $F_n + L_n = 2F_{n+1}$, so $2F_{n+1} - F_n = L_n$. Substituting this into Property 13.5 we find that

$$(2F_{n+1} - F_n)^2 - 5F_n^2 = 4(-1)^n$$
$$(2F_{n+1} - F_n)^2 = 5F_n^2 + 4(-1)^n$$
$$2F_{n+1} - F_n = \pm\sqrt{5F_n^2 + 4(-1)^n}$$
$$F_{n+1} = \frac{F_n \pm \sqrt{5F_n^2 + 4(-1)^n}}{2}.$$

Since $F_n \geq 0$, it follows that

$$F_{n+1} = \frac{F_n + \sqrt{5\,F_n^2 + 4(-1)^n}}{2}. \qquad \square$$

Our last theorem for this chapter provides a generalization of the Cassini identity of Property 4.9. It uses a result that readily follows from Property 13.5. This generalization was established in 1879 by the Belgian mathematician Eugène Charles

Catalan (1814–1894). We will meet with this mathematician once again in the second part of this book.

Example 13.5: Before we state the generalization of Cassini's identity, let us consider the following:

For $n \geq 0$, using the Binet form for the Lucas numbers, we see that

$$L_n^2 = (\alpha^n + \beta^n)^2 = \alpha^{2n} + \beta^{2n} + 2(\alpha\beta)^n$$
$$= L_{2n} + 2(-1)^n, \quad \text{since } \alpha\beta = -1.$$

Then Property 13.5 leads us to $5F_n^2 = L_n^2 - 4(-1)^n = [L_{2n} + 2(-1)^n] - 4(-1)^n = L_{2n} - 2(-1)^n$ or

$$5F_n^2 + 2(-1)^n = L_{2n}.$$

This will now help us to establish the following:

Theorem 13.2: (**Catalan's Generalization of the Cassini Identity**): For $n \geq r > 0$,

$$F_{n+r}F_{n-r} - F_n^2 = (-1)^{n+r+1}F_r^2.$$

Proof: Using the Binet form for the Fibonacci numbers, we see that

$$F_{n+r}F_{n-r} - F_n^2$$
$$= \frac{\alpha^{n+r} - \beta^{n+r}}{\alpha - \beta} \frac{\alpha^{n-r} - \beta^{n-r}}{\alpha - \beta} - \left(\frac{\alpha^n - \beta^n}{\alpha - \beta}\right)^2$$
$$= \frac{\alpha^{2n} - \alpha^{n+r}\beta^{n-r} - \alpha^{n-r}\beta^{n+r} + \beta^{2n}}{5} - \frac{\alpha^{2n} - 2(\alpha\beta)^n + \beta^{2n}}{5},$$

because $(\alpha - \beta)^2 = (\sqrt{5})^2 = 5$. Since $\alpha\beta = -1$, we have $\alpha^{-1} = -\beta$ and $\beta^{-1} = -\alpha$, and this takes us to

$$\frac{-(\alpha\beta)^n\alpha^r\beta^{-r} - (\alpha\beta)^n\alpha^{-r}\beta^r + 2(-1)^n}{5}$$
$$= \frac{-(-1)^n\alpha^r(\beta^{-1})^r - (-1)^n(\alpha^{-1})^r\beta^r + 2(-1)^n}{5}$$
$$= \frac{-(-1)^n\alpha^r(-\alpha)^r - (-1)^n(-\beta)^r\beta^r + 2(-1)^n}{5}$$
$$= \frac{-(-1)^n(-1)^r\alpha^{2r} - (-1)^n(-1)^r\beta^{2r} + 2(-1)^n}{5}$$
$$= \frac{(-1)^{n+r+1}[\alpha^{2r} + \beta^{2r}] + 2(-1)^n}{5}$$

$$= \frac{(-1)^{n+r+1} L_{2r} + 2(-1)^n}{5}$$

$$= \frac{(-1)^{n+r+1} [5 F_r^2 + 2(-1)^r] + 2(-1)^n}{5}, \qquad \text{from Example 13.5}$$

$$= (-1)^{n+r+1} F_r^2 + \left(\frac{2}{5}\right) [(-1)^{n+2r+1} + (-1)^n]$$

$$= (-1)^{n+r+1} F_r^2 + \left(\frac{2}{5}\right) [(-1)^{n+1} + (-1)^n] = (-1)^{n+r+1} F_r^2.$$

$\square$

EXERCISES FOR CHAPTER 13

1. (a) Cinda and Katelyn have a 1×25 circular chessboard made up of 25 curved squares, numbered 1, 2, ..., 25—with curved square 1 following curved square 25. In how many ways can they tile this chessboard with curved 1×1 square tiles and curved 1×2 rectangular tiles?

   (b) In how many ways can they place nontaking kings on their 1×25 circular chessboard?

2. Let A be the set of all positive and negative (integer) divisors of 6. Define the relation $\mathcal{R}$ on A by $x \mathcal{R} y$ if x (exactly) divides y. Is $\mathcal{R}$ a partial order for A? Why?

3. For $n \geq 0$, prove that $F_{2n} = F_n L_n$.

4. Use the Binet forms for the Fibonacci and Lucas numbers to prove that for $n \geq 1$, $F_{n-1} + F_{n+1} = L_n$.

5. For $n \geq 1$, prove that $L_{n-1} + L_{n+1} = 5F_n$.

6. For $n \geq 0$, prove that $2F_{n+1} = F_n + L_n$.

7. For $n \geq 0$, prove that $2L_{n+1} = 5F_n + L_n$.

8. For $n \geq 0$, prove that $L_n^2 - 5F_n^2 = 4(-1)^n$.

9. For $m \geq n \geq 0$, prove that $L_{2m} L_{2n} = L_{m+n}^2 + 5F_{m-n}^2$.

10. For $n \geq 0$, prove that $\sum_{k=0}^{n} \binom{n}{k} F_k F_{n-k} = (1/5)(2^n L_n - 2)$.

11. For $n \geq 0$, prove that $\sum_{k=0}^{2n} (-1)^k \binom{2n}{k} 2^{k-1} L_k = 5^n$. (J. L. Brown, Jr., 1967) [10].

12. For $n \geq 0$, prove that $L_{n+1} = \left(L_n + \sqrt{5 \left[L_n^2 - 4(-1)^n \right]} \right) / 2$.

13. For $n \geq 0$, prove that

$$\sum_{k=0}^{n} \binom{n}{k} F_k L_{n-k} = 2^n F_n.$$

14. For $m, n \geq 0$, prove that $2L_{m+n} = 5F_m F_n + L_m L_n$.

15. For $n \geq 2$, prove that $5F_{2n+3} F_{2n-3} = L_{4n} + 18$. (Blazej, 1975) [4].

16. For $m, n \geq 0$, prove that $F_m L_n + F_n L_m = 2F_{m+n}$. (Blazej, 1975) [5].

17. For $m, n \geq 0$, prove that $L_{2m+n} - (-1)^m L_n = 5F_m F_{m+n}$.

18. For $m, n \geq 0$, prove that $F_{2m+n} + (-1)^m F_n = F_{m+n} L_m$.

19. For $m, n \geq 0$, prove that $F_{2m+n} - (-1)^m F_n = F_m L_{m+n}$.

20. For $m \geq 0, n \geq 1$, prove that $L_{m+n} = F_{m+1} L_n + F_m L_{n-1}$. (This is comparable to Property 7.1.)

21. (a) For all $n \geq 1$, prove that $232\, L_n + 144\, L_{n-1} = (F_{13} - 1)L_n + F_{12} L_{n-1}$ is divisible by 10.

    (b) In Exercise 15 of Chapter 7, we learned that the sequence of units digits of the Fibonacci numbers has period 60. Determine, with proof, the period for the sequence of units digits of the Lucas numbers.

22. (a) For $n \geq 0$, prove that $2L_n + L_{n+1} = 5F_{n+1}$.

    (b) For $n \geq 0$, prove that $4nL_{n+1} - 2F_n$ is a multiple of 10. [An equivalent way to express this result is as follows: for $n \geq 0$, $4nL_{n+1} - 2F_n \equiv 0 \pmod{10}$.]

23. For $n \geq 1$, prove that $3^{n-1} - L_n$ is divisible by 5.

24. The *generalized Fibonacci numbers* are defined by

$$G_0 = (b - a), \quad G_1 = a, \quad \text{and} \quad G_n = G_{n-1} + G_{n-2}, \quad \text{for } n \geq 2.$$

    (a) Prove that for $n \geq 2$, $G_n = aF_{n-2} + bF_{n-1}$.

    (b) Which sequence arises when $a = 1$ and $b = 3$?

25. For $n \geq 0$, prove that

$$\sum_{r=0}^{n} G_r = G_{n+2} - a.$$

26. For $n \geq 1$, prove that

$$\sum_{r=1}^{n} G_{2r-1} = G_{2n} + a - b.$$

27. For $n \geq 0$, prove that

$$\sum_{r=0}^{n} G_{2r} = G_{2n+1} + b - 2a.$$

28. For $n \geq 0$, prove that

$$\sum_{r=0}^{n} G_r^2 = G_n G_{n+1} + 2a^2 - 3ab + b^2.$$

29. Fix $m \geq 2$. For $n \geq 0$, prove that

$$\sum_{r=0}^{n} G_{m+r} = G_{m+n+2} - G_{m+1}.$$

30. When dealing with the generalized Fibonacci numbers, let

$$c = a + (a - b)\beta \quad \text{and} \quad d = a + (a - b)\alpha.$$

For $n \geq 0$, prove that

$$G_n = \frac{c\alpha^n - d\beta^n}{\alpha - \beta}.$$

(This is called the *Binet form* for G_n.)

31. Determine (a) $\lim_{n \to \infty}(G_n/F_n)$ and (b) $\lim_{n \to \infty}(G_n/L_n)$.
32. For $n \geq 1$, prove that $G_{n+1}G_{n-1} - G_n^2 = (-1)^n(a^2 + ab - b^2)$. (This is analogous to the Cassini identity for the Fibonacci numbers.)
33. (a) (i) Write α, α^2, α^3, and α^4 in the form $(a + b\sqrt{5})/2$, where a and b are integers. (ii) For $n \geq 1$, how is α^n related to F_n and L_n?
    (b) (i) Write β, β^2, β^3, and β^4 in the form $(a - b\sqrt{5})/2$, where a and b are integers. (ii) For $n \geq 1$, how is β^n related to F_n and L_n?

CHAPTER 14

Matrices, The Inverse Tangent Function, and an Infinite Sum

Now the time has come to introduce some fundamental ideas from linear algebra into our discussion.

Let us start with the matrix

$$\mathbf{Q} = \begin{bmatrix} 1 & 1 \\ 1 & 0 \end{bmatrix}.$$

Properties of this matrix were investigated in 1960 by Charles H. King in his Master's thesis at what was, at that time, the San Jose State College in California.

Computing $\mathbf{Q}^n$ for $n = 2, 3, 4,$ and 5, we arrive at the following:

$$\mathbf{Q}^2 = \begin{bmatrix} 1 & 1 \\ 1 & 0 \end{bmatrix} \begin{bmatrix} 1 & 1 \\ 1 & 0 \end{bmatrix} = \begin{bmatrix} 2 & 1 \\ 1 & 1 \end{bmatrix}$$

$$\mathbf{Q}^3 = \mathbf{Q}\mathbf{Q}^2 = \begin{bmatrix} 1 & 1 \\ 1 & 0 \end{bmatrix} \begin{bmatrix} 2 & 1 \\ 1 & 1 \end{bmatrix} = \begin{bmatrix} 3 & 2 \\ 2 & 1 \end{bmatrix}$$

$$\mathbf{Q}^4 = \mathbf{Q}\mathbf{Q}^3 = \begin{bmatrix} 1 & 1 \\ 1 & 0 \end{bmatrix} \begin{bmatrix} 3 & 2 \\ 2 & 1 \end{bmatrix} = \begin{bmatrix} 5 & 3 \\ 3 & 2 \end{bmatrix}$$

$$\mathbf{Q}^5 = \mathbf{Q}\mathbf{Q}^4 = \begin{bmatrix} 1 & 1 \\ 1 & 0 \end{bmatrix} \begin{bmatrix} 5 & 3 \\ 3 & 2 \end{bmatrix} = \begin{bmatrix} 8 & 5 \\ 5 & 3 \end{bmatrix}$$

It appears that the entries in these matrices are Fibonacci numbers, since

$$\mathbf{Q} = \begin{bmatrix} F_2 & F_1 \\ F_1 & F_0 \end{bmatrix} \quad \mathbf{Q}^2 = \begin{bmatrix} F_3 & F_2 \\ F_2 & F_1 \end{bmatrix} \quad \mathbf{Q}^3 = \begin{bmatrix} F_4 & F_3 \\ F_3 & F_2 \end{bmatrix}$$

$$\mathbf{Q}^4 = \begin{bmatrix} F_5 & F_4 \\ F_4 & F_3 \end{bmatrix} \quad \mathbf{Q}^5 = \begin{bmatrix} F_6 & F_5 \\ F_5 & F_4 \end{bmatrix}$$

Fibonacci and Catalan Numbers: An Introduction, First Edition. Ralph P. Grimaldi.
© 2012 John Wiley & Sons, Inc. Published 2012 by John Wiley & Sons, Inc.

and these results suggest the following.

Theorem 14.1:

For $\mathbf{Q} = \begin{bmatrix} 1 & 1 \\ 1 & 0 \end{bmatrix}$ and $n \geq 1$,

$$\mathbf{Q}^n = \begin{bmatrix} F_{n+1} & F_n \\ F_n & F_{n-1} \end{bmatrix}.$$

This theorem can be established by the Principle of Mathematical Induction. We leave this for the exercises at the end of this chapter.

Our next results will follow from the following properties of determinants.

For any square matrix A, we let $\det(A)$ denote the determinant of A. Then if A and B are two $k \times k$ matrices, we have

(1) $\det(AB) = \det(A)\det(B)$
(2) $\det(A^n) = [\det(A)]^n$
(3) $\det(A^{m+n}) = \det(A^m A^n) = \det(A^m)\det(A^n)$

With these properties in hand, start by observing that $\det(\mathbf{Q})$, the determinant of our matrix $\mathbf{Q}$, is

$$\begin{vmatrix} F_2 & F_1 \\ F_1 & F_0 \end{vmatrix} = \begin{vmatrix} 1 & 1 \\ 1 & 0 \end{vmatrix} = 1 \cdot 0 - 1 \cdot 1 = -1.$$

Consequently, from (2) we see that for $n \geq 1$,

$$F_{n+1}F_{n-1} - F_n^2 = \begin{vmatrix} F_{n+1} & F_n \\ F_n & F_{n-1} \end{vmatrix} = \det(\mathbf{Q}^n) = [\det(\mathbf{Q})]^n = \left(\begin{vmatrix} 1 & 1 \\ 1 & 0 \end{vmatrix} \right)^n$$
$$= (-1)^n,$$

or

$$F_{n+1}F_{n-1} - F_n^2 = (-1)^n. \quad \text{(Cassini's identity, seen earlier in Property 4.9.)}$$

This is now the third time that we have obtained Cassini's identity. (The first time was in Property 4.9, where the method of proof suggested was the Principle of Mathematical Induction. Then again, following the introduction of the Binet form for F_n—just prior to Property 10.1.) So we have obtained Cassini's identity by three different approaches.

When n is odd, say $n = 2m + 1$, Cassini's identity takes the form

$$F_{2m+2}F_{2m} - F_{2m+1}^2 = (-1)^{2m+1} = -1$$

or

$$F_{2m+2}F_{2m} = F_{2m+1}^2 - 1.$$

Now at this point—and rightfully so—the reader may be wondering where would one use Cassini's identity, in one form or another. To answer this, we introduce a new identity involving the Fibonacci numbers. This time the inverse tangent function comes into play, as we discover in the next example.

Example 14.1:   We want to verify the identity

$$\arctan\left(\frac{1}{F_{2m+1}}\right) + \arctan\left(\frac{1}{F_{2m+2}}\right) = \arctan\left(\frac{1}{F_{2m}}\right), \quad m \geq 1.$$

To do so, we need to use the trigonometric identity

$$\tan(x + y) = \frac{\tan x + \tan y}{1 - \tan x \tan y}$$

We shall start by taking the tangents of the angles on the two sides of the identity we want to establish. So we have

$$\tan\left(\arctan\left(\frac{1}{F_{2m+1}}\right) + \arctan\left(\frac{1}{F_{2m+2}}\right)\right) = \tan\left(\arctan\left(\frac{1}{F_{2m}}\right)\right).$$

Substituting $\arctan\left(\frac{1}{F_{2m+1}}\right)$ for x and $\arctan\left(\frac{1}{F_{2m+2}}\right)$ for y in the identity for $\tan(x + y)$, since $\tan(\arctan \theta) = \theta$, we now find that

$$\frac{(1/F_{2m+1}) + (1/F_{2m+2})}{1 - (1/F_{2m+1})(1/F_{2m+2})} = (1/F_{2m}).$$

From this it follows that

$$\left(\frac{1}{F_{2m+1}}\right) + \left(\frac{1}{F_{2m+2}}\right) = \left(\frac{1}{F_{2m}}\right) - \left(\frac{1}{F_{2m}F_{2m+1}F_{2m+2}}\right)$$

and so

$$F_{2m}F_{2m+2} + F_{2m}F_{2m+1} = F_{2m+1}F_{2m+2} - 1.$$

But this implies that

$$
\begin{aligned}
F_{2m} F_{2m+2} &= F_{2m+1}(F_{2m+2} - F_{2m}) - 1 \\
&= F_{2m+1} F_{2m+1} - 1 \\
&= F_{2m+1}^2 - 1,
\end{aligned}
$$

the result obtained when we replaced n by $2m + 1$ in Cassini's identity. Working backward from this identity, as all of our steps are reversible, we obtain the identity

$$
\arctan\left(\frac{1}{F_{2m+1}}\right) + \arctan\left(\frac{1}{F_{2m+2}}\right) = \arctan\left(\frac{1}{F_{2m}}\right), \ m \geq 1.
$$

Using the identity established in Example 14.1, now it is time to determine

$$
\sum_{m=1}^{\infty} \arctan\left(\frac{1}{F_{2m+1}}\right).
$$

Example 14.2:   The following result was discovered by the American mathematician Derrick H. Lehmer (1905–1991) in 1936, when he was a professor at Lehigh University. For any fixed positive integer K,

$$
\begin{aligned}
\sum_{m=1}^{K} \arctan\left(\frac{1}{F_{2m+1}}\right) &= \sum_{m=1}^{K} \left[\arctan\left(\frac{1}{F_{2m}}\right) - \arctan\left(\frac{1}{F_{2m+2}}\right)\right] \\
&= \left[\arctan\left(\frac{1}{F_2}\right) - \arctan\left(\frac{1}{F_4}\right)\right] + \left[\arctan\left(\frac{1}{F_4}\right) - \arctan\left(\frac{1}{F_6}\right)\right] \\
&\quad + \left[\arctan\left(\frac{1}{F_6}\right) - \arctan\left(\frac{1}{F_8}\right)\right] + \cdots \\
&\quad + \left[\arctan\left(\frac{1}{F_{2K}}\right) - \arctan\left(\frac{1}{F_{2K+2}}\right)\right] \\
&= \arctan\left(\frac{1}{F_2}\right) - \arctan\left(\frac{1}{F_{2K+2}}\right).
\end{aligned}
$$

Consequently,

$$
\begin{aligned}
\sum_{m=1}^{\infty} \arctan\left(\frac{1}{F_{2m+1}}\right) &= \lim_{K \to \infty} \sum_{m=1}^{K} \arctan\left(\frac{1}{F_{2m+1}}\right) \\
&= \lim_{K \to \infty} \left[\arctan\left(\frac{1}{F_2}\right) - \arctan\left(\frac{1}{F_{2K+2}}\right)\right] \\
&= \arctan\left(\frac{1}{F_2}\right) - \arctan 0 = \arctan\left(\frac{1}{1}\right) - 0 \\
&= \arctan 1 = \pi/4.
\end{aligned}
$$

Now we shall use $\mathbf{Q}$ once again to derive another property for the Fibonacci numbers. Turn your attention to property (3) for determinants, as given previously—following Theorem 14.1.

Upon replacing A by $\mathbf{Q}$, from $\mathbf{Q}^{m+n} = \mathbf{Q}^m \mathbf{Q}^n$ we see that

$$
\begin{vmatrix} F_{m+n+1} & F_{m+n} \\ F_{m+n} & F_{m+n-1} \end{vmatrix} = \det(\mathbf{Q}^{m+n}) = \det(\mathbf{Q}^m \mathbf{Q}^n) = \det(\mathbf{Q}^m)\det(\mathbf{Q}^n)
$$

$$
= \begin{vmatrix} F_{m+1} & F_m \\ F_m & F_{m-1} \end{vmatrix} \begin{vmatrix} F_{n+1} & F_n \\ F_n & F_{n-1} \end{vmatrix}.
$$

When these determinants are expanded, we obtain one more property satisfied by the Fibonacci numbers.

Property 14.1: For $m \geq 1$, $n \geq 1$,

$$
F_{m+n+1} F_{m+n-1} - F_{m+n}^2 = (F_{m+1} F_{m-1} - F_m^2)(F_{n+1} F_{n-1} - F_n^2).
$$

In closing this chapter, we want to make one more observation about $\mathbf{Q}$. In so doing, we shall find ourselves confronted with numbers we have seen several times before. So let us compute what are called the *eigenvalues* of $\mathbf{Q}$. These are the values of λ such that

$$
\det(\mathbf{Q} - \lambda \mathbf{I}) = 0,
$$

where $\mathbf{I} = \begin{bmatrix} 1 & 0 \\ 0 & 1 \end{bmatrix}$, the 2×2 multiplicative identity. We find that

$$
0 = \det(\mathbf{Q} - \lambda \mathbf{I}) = \det\left(\begin{bmatrix} 1 & 1 \\ 1 & 0 \end{bmatrix} - \lambda \begin{bmatrix} 1 & 0 \\ 0 & 1 \end{bmatrix} \right)
$$

$$
= \begin{vmatrix} 1 - \lambda & 1 \\ 1 & -\lambda \end{vmatrix} = (1 - \lambda)(-\lambda) - 1 = \lambda^2 - \lambda - 1,
$$

$$
\text{so } \lambda = \frac{-(-1) \pm \sqrt{(-1)^2 - 4(1)(-1)}}{2(1)} = \frac{1 \pm \sqrt{5}}{2}.
$$

Consequently, the eigenvalues of $\mathbf{Q}$ are the familiar constants α and β.

EXERCISES FOR CHAPTER 14

1. If

$$\mathbf{Q} = \begin{bmatrix} 1 & 1 \\ 1 & 0 \end{bmatrix},$$

prove that for $n \geq 1$,

$$\mathbf{Q}^n = \begin{bmatrix} F_{n+1} & F_n \\ F_n & F_{n-1} \end{bmatrix}.$$

2. For $\mathbf{Q}$ as in Exercise 1, prove that $\mathbf{Q}^2 - \mathbf{Q} - \mathbf{I} = \mathbf{Z}$, where $\mathbf{I} = \begin{bmatrix} 1 & 0 \\ 0 & 1 \end{bmatrix}$, the 2×2 multiplicative identity, and $\mathbf{Z} = \begin{bmatrix} 0 & 0 \\ 0 & 0 \end{bmatrix}$. [This shows that $\mathbf{Q}$ satisfies its characteristic equation—namely, $\det(\mathbf{Q} - \lambda \mathbf{I}) = \lambda^2 - \lambda - 1 = 0$. In general, any square matrix satisfies its characteristic equation. This is the celebrated *Cayley–Hamilton theorem* of matrix theory, named after the English mathematician Arthur Cayley (1821–1895) and the Irish mathematician William Rowan Hamilton (1805–1865).]

3. For $\mathbf{Q}$ as in Exercise 1, determine

$$\lim_{n \to \infty} \frac{\mathbf{Q}^n}{F_{n-1}}.$$

4. For $\mathbf{Q}$ and $\mathbf{I}$ as in Exercise 2, prove that

$$\mathbf{Q}^n = F_n \mathbf{Q} + F_{n-1} \mathbf{I}, \quad n \geq 1.$$

(This result is comparable to those given in Exercise 18 for Chapter 10.)

5. Prove that $\arctan 2 = 2 \arctan |\beta|$.

6. Prove that
   (a) $\arcsin |\beta| = \arccos \left(1/\sqrt{\alpha} \right) = \arccos \sqrt{-\beta} = \arccos \sqrt{|\beta|}$
   (b) $\arccos |\beta| = \arcsin \left(1/\sqrt{\alpha} \right) = \arcsin \sqrt{-\beta} = \arcsin \sqrt{|\beta|}$.

7. Comparable to the matrix $\mathbf{Q}$ in Exercise 1, we have the matrix

$$M = \begin{bmatrix} 1 & 1 \\ 1 & 2 \end{bmatrix}.$$

(This matrix was investigated in 1983 by Sam Moore of the Community College of Allegheny County in Pennsylvania.)

(a) Compute M^2, M^3, M^4, and M^5.

(b) Conjecture a general formula for M^n, where $n \geq 1$, and establish your conjecture by the Principle of Mathematical Induction.

(c) Determine

$$\lim_{n \to \infty} \frac{M^n}{F_{2n-1}}.$$

8. For M as in Exercise 7, determine the eigenvalues of the matrix M—that is, determine the values of λ such that $\det(M - \lambda\mathbf{I}) = 0$, where $\mathbf{I} = \begin{bmatrix} 1 & 0 \\ 0 & 1 \end{bmatrix}$, the 2×2 multiplicative identity.

9. (a) Determine the value of

$$\begin{vmatrix} F_1 & F_2 & F_3 \\ F_2 & F_3 & F_4 \\ F_3 & F_4 & F_5 \end{vmatrix}.$$

(b) Determine the value of

$$\begin{vmatrix} F_1 & F_2 & F_3 & F_4 \\ F_2 & F_3 & F_4 & F_5 \\ F_3 & F_4 & F_5 & F_6 \\ F_4 & F_5 & F_6 & F_7 \end{vmatrix}.$$

(c) Generalize the results from parts (a) and (b).

10. For $n \geq 1$, prove that the value of the $n \times n$ determinant

$$\begin{vmatrix} 3 & i & 0 & 0 & \cdots & 0 & 0 \\ i & 1 & i & 0 & \cdots & 0 & 0 \\ 0 & i & 1 & i & \cdots & 0 & 0 \\ 0 & 0 & i & 1 & \cdots & 0 & 0 \\ \vdots & \vdots & \vdots & \vdots & \vdots & \vdots & \vdots \\ 0 & 0 & 0 & 0 & \cdots & 1 & i \\ 0 & 0 & 0 & 0 & \cdots & i & 1 \end{vmatrix}$$

is the Lucas number L_{n+1}. Here $i = \sqrt{-1}$. (P. F. Byrd, 1963) [13].

11. For $n \geq 1$, let A_n be the $n \times n$ matrix

$$\begin{bmatrix} 3 & -1 & 0 & 0 & 0 & \dots & 0 & 0 \\ -1 & 3 & -1 & 0 & 0 & \dots & 0 & 0 \\ 0 & -1 & 3 & -1 & 0 & \dots & 0 & 0 \\ \vdots & \vdots & \vdots & \vdots & \vdots & & \vdots & \vdots \\ 0 & 0 & 0 & 0 & 0 & \dots & -1 & 3 \end{bmatrix}.$$

(a) Evaluate the determinants for A_1, A_2, A_3, and A_4.

(b) Conjecture a formula for the determinant of A_n, $n \geq 1$. Prove your conjecture.

CHAPTER 15

The gcd Property for the Fibonacci Numbers

First and foremost, the reader may be puzzling over what we mean by the *gcd Property*. Instead of simply stating it, as if it comes out of nowhere, let us present some motivation for this amazing property. So consider the following two examples:

(i) $\gcd(F_8, F_{12}) = \gcd(21, 144) = \gcd(3 \cdot 7, 2^4 \cdot 3^2) = 3 = F_4 = F_{\gcd(8,12)}$

(ii) $\gcd(F_{12}, F_{18}) = \gcd(144, 2584) = \gcd(2^4 \cdot 3^2, 2^3 \cdot 17 \cdot 19) = 2^3$
$$= 8 = F_6 = F_{\gcd(12,18)}$$

These results suggest the following:

Property 15.1: For $m \geq n \geq 1$,

$$\gcd(F_m, F_n) = F_{\gcd(m,n)}.$$

To establish this property, we return to the matrix $\mathbf{Q}$ of Chapter 14.

For $m, n \geq 1$, we know that $\mathbf{Q}^{m+n} = \mathbf{Q}^m \mathbf{Q}^n$. Previously we learned from Theorem 14.1 that $\mathbf{Q}^k = \begin{bmatrix} F_{k+1} & F_k \\ F_k & F_{k-1} \end{bmatrix}$ for $k \geq 1$. So $\mathbf{Q}^{m+n} = \mathbf{Q}^m \mathbf{Q}^n$ now becomes

$$\begin{bmatrix} F_{m+n+1} & F_{m+n} \\ F_{m+n} & F_{m+n-1} \end{bmatrix} = \begin{bmatrix} F_{m+1} & F_m \\ F_m & F_{m-1} \end{bmatrix} \begin{bmatrix} F_{n+1} & F_n \\ F_n & F_{n-1} \end{bmatrix} \quad \text{or}$$

$$\begin{bmatrix} F_{m+n+1} & F_{m+n} \\ F_{m+n} & F_{m+n-1} \end{bmatrix} = \begin{bmatrix} F_{m+1}F_{n+1} + F_m F_n & F_{m+1}F_n + F_m F_{n-1} \\ F_m F_{n+1} + F_{m-1} F_n & F_m F_n + F_{m-1} F_{n-1} \end{bmatrix}.$$

Fibonacci and Catalan Numbers: An Introduction, First Edition. Ralph P. Grimaldi.
© 2012 John Wiley & Sons, Inc. Published 2012 by John Wiley & Sons, Inc.

When two matrices are equal, their corresponding entries must be equal. Consequently, we arrive at the following four results. For m, $n \geq 1$,

$$(1) \quad F_{m+n+1} = F_{m+1}F_{n+1} + F_m F_n$$
$$(2) \quad F_{m+n} = F_{m+1}F_n + F_m F_{n-1}$$
$$(3) \quad F_{m+n} = F_m F_{n+1} + F_{m-1} F_n$$
$$(4) \quad F_{m+n-1} = F_m F_n + F_{m-1} F_{n-1}.$$

Result (3) is precisely Property 7.1 of Example 7.3, which we established by a combinatorial argument involving the tiling of a $1 \times (n + m - 1)$ chessboard. Results (1), (2), and (4) are simply variations of Property 7.1. The matrix $\mathbf{Q}$ has now provided another way to obtain Property 7.1, as well as these three other related results.

In order to establish the gcd Property for the Fibonacci numbers, we need the following four lemmas:

Lemma 15.1:   For $m \geq 1, n \geq 1$, F_m divides F_{mn}.

Proof: This result was given earlier as Property 7.2, where the Principle of Mathematical Induction was suggested as a method of proof. For the sake of completeness in deriving the gcd Property, at this point we shall leave no stone unturned and provide a complete proof for this first lemma.

We keep m fixed but arbitrary. When $n = 1$, the result certainly follows since F_m divides itself. This establishes the basis step for our inductive proof. Next we assume the result true for some $n = k$ (≥ 1)—that is, we assume that F_m divides F_{mk}. Now for $n = k + 1$ (≥ 2), we have

$$F_{m(k+1)} = F_{mk+m} = F_{mk}F_{m+1} + F_{mk-1}F_m,$$

by result (3) from above (or Property 7.1). Since F_m divides itself and F_m divides F_{mk} (by the induction hypothesis), it follows that F_m divides $F_{m(k+1)}$. So by the Principle of Mathematical Induction, for $m \geq 1, n \geq 1$,

$$F_m \text{ divides } F_{mn}. \qquad \square$$

This first lemma will immediately help us with the following.

Lemma 15.2:   For $q \geq 1, n \geq 1$, $\gcd(F_{qn-1}, F_n) = 1$.

Proof: Let $d = \gcd(F_{qn-1}, F_n)$. Then d divides F_{qn-1} and d divides F_n. Since d divides F_n, it follows from Lemma 15.1 that d divides F_{qn}. Consequently, d divides the consecutive Fibonacci numbers F_{qn-1} and F_{qn}. But from Property 4.1 we know that the greatest common divisor of any two consecutive Fibonacci numbers is 1. Consequently, $\gcd(F_{qn-1}, F_n) = d = 1$. $\qquad \square$

The next lemma establishes a very useful property about greatest common divisors. Its proof uses the following important property of the greatest common divisor.

For positive integers a and b, the greatest common divisor d of a and b is the smallest positive integer that can be expressed as a linear combination of a and b—that is, there exist integers x and y so that $d = ax + by$, and no positive integer smaller than d can be expressed, in terms of a and b, in this way. Now let us see how this comes into play!

Lemma 15.3:   Let a, b, and c be positive integers with $\gcd(a, c) = 1$. Then $\gcd(ab, c) = \gcd(b, c)$.

Proof: Let $d_1 = \gcd(b, c)$ and $d_2 = \gcd(ab, c)$. Now $d_1 = \gcd(b, c) \Rightarrow [(d_1$ divides $b)$ and $(d_1$ divides $c)]$. Consequently, d_1 divides ab, and so it follows that d_1 divides d_2.

With $\gcd(a, c) = 1$, there exist integers x and y so that $ax + cy = 1$. Therefore, $abx + bcy = b$. Now $d_2 = \gcd(ab, c) \Rightarrow [(d_2$ divides $ab)$ and $(d_2$ divides $c)]$, so d_2 divides $(ab)x + b(c)y = b$. Then $[(d_2$ divides $b)$ and $(d_2$ divides $c)] \Rightarrow [d_2$ divides $d_1]$.

Finally, since d_1 divides d_2, and d_2 divides d_1, and d_1, d_2 are positive, we have $d_1 = d_2$—or $\gcd(ab, c) = \gcd(b, c)$.   □

Lemma 15.4:   Let m, n be positive integers with $m \geq n$. Suppose that $m = qn + r$, where q is a positive integer and $0 \leq r < n$. Then $\gcd(F_m, F_n) = \gcd(F_n, F_r)$.

Proof: First we see that

$$\gcd(F_m, F_n) = \gcd(F_{qn+r}, F_n) = \gcd(F_{qn}F_{r+1} + F_{qn-1}F_r, F_n),$$

from result (3) prior to Lemma 15.1 (or Property 7.1). Let $h = \gcd(F_{qn}F_{r+1} + F_{qn-1}F_r, F_n)$ and $k = \gcd(F_{qn-1}F_r, F_n)$. Then h divides $F_{qn}F_{r+1} + F_{qn-1}F_r$ and h divides F_n. Since h divides F_n, by Lemma 15.1 h divides F_{qn}. Consequently, h divides $F_{qn}F_{r+1}$. Then, as h divides both $F_{qn}F_{r+1}$ and $F_{qn}F_{r+1} + F_{qn-1}F_r$, it follows that h divides $F_{qn-1}F_r$. So h divides k.

Likewise, k divides $F_{qn-1}F_r$ and k divides F_n. Since $[k$ divides $F_n] \Rightarrow [k$ divides $F_{qn}] \Rightarrow [k$ divides $F_{qn}F_{r+1}]$, we see that k divides $F_{qn}F_{r+1} + F_{qn-1}F_r$. So k divides h. Since h, k are positive, it follows that $h = k$.

Consequently, $\gcd(F_m, F_n) = \gcd(F_{qn}F_{r+1} + F_{qn-1}F_r, F_n) = h = k = \gcd(F_{qn-1}F_r, F_n)$. From Lemma 15.3 (with $a = F_{qn-1}$, $b = F_r$, and $c = F_n$), it follows that $\gcd(F_{qn-1}F_r, F_n) = \gcd(F_r, F_n) = \gcd(F_n, F_r)$. So $\gcd(F_m, F_n) = \gcd(F_n, F_r)$.   □

And now it is time to establish the gcd Property. Before we do so, however, it should be pointed out that the results in Lemmas 15.1–15.4 are general results that one encounters in studying number theory and abstract algebra. These lemmas are not just used to help us prove the following theorem.

Theorem 15.1:   Let m, n be positive integers, with $m \geq n$. Then

$$\gcd(F_m, F_n) = F_{\gcd(m,n)}.$$

Proof: Once again we use the Euclidean algorithm and find that

$$\begin{aligned}
m &= q_1 n + r_1, & 0 \leq r_1 < n \\
n &= q_2 r_1 + r_2, & 0 \leq r_2 < r_1 \\
r_1 &= q_3 r_2 + r_3, & 0 \leq r_3 < r_2 \\
&\cdots \qquad \cdots & \cdots \\
r_{k-2} &= q_k r_{k-1} + r_k, & 0 \leq r_k < r_{k-1} \\
r_{k-1} &= q_{k+1} r_k.
\end{aligned}$$

So $r_k = \gcd(m, n)$.

From Lemma 15.4 we learn that

$$\begin{aligned}
\gcd(F_m, F_n) &= \gcd(F_{q_1 n + r_1}, F_n) \\
&= \gcd(F_n, F_{r_1}) \\
&= \gcd(F_{q_2 r_1 + r_2}, F_{r_1}) \\
&= \gcd(F_{r_1}, F_{r_2}) \\
&= \gcd(F_{q_3 r_2 + r_3}, F_{r_2}) \\
&= \gcd(F_{r_2}, F_{r_3}) \\
&\quad \vdots \\
&= \gcd(F_{q_k r_{k-1} + r_k}, F_{r_{k-1}}) \\
&= \gcd(F_{r_{k-1}}, F_{r_k}).
\end{aligned}$$

Since r_k divides r_{k-1}, it follows from Lemma 15.1 (or Property 7.2) that F_{r_k} divides $F_{r_{k-1}}$. Consequently,

$$\gcd(F_m, F_n) = \gcd(F_{r_{k-1}}, F_{r_k}) = F_{r_k} = F_{\gcd(m,n)}. \qquad \square$$

Now we shall close this chapter by illustrating how the gcd Property for the Fibonacci numbers can reduce the amount of computation needed to determine the greatest common divisor of two Fibonacci numbers.

Example 15.1:   Consider the following Fibonacci numbers along with their prime factorizations:

$$\begin{aligned}
F_{30} &= 832,040 = 2^3 \cdot 5 \cdot 11 \cdot 31 \cdot 61 \text{ and} \\
F_{70} &= 190,392,490,709,135 = 5 \cdot 11 \cdot 13 \cdot 29 \cdot 71 \cdot 911 \cdot 141961
\end{aligned}$$

Having invested the time to determine these prime factorizations, we can then readily write

$$\gcd(F_{30}, F_{70}) = 5 \cdot 11 = 55.$$

But we can bypass the effort needed to find the prime factorizations because from Theorem 15.1 we now know that

$$\gcd(F_{30}, F_{70}) = F_{\gcd(30,70)} = F_{10} = 55.$$

EXERCISES FOR CHAPTER 15

1. Find $\gcd(F_{57}, F_{75})$.
2. If p and q are relatively prime integers—that is, $\gcd(p, q) = 1$—what is $\gcd(F_p, F_q)$?
3. Does the gcd Property hold for the Lucas numbers? Why?
4. For positive integers m, n, the least common multiple of m and n is the smallest positive integer that is a multiple of both m and n. It is denoted by $\text{lcm}(m, n)$. If $m = p_1^{e_1} p_2^{e_2} p_3^{e_3} \ldots p_r^{e_r}$ and $n = p_1^{f_1} p_2^{f_2} p_3^{f_3} \ldots p_r^{f_r}$, where p_1, p_2, p_3, ..., p_r are distinct primes, and $e_i \geq 0$, $f_i \geq 0$, for $1 \leq i \leq r$, then $\text{lcm}(m, n) = p_1^{k_1} p_2^{k_2} p_3^{k_3} \ldots p_r^{k_r}$, where $k_i = \max\{e_i, \ f_i\}$, for $1 \leq i \leq r$. For example, if $m = 440 = 2^3 5^1 11^1$ and $n = 525 = 3^1 5^2 7^1$, then

$$\begin{aligned}
\text{lcm}(440, 525) &= \text{lcm}(2^3 5^1 11^1, 3^1 5^2 7^1) \\
&= \text{lcm}(2^3 3^0 5^1 7^0 11^1, 2^0 3^1 5^2 7^1 11^0) \\
&= 2^3 3^1 5^2 7^1 11^1 = 46,200.
\end{aligned}$$

   (a) Is there an lcm Property for the Fibonacci numbers, comparable to the gcd Property? That is, is it true that for all $m, n \geq 0$, $\text{lcm}(F_m, F_n) = F_{\text{lcm}(m, n)}$? Why?

CHAPTER 16

Alternate Fibonacci Numbers

In this chapter, we shall provide a variety of examples that are counted by alternate Fibonacci numbers—that is, subsequences of the Fibonacci numbers, such as $1 \ (= F_1), 2 \ (= F_3), 5 \ (= F_5), 13 \ (= F_7), \ldots$, or $1 \ (- F_2), 3 \ (- F_4), 8 \ (= F_6),$ $21 \ (= F_8), \ldots$.

Example 16.1: Our first example deals with the undirected graph G_n shown in Fig. 16.1. This graph has the $n + 1$ vertices: $v_0, v_1, v_2, v_3, \ldots, v_{n-2}, v_{n-1}, v_n$, and the $n + (n - 1) = 2n - 1$ edges:

$$v_0 v_1, v_0 v_2, v_0 v_3, \ldots, v_0 v_{n-2}, v_0 v_{n-1}, \ v_0 v_n, v_1 v_2, v_2 v_3, \ldots, v_{n-2} v_{n-1}, v_{n-1} v_n.$$

It is called the *fan* on n vertices.

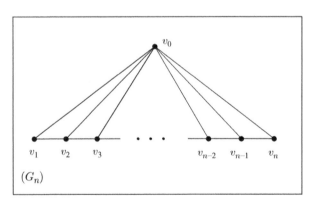

FIGURE 16.1

For $n = 2$, the undirected graph G_2 is shown in Fig. 16.2 (a). In Fig. 16.2 (b) there are three subgraphs of G_2. Note that each of these subgraphs contains all three vertices of G_2, contains no loops or cycles, and is connected. These are the three *spanning trees* of G_2.

Fibonacci and Catalan Numbers: An Introduction, First Edition. Ralph P. Grimaldi.
© 2012 John Wiley & Sons, Inc. Published 2012 by John Wiley & Sons, Inc.

The fan for $n = 3$ is shown in Fig. 16.3 (a) on p. 128. In Fig. 16.3 (b) we have three of the eight spanning trees for G_3. Note how these three spanning trees can be obtained from the three spanning trees in Fig. 16.2 (b) by adding the edge v_0v_3. Part (c) of Fig. 16.3 provides three more of the spanning trees for G_3. These can be obtained from the trees in Fig. 16.2 (b) by adding the edge v_2v_3. Finally, the remaining two spanning trees for G_3 are shown in Fig. 16.3 (d). These can be obtained from the first two spanning trees in Fig. 16.2 (b) by removing the edge v_0v_2 and then adding the edges v_0v_3 and v_2v_3, so that in either case no cycle is created.

We shall let t_n count the number of spanning trees that exist for the fan G_n, where $n \geq 1$. To determine t_n, we realize that we want to consider spanning trees of smaller fans and somehow account for the various ways in which we can include the vertex v_n. Consequently, we find that

$$t_n = t_{n-1} + t_{n-1} + (t_{n-1} - t_{n-2}),$$

where

(1) the first summand of t_{n-1} accounts for when we add the edge v_0v_n to a spanning tree for G_{n-1} (but not the edge $v_{n-1}v_n$);

(2) the second summand of t_{n-1} accounts for when we add the edge $v_{n-1}v_n$ to a spanning tree for G_{n-1} (but not the edge v_0v_n); and,

(3) the summand $(t_{n-1} - t_{n-2})$ accounts for the remaining spanning trees. For here we are asking ourselves for the number of spanning trees of G_{n-1} that do not contain the edge v_0v_{n-1}. That would be the total number (namely, t_{n-1}) minus the t_{n-2} spanning trees of G_{n-2} (to which we would add the edge $v_{n-2}v_{n-1}$). But now we remove the edge $v_{n-2}v_{n-1}$ and add the edges v_0v_n and $v_{n-1}v_n$ to get the remaining spanning trees for G_n.

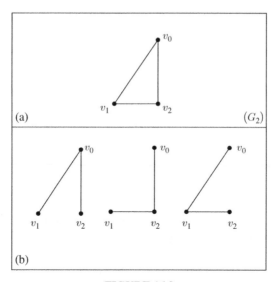

FIGURE 16.2

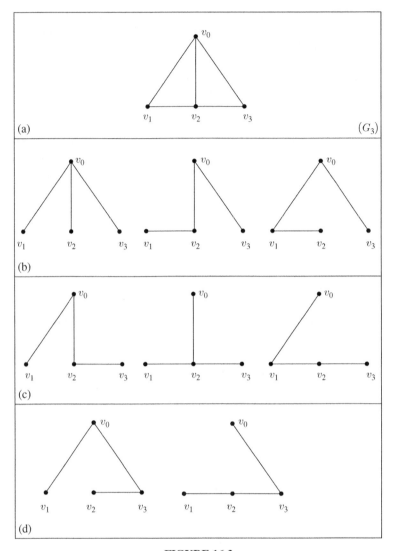

FIGURE 16.3

To solve the initial value problem

$$t_n = 3t_{n-1} - t_{n-2}, \quad n \geq 3, \quad t_1 = 1, \quad t_2 = 3,$$

we return to the method we developed in Chapter 10. Substituting Ar^n for t_n, where $A \neq 0$ and $r \neq 0$, we obtain the equation

$$Ar^n = 3Ar^{n-1} - Ar^{n-2}.$$

Dividing through by Ar^{n-2} leads us to the characteristic equation

$$r^2 - 3r + 1 = 0.$$

Upon solving this equation, we learn that the characteristic roots are

$$r = \frac{-(-3) \pm \sqrt{(-3)^2 - 4(1)(1)}}{2(1)} = \frac{3 \pm \sqrt{5}}{2}.$$

We learned that $\alpha^2 = ((1 + \sqrt{5})/2)^2 = (3 + \sqrt{5})/2$ and $\beta^2 = ((1 - \sqrt{5})/2)^2 = (3 - \sqrt{5})/2$ in Exercise 33 of Chapter 13. Consequently, the solution of the recurrence relation for t_n has the form

$$t_n = c_1(\alpha^2)^n + c_2(\beta^2)^n.$$

From $1 = t_1 = c_1\alpha^2 + c_2\beta^2$ and $3 = t_3 = c_1\alpha^4 + c_2\beta^4$, we learn that $c_1 = 1/\sqrt{5}$ and $c_2 = -1/\sqrt{5}$. Recall that $\alpha - \beta = \sqrt{5}$, so for $n \geq 1$,

$$t_n = \frac{\alpha^{2n} - \beta^{2n}}{\alpha - \beta}, \qquad \text{the Binet form for } F_{2n}.$$

Example 16.2: In Example 6.1, we learned that a positive integer n has 2^{n-1} compositions. However, if one can use only 1's and 2's as summands, in Example 6.2 we then found that n has only F_{n+1} compositions. Now suppose there are no restrictions on the positive integers we can use in forming our compositions, and, in addition, two different kinds of ones are available—say 1 and $1'$. This was proposed in Reference [19].

(One can also think of this in terms of tiling a $1 \times n$ chessboard using $1 \times k$ tiles, for $1 \leq k \leq n$. Here the 1×1 tiles come in two colors, while for $2 \leq k \leq n$, the $1 \times k$ tiles come in only one color.)

If we let a_n count the number of compositions of n, where we have two different kinds of ones, we find that

(1) $a_1 = 2$, for the compositions 1 and $1'$, and
(2) $a_2 = 5$, for the compositions 2, $1 + 1$, $1 + 1'$, $1' + 1$, and $1' + 1'$.

To determine the general formula for a_n, we consider the recurrence relation

$$a_n = 2a_{n-1} + (a_{n-1} - a_{n-2}).$$

Here the first summand accounts for the fact that we can obtain the compositions of n that end in 1 or $1'$ from the compositions of $n - 1$ by simply appending "+1" or "+1'" to the end of each of these compositions (of $n - 1$). For the other summand, we take the compositions of $n - 1$ and increase the last summand in the composition by 1. Unfortunately, this means that if the last summand in a composition of $n - 1$ is either a 1 or a 1', then the last summand in the composition of n is a 2 in both cases. To account for this overcounting, we remove the a_{n-2} compositions of $n - 1$ that end with 1'.

As in the previous example, we find that

$$a_n = c_1 \left(\frac{3 + \sqrt{5}}{2} \right)^n + c_2 \left(\frac{3 - \sqrt{5}}{2} \right)^n.$$

But here, with $a_1 = 2$ and $a_2 = 5$, it follows that

$$c_1 = \frac{5 + \sqrt{5}}{10} \text{ and } c_2 = \frac{5 - \sqrt{5}}{10},$$

so for $n \geq 1$, we have

$$\begin{aligned}
a_n &= \left(\frac{5 + \sqrt{5}}{10} \right) \left(\frac{3 + \sqrt{5}}{2} \right)^n + \left(\frac{5 - \sqrt{5}}{10} \right) \left(\frac{3 - \sqrt{5}}{2} \right)^n \\
&= \left(\frac{1}{\sqrt{5}} \right) \left(\frac{1 + \sqrt{5}}{2} \right) \left(\frac{1 + \sqrt{5}}{2} \right)^{2n} - \left(\frac{1}{\sqrt{5}} \right) \left(\frac{1 - \sqrt{5}}{2} \right) \left(\frac{1 - \sqrt{5}}{2} \right)^{2n} \\
&= \left(\frac{1}{\sqrt{5}} \right) \left(\frac{1 + \sqrt{5}}{2} \right)^{2n+1} - \left(\frac{1}{\sqrt{5}} \right) \left(\frac{1 - \sqrt{5}}{2} \right)^{2n+1} \\
&= \frac{\alpha^{2n+1} - \beta^{2n+1}}{\alpha - \beta} = F_{2n+1}.
\end{aligned}$$

(This result gives us those Fibonacci numbers where the subscripts are odd. The result in Example 16.1 had the subscripts even.)

Example 16.3: This example will provide us with an opportunity to use one of the properties of the Fibonacci numbers that we established earlier—namely, Property 4.7. The example was proposed in 1970 by Reuben C. Drake of the North Carolina A & T University at Greensboro. The solution given here is based on the one given by Leonard Carlitz of Duke University.

For $n \geq 0$, we shall let p_n count the number of paths from $(0, 0)$ to $(n, 0)$, using the following four types of steps:

(1) $(x, 0) \uparrow (x, 1)$
(2) $(x, 0) \rightarrow (x + 1, 0)$
(3) $(x, 1) \rightarrow (x + 1, 1)$
(4) $(x, 1) \searrow (x + 1, 0)$

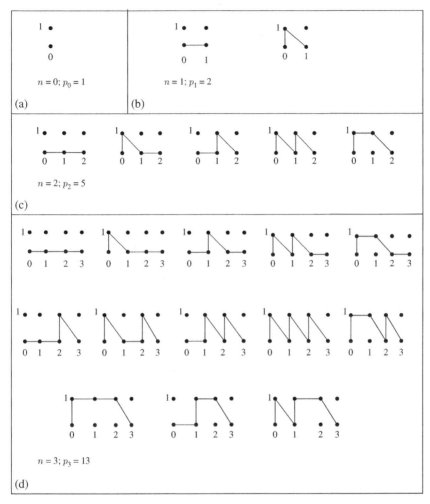

FIGURE 16.4

Part (a) of Fig. 16.4 shows us that when $n = 0$, we stay put. Since there is only one way to do this, we assign p_0 the value 1 ($= F_1$). There are two ways to go from $(0, 0)$ to $(1, 0)$, as shown in Fig. 16.4 (b). Consequently, we have $p_1 = 2$ ($= F_3$). Part (c) of Fig. 16.4 reveals that there are five ways to go from $(0, 0)$ to $(2, 0)$, so $p_2 = 5$ ($= F_5$). Now let us consider the paths from $(0, 0)$ to $(3, 0)$, as shown in Fig. 16.4 (d). Note how we can obtain the first five of these paths from those for $n = 2$ by appending the move $(2, 0) \rightarrow (3, 0)$. Likewise, the second set of five paths result when we append the moves $(2, 0) \uparrow (2, 1) \searrow (3, 0)$ to each of the paths for $n = 2$. The remaining three paths are determined by the last occurrence of a move of the type $(x, 0) \uparrow (x, 1)$.

(1) For $x = 0$, we have one ($= p_0$) path. From here—namely, $(0, 0)$—there is only one move: $(0, 0) \uparrow (0, 1)$. Following this, we have the two moves:

$(0, 1) \to (1, 1)$ and $(1, 1) \to (2, 1)$. Finally, the move $(2, 1) \searrow (3, 0)$ completes the path.

(2) When $x = 1$, there are two $(= p_1)$ paths from $(0, 0)$ to $(1, 0)$. This is then followed by the move $(1, 0) \uparrow (1, 1)$. From $(1, 1)$, we get to $(3, 0)$ by $(1, 1) \to (2, 1) \searrow (3, 0)$. This accounts for the other two paths in the third row of paths for $n = 3$. Consequently, we see that

$$p_3 = p_2 + p_2 + (p_0 + p_1).$$

To generalize this result, consider $n \geq 1$ and how p_{n+1} is related to p_n. Of the p_{n+1} paths from $(0, 0)$ to $(n + 1, 0)$, we know that p_n of these paths have the final move $(n, 0) \to (n + 1, 0)$. Another p_n paths (among the p_{n+1} paths) end with the moves $(n, 0) \uparrow (n, 1) \searrow (n + 1, 0)$. Finally, to determine the remaining paths from $(0, 0)$ to $(n + 1, 0)$, consider the paths where $(k, 0) \uparrow (k, 1)$ is the last move of this type, for $0 \leq k \leq n - 1$. Now there are p_k paths from $(0, 0)$ to $(k, 0)$. Each of these paths is then followed by the move: $(k, 0) \uparrow (k, 1)$, and then by the $n - k$ moves: $(k, 1) \to (k + 1, 1) \to \cdots \to (n, 1)$. The final move—namely, $(n, 1) \searrow (n + 1, 0)$—then completes the path [from $(0, 0)$ to $(n + 1, 0)$]. So now we have the recurrence relation

$$p_{n+1} = p_n + p_n + (p_0 + p_1 + \cdots + p_{n-1}).$$

This is quite different from any other recurrence relation we have encountered. To solve this for p_n, let us review what we know about the values of p_n for $n = 0, 1, 2,$ and 3. We have

$$p_0 = 1 = F_1, \quad p_1 = 2 = F_3, \quad p_2 = 5 = F_5, \quad p_3 = 13 = F_7.$$

These results suggest the pattern

$$p_n = F_{2n+1}.$$

To show that this is true in general, we shall use the Alternative, or Strong, form of the Principle of Mathematical Induction. We know that the result is true for $n = 0$ and $n = 1$ (as well as for $n = 2$ and $n = 3$). This establishes the basis step for our inductive proof. So now we assume the result true for $n = 0, 1, 2, \ldots, k \ (\geq 1)$—that is, we assume that $p_n = F_{2n+1}$, for all $0 \leq n \leq k$. Then for the case where $n = k + 1 \ (\geq 2)$, we have

$$\begin{aligned}
p_{k+1} &= p_k + p_k + (p_0 + p_1 + \cdots + p_{k-1}) \\
&= F_{2k+1} + F_{2k+1} + (F_1 + F_3 + \cdots + F_{2(k-1)+1}) \\
&= F_{2k+1} + (F_1 + F_3 + \cdots + F_{2k-1} + F_{2k+1}).
\end{aligned}$$

But Property 4.7 tells us that

$$F_1 + F_3 + \cdots + F_{2k-1} + F_{2k+1} = \sum_{i=0}^{k} F_{2i+1} = F_{2k+2}.$$

So

$$p_{k+1} = F_{2k+1} + F_{2k+2} = F_{2k+3} = F_{2(k+1)+1},$$

and the result is true for all $n \geq 0$, by the Alternative, or Strong, form of the Principle of Mathematical Induction.

The next example is different in that it came out of a study dealing with the location of sewage treatment plants.

Example 16.4:   The following was studied in 1974 in Reference [17] by Rolf A. Deininger at the University of Michigan in Ann Arbor.

Suppose that there are n cities along a river and that these cities pollute this river by discharging untreated sewage directly into the water. To stop this environmental problem, the city managers of these n cities have met to discuss plans to build one or more treatment plants to deal with the sewage. The objective is for each city to have a treatment plant of its own or to be upstream or downstream from a city that has such a plant. However, there is one restriction. If a city transports its untreated sewage (through a connecting sewer) to a neighboring city, it must transport all of the sewage to that one city. It cannot split the sewage and transport part of it upstream and part of it downstream. These considerations will then place an upper bound on the number of feasible plans to be investigated—but the city managers will still have to determine the most economical plan, or plans, by also considering such costs as labor, materials, and the allocation of materials.

To determine this upper bound, let us start with the case of $n = 2$ cities. In Fig. 16.5, we see the three feasible plans.

(1) Treatment plants (T) are built at each city. The "0" at the bottom of Fig. 16.5 (a) indicates that no sewage is transported from either of these two cities.

(2) Here a treatment plant (T) is built at city 2. Then the sewage from city 1 is transported to city 2 to be processed. The "1" at the bottom of Fig. 16.5 (b) indicates the transport is upstream.

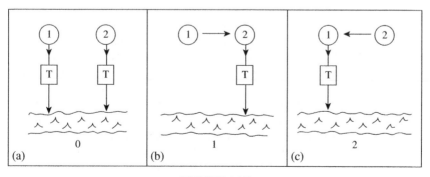

FIGURE 16.5

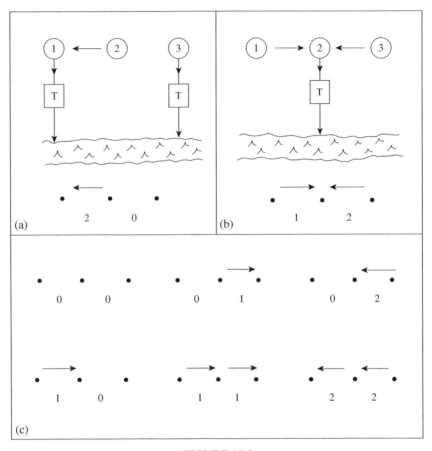

FIGURE 16.6

(3) In this last feasible solution, a treatment plant (T) is built at city 1. Now the sewage from city 2 is transported downstream to city 1. The "2" at the bottom of Fig. 16.5 (c) indicates the downstream transport.

For the case of $n = 3$ cities, we find that there are eight feasible solutions to consider. In Fig. 16.6 (a) we find the feasible solution where treatment plants are built at cities 1 and 3, and sewage is transported downstream from city 2. This is represented by the ternary string "20," as shown at the bottom of Fig. 16.6 (a). The plan for Fig. 16.6 (b) calls for a plant at city 2 with city 1 transporting its sewage upstream and city 3 doing so downstream. Here, the ternary string "12" represents the situation. The other six feasible solutions are represented in Fig. 16.6 (c). Although there are $3 \cdot 3 = 9$ ternary strings of length 2, we only have eight of them here, because the string "21" indicates sewage being transported from city 2 downstream to city 1 and upstream to city 3. But no city is allowed to split the transport of its sewage.

If there are four such cities along this river, the feasible solutions can be represented by ternary strings of length 3 that avoid the substring "21." For example, the ternary

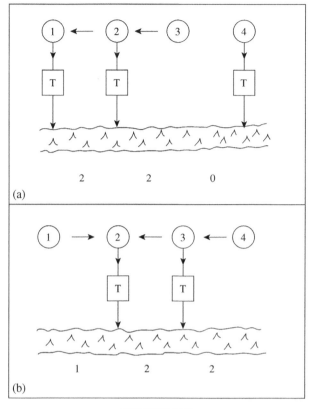

(a)

(b)

FIGURE 16.7

string for the feasible solution in Fig. 16.7 (a) is "220." The feasible solution in Fig. 16.7 (b) is represented by the ternary string "122." Now there are $3^3 = 27$ possible ternary strings of length 3, but each of the strings

$$210, 211, 212, 021, 121, \text{ and } 221$$

contain the substring "21." Consequently, the $27 - 6 = 21$ feasible solutions for four cities can be represented by the 21 ternary strings shown in Table 16.1.

TABLE 16.1

000	010	020	101	111	122	202
001	011	022	102	112	200	220
002	012	100	110	120	201	222

In order to determine an upper bound on the number of feasible solutions for n cities, let s_n count the number of such feasible solutions. At the same time, let a_n count the number of ternary strings of length n that do not contain the substring

"21." Then we find that

$$s_1 = 1 = a_0$$
$$s_2 = 3 = a_1$$
$$s_3 = 8 = a_2$$
$$s_4 = 21 = a_3.$$

To determine a_n, we consider the recurrence relation

$$a_n = 3a_{n-1} - a_{n-2}, \quad n \geq 3, \quad a_1 = 3, \quad a_2 = 8.$$

Upon appending any of the symbols 0, 1, or 2 to the right of each of the a_{n-1} strings of length $n - 1$, the only resulting strings of length n that contain the substring "21" are the ternary strings of length n that have "21" in positions $n - 1$ and n, respectively. There are a_{n-2} such ternary strings of length n. Consequently, arguing as we did in Example 16.1, we find that

$$a_n = F_{2n+2}, \quad n \geq 0,$$

and so

$$s_n = a_{n-1} = F_{2(n-1)+2} = F_{2n}, \quad n \geq 1.$$

(More on ternary strings that do not contain the substring "21" can be found in Reference [28].)

Now for one final example!

Example 16.5: Let us return to the compositions in Example 6.2. There we learned that the number of compositions of n, where the only summands allowed are 1's and 2's, is F_{n+1}. Before we state the result we are seeking here, let us examine two specific examples to illustrate how one might conjecture the general result.

(i) Consider all four compositions of 3 as shown in Table 16.2.

TABLE 16.2

Composition of 3	Product of Summands in the Composition
3	3
$2 + 1$	$2 \cdot 1 = 2$
$1 + 2$	$1 \cdot 2 = 2$
$1 + 1 + 1$	$1 \cdot 1 \cdot 1 = 1.$

The sum of these four products is

$$3 + 2 + 2 + 1 = 8 \, (= F_6)$$

or

$$\sum_{\substack{a_1 + \cdots + a_m = 3 \\ a_i \geq 1,\, i = 1,\ldots,\, m}} a_1 \cdots a_m = 8 \ (= F_6).$$

(ii) Now examine the eight compositions of 4 as shown in Table 16.3.

TABLE 16.3

Composition of 4	Product of Summands in the Composition
4	4
$3 + 1$	$3 \cdot 1 = 3$
$2 + 2$	$2 \cdot 2 = 4$
$2 + 1 + 1$	$2 \cdot 1 \cdot 1 = 2$
$1 + 3$	$1 \cdot 3 = 3$
$1 + 2 + 1$	$1 \cdot 2 \cdot 1 = 2$
$1 + 1 + 2$	$1 \cdot 1 \cdot 2 = 2$
$1 + 1 + 1 + 1$	$1 \cdot 1 \cdot 1 \cdot 1 = 1$

Here the sum of the eight products is

$$4 + 3 + 4 + 2 + 3 + 2 + 2 + 1 = 21 \ (= F_8)$$

or

$$\sum_{\substack{a_1 + \cdots + a_m = 4 \\ a_i \geq 1,\, i = 1,\ldots,\, m}} a_1 \cdots a_m = 21 \ (= F_8).$$

With these two examples to lead the way, let us now consider a composition of a slightly larger integer like 12. So consider

$$3 + 7 + 2 = 12.$$

The following idea, developed by Ira Gessel, will help us to set up a correspondence between all compositions of 12 and the compositions of 23 $[= 2(12) - 1]$, where only 1's and 2's are allowed as summands.

Place 12 dots in a line as shown in Fig. 16.8 on p. 138, then insert at most one (separating) bar between any two adjacent dots. Here we have inserted two such separating bars. Now select and circle one of the three dots that appear before the first bar, then one dot between the first and second bars, and finally one dot following the second bar.

Having made the above selection, at this point we generate a composition of 23, using only 1's and 2's as summands, by doing the following: (i) Replace each circled dot with a 1. (ii) Replace each bar with a 1. (iii) Replace each uncircled dot with a 2.

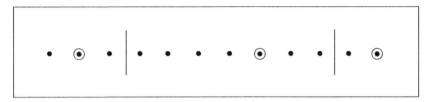

FIGURE 16.8

The result is the following composition of 23.

$$2 + 1 + 2 + 1 + 2 + 2 + 2 + 2 + 1 + 2 + 2 + 1 + 2 + 1 = 23 = 2(12) - 1.$$

Note that in this example there were three choices for circling the first dot, seven choices for circling the second dot, and two choices for circling the third dot. So from this one composition, $3 + 7 + 2 = 12$, we get $3 \cdot 7 \cdot 2 \, (= 42)$ of the compositions of 23 where the only summands allowed are 1's and 2's.

If we do the same thing with the composition $3 + 3 + 4 + 2 = 12$, as shown in Fig. 16.9, we obtain another composition of 23, using only 1's and 2's as summands—namely,

$$2 + 1 + 2 + 1 + 1 + 2 + 2 + 1 + 2 + 1 + 2 + 2 + 1 + 2 + 1 = 23 = 2(12) - 1.$$

Note how adding another bar brings about another circled dot. The uncircled dot that was originally replaced by a 2 is now replaced by a 1, but the additional bar provides another 1, so the sum is still 23. Also note that this one composition of 12 provides us with $3 \cdot 3 \cdot 4 \cdot 2 \, (= 72)$ more of the compositions of 23, using only 1's and 2's as summands.

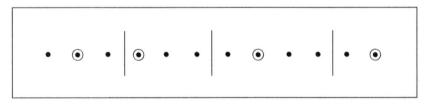

FIGURE 16.9

We also realize that this process is reversible. For example, given the composition

$$1 + 1 + 1 + 1 + 2 + 2 + 2 + 1 + 2 + 1 + 2 + 2 + 2 + 1 + 1 + 1 = 23 = 2(12) - 1,$$

we determine the corresponding string of dots, circled dots, and bars that generate it, as follows: (i) Replace each 2 with an uncircled dot. (ii) Scanning the composition from left to right, replace the kth 1 by a circled dot, when k is odd, and by a bar, when

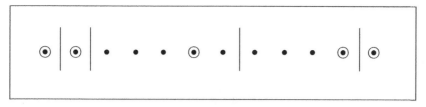

FIGURE 16.10

k is even. This gives us the result in Fig. 16.10, which shows us that this composition of 23 (where the only summands are 1's and 2's) comes about from the composition

$$1 + 1 + 5 + 4 + 1 = 12.$$

This composition of 12 accounts for $1 \cdot 1 \cdot 5 \cdot 4 \cdot 1 = 20$ compositions of 23 using only 1's and 2's as summands. Consequently, we see that the total number of compositions of 23, using only 1's and 2's as summands, equals

$$\sum_{\substack{a_1+a_2+\cdots+a_m = 12 \\ a_i \geq 1,\ i=1,\ldots,m}} a_1 a_2 \cdots a_m.$$

In Example 6.2, we learned that the number of compositions of n using only 1's and 2's as summands is F_{n+1}. Therefore, it now follows that

$$\sum_{\substack{a_1+a_2+\cdots+a_m=12 \\ a_i \geq 1,\ i=1,\ldots,m}} a_1 a_2 \cdots a_m = F_{(2(12)-1)+1} = F_{24} = F_{2(12)}.$$

The above argument generalizes to tell us that for $n \geq 1$,

$$\sum_{\substack{a_1+a_2+\cdots+a_m = n \\ a_i \geq 1,\ i=1,\ldots,m}} a_1 a_2 \cdots a_m = F_{(2n-1)+1} = F_{2n}.$$

EXERCISE FOR CHAPTER 16

1. For $n \geq 1$, let a_n count the number of ternary strings of length n that avoid the substring "21" and also avoid the initial symbol "0." For example, $a_1 = 2$ for the ternary strings 1 and 2 and $a_2 = 5$ for the ternary strings 10, 20, 11, 22, and 12. Set up and solve a recurrence relation for a_n.

CHAPTER 17

One Final Example?

Why the question mark at the end of this chapter title. Is it a misprint? Let us see.

At this point, we want to make sure that the reader understands and appreciates the pains we have gone through in always setting up the somewhat repetitious recurrence relation that arose in the examples where the solution was F_n (or F_{n-1}, F_{n+1}, F_{n+2}, F_{2n}, or F_{2n+1}). When we are proving a theorem, we cannot draw any general conclusions from a few (or even, perhaps, many) particular instances where the result stated in the theorem happens to be true. The same is true here, and the following example should serve to drive this point home!

Example 17.1: We start with n identical circles and let a_n count the number of ways we can arrange these circles—*contiguous* in each row—with each circle above the bottom row tangent to two circles in the row below it. In Fig. 17.1, we see the possible ways to so arrange the n circles for $1 \leq n \leq 6$. It follows from this that

$$a_1 = 1 \quad a_2 = 1 \quad a_3 = 2$$
$$a_4 = 3 \quad a_5 = 5 \quad a_6 = 8.$$

These results definitely *suggest* that we have encountered another instance where the Fibonacci numbers arise. But before we write $a_n = F_n$, for all $n \geq 1$, remember that we have not presented a general argument or set up a recurrence relation.

Unfortunately, we have been led astray in this instance. Here one finds, for example, that

$$a_7 = 12 \neq 13 = F_7$$
$$a_8 = 18 \neq 21 = F_8$$
$$a_9 = 26 \neq 34 = F_9.$$

The sequence of numbers a_1, a_2, a_3, ... was studied in Reference [2]. Other such counterexamples can be found in Reference [30].

Fibonacci and Catalan Numbers: An Introduction, First Edition. Ralph P. Grimaldi.
© 2012 John Wiley & Sons, Inc. Published 2012 by John Wiley & Sons, Inc.

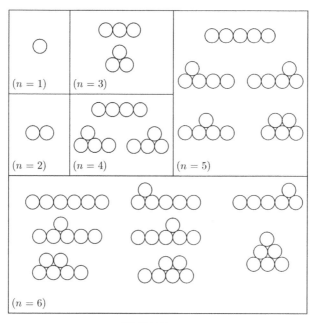

FIGURE 17.1

EXERCISE FOR CHAPTER 17

1. Consider Example 17.1 where we arranged n circles. This time, however, we specify the following conditions for the arrangements, where the second and fourth conditions are new: (1) The circles in the bottom row must still be contiguous. (2) Now at most two rows are allowed. (3) Each circle in the upper row must still rest on two circles below it. (4) The circles in the upper row need not be contiguous. For $1 \leq n \leq 5$, the ways to so arrange n circles in this way are precisely as shown in Fig. 17.1.

   (a) Set up and solve a recurrence relation with initial conditions to show that the pattern continues in this case.

   (b) If $n = 10$, how many arrangements have (i) no circles in the upper row; (ii) one circle in the upper row; (iii) two circles in the upper row; (iv) three circles in the upper row; and, (v) four circles in the upper row.

   (c) Determine the sum of the results in parts (i)–(v) of part (b).

   (d) Generalize the results in parts (b) and (c).

   (This exercise is due to Gary Stevens of Hartwick College.)

REFERENCES

1. Alladi, Krishnaswami, and Hoggatt, Jr., Verner E. Compositions with Ones and Twos. *The Fibonacci Quarterly*, Volume 13, Issue 3, October, 1975, Pp. 233 - 239.
2. Auluck, F. C. On Some New Types of Partitions Associated with Generalized Ferrers Graphs. *Proceedings of the Cambridge Philosophical Society*, Volume 47, 1951, Pp. 679 - 685.
3. Benjamin, Arthur T., and Quinn, Jennifer J. *Proofs That Really Count (The Art of Combinatorial Proof)*. Washington, DC: The Mathematical Association of America, 2003.
4. Blazej, Richard. Problem B - 298. *The Fibonacci Quarterly*, Volume 13, Issue 1, February, 1975, P. 94.
5. Blazej, Richard. Problem B - 294. *The Fibonacci Quarterly*, Volume 13, Issue 3, October, 1975, P. 375.
6. Brigham, Robert C., Caron, Richard M., Chinn, Phyllis Z., Grimaldi, Ralph P. A Tiling Scheme for the Fibonacci Numbers. *The Journal of Recreational Mathematics*, Volume 28, Issue 1, 1996-97, Pp. 10 - 16.
7. Brigham, Robert C., Carrington, Julie R., Jeong, Dal Y., Vitray, Richard P., and Yellen, Jay. Domination in Fibonacci Trees. *The Fibonacci Quarterly*, Volume 43, Issue 2, May, 2005, Pp. 157 - 165.
8. Brousseau, Brother Alfred. Problem H - 29. *The Fibonacci Quarterly*, Volume 2, Issue 4, December, 1964, P. 305.
9. Brousseau, Brother Alfred. Fibonacci-Lucas Infinite Series — Research Topic. *The Fibonacci Quarterly*, Volume 7, Issue 2, April, 1968, Pp. 211 - 217.
10. Brown, Jr., John L. Problem H - 71. *The Fibonacci Quarterly*, Volume 5, Issue 2, April, 1967, Pp. 166 - 167.
11. Brualdi, Richard A. *Introductory Combinatorics*, fifth edition. Upper Saddle River, New Jersey: Pearson Prentice-Hall, 2010.
12. Butchart, J. H. Problem B - 124. *The Fibonacci Quarterly*, Volume 6, Issue 4, October, 1968, Pp. 289 - 290.
13. Byrd, Paul F. Problem B - 12. *The Fibonacci Quarterly*, Volume 1, Issue 2, April, 1963, P. 86.
14. Chinn, Phyllis Z., and Heubach, Silvia P. Compositions of n with No Occurrence of k. *Congressus Numerantium*, Volume 164, 2003, Pp. 33 - 51.
15. Conway, John H., and Guy, Richard K. *The Book of Numbers*. New York: Springer-Verlag, 1996.
16. Davis, Basil. Fibonacci Numbers in Physics. *The Fibonacci Quarterly*, Volume 10, Issue 6, December, 1972, Pp. 659 - 660, 662.
17. Deininger, Rolf A. Fibonacci Numbers and Water Pollution Control. *The Fibonacci Quarterly*, Volume 10, Issue 43, April, 1972, Pp. 299 - 300, 302.
18. Dence, T. P. Problem B - 129. *The Fibonacci Quarterly*, Volume 6, Issue 4, October, 1968, P. 296.
19. Deutsch, Emeric. Problem H - 641. *The Fibonacci Quarterly*, Volume 44, Issue 2, May, 2006, P. 188.
20. Drake, Reuben C. Problem B - 180. *The Fibonacci Quarterly*, Volume 8, Issue 5, December, 1970, Pp. 547 - 548.
21. Freitag, Herta Taussig. Problem B - 286. *The Fibonacci Quarterly*, Volume 13, Issue 3, October, 1975, P. 286.
22. Gardner, Martin. About phi, an Irrational Number that Has Some Remarkable Geometrical Expressions. *Scientific American*, Volume 201, No. 2, August, 1959, Pp. 128 - 134.

23. Gordon, M., and Davison, W. H. T. Theory of Resonance Topology of Fully Aromatic Hydrocarbons. I. *Journal of Chemical Physics*. Volume 20, Number 3, March, 1952, Pp. 428 - 435.

24. Gould, Henry W. Problem B - 7. *The Fibonacci Quarterly*, Volume 1, Issue 3, October, 1963, P. 80.

25. Grimaldi, Ralph P. *Discrete and Combinatorial Mathematics*, fifth edition. Boston, Massachusetts: Pearson Addison Wesley, 2004.

26. Grimaldi, Ralph P. Generating Sets and the Fibonacci Numbers, *Congressus Numerantium*, Volume 110, 1995, Pp. 129 - 136.

27. Grimaldi, Ralph P. Properties of Fibonacci Trees. *Congressus Numerantium*. Volume 84, 1991, Pp. 21 - 32.

28. Grimaldi, Ralph P. Ternary Strings that Avoid the Substring '21', *Congressus Numerantium*, Volume 170, 2004, Pp. 33 - 49.

29. Gutman, Ivan, Vukičević, Damir, Graovac, Ante, and Randić, Milan. Algebraic Kekulé Structures of Benzenoid Hydrocarbons, *Journal of Chemical Information and Computer Science*, Volume 44, 2004, Pp. 296 - 299.

30. Guy, Richard K. The Second Strong Law of Large Numbers. *The Mathematics Magazine*, Volume 63, No. 1, February, 1990, Pp. 3 - 20.

31. Hoggatt, Jr., Verner E., and Lind, D. A. The Heights of Fibonacci Polynomials and an Associated Function. *The Fibonacci Quarterly*, Volume 5, Issue 2, April, 1967, Pp. 141 - 152

32. Hope-Jones, W. The Bee and the Regular Pentagon. *The Mathematical Gazette*, Volume X, No. 150, January, 1921, P. 206. [Reprinted in Volume LV, No. 392, March, 1971, P. 220.]

33. Hunter, J. A. H. Triangle Inscribed in Rectangle. *The Fibonacci Quarterly*, Volume 1, Issue 3, October, 1963, P. 66.

34. Jean, Roger V. *A Mathematical Approach to Pattern and Form in Plant Growth*. New York: John Wiley & Sons, Inc., 1984.

35. King, Charles H. Some Properties of the Fibonacci Numbers. Master's Thesis. San Jose State College, San Jose, California, June 1960.

36. Knisley, J., Wallis, Charles, and Domke, Gayla. The Partitioned Graph Isomorphism Problem. *Congressus Numerantium*. Volume 89, 1992, Pp. 39 - 44.

37. Koshy, Thomas. New Fibonacci and Lucas Identities. *The Mathematical Gazette*, Volume 82, November, 1998, Pp. 481 - 484.

38. Koshy, Thomas. *Fibonacci and Lucas Numbers with Applications*. New York: John Wiley & Sons, Inc., 2001.

39. Mana, Phil. Problem B - 152. *The Fibonacci Quarterly*, Volume 7, Issue 3, October, 1969, P. 336.

40. McNabb, Sister Mary DeSales. Phyllotaxis. *The Fibonacci Quarterly*, Volume 1, Issue 4, December, 1963, Pp. 57 - 60.

41. Moore, Sam. Fibonacci Matrices. *The Mathematical Gazette*. Volume 67, 1983, Pp. 56 - 57.

42. Moser, Leo, and Brown Jr., John L. Some Reflections. *The Fibonacci Quarterly*, Volume 1, Issue 1, December, 1963, P. 75.

43. Moser, Leo, and Wyman, Max. Problem B - 6. *The Fibonacci Quarterly*, Volume 1, Issue 1, February, 1963, P. 74.

44. Pauling, Linus. *The Nature of the Chemical Bond*, third edition. Ithaca: Cornell University Press, 1960.

45. Rao, K. S. Some Properties of Fibonacci Numbers. *The American Mathematical Monthly*, Volume 60, May, 1953, Pp. 680 - 684.
46. Recke, Klaus Gunther. Problem B - 153. *The Fibonacci Quarterly*, Volume 7, Issue 3, October, 1969, P. 276.
47. Rosen, Kenneth H. *Discrete Mathematics and Its Applications*, sixth edition. New York: McGraw-Hill, 2007.
48. Schröder, Bernd S. W. *Ordered Sets, An Introduction*. Boston: Birkhauser, 2003.
49. Sharpe, B. On Sums $F_x^2 \pm F_y^2$. *The Fibonacci Quarterly*, Volume 3, Issue 1, February, 1965, P. 63.
50. Sloane, Neil James Alexander. The On-Line Encyclopedia of Integer Sequences. http://www.research.att.com/~njas/sequences/
51. Stevens, Gary E. The Bishop's Tale: A Combinatorial Proof of $F_n^2 = 2(F_{n-1}^2 + F_{n-2}^2) - F_{n-3}^2$. *The Fibonacci Quarterly*, Volume 45, Issue 4, November, 2007, Pp. 319 - 321.
52. Swamy, M. N. S. Problem B - 83. *The Fibonacci Quarterly*, Volume 4, Issue 4, December, 1966, P. 375.
53. Tadlock, Sheryl B. Products of Odds. *The Fibonacci Quarterly*, Volume 3, Issue 1, February, 1965, Pp. 54 - 56.
54. Tanton, James S. Fibonacci Numbers, Generating Sets, and Hexagonal Properties. *The Fibonacci Quarterly*, Volume 38, Issue 4, August, 2000, Pp. 299 - 309.
55. Tošić, Ratko, and Stojmenović, Ivan. Fibonacci Numbers and the Numbers of Perfect Matchings of Square, Pentagonal, and Hexagonal Chains. *The Fibonacci Quarterly*, Volume 30, Issue 4, November, 1992, Pp. 315 - 321.
56. Walser, Hans. *The Golden Section*. Washington, D. C.: The Mathematical Association of America, 2001.
57. Williams, R. F. *The Shoot Apex and Leaf Growth*. New York: Cambridge University Press, 1974.

PART TWO

THE CATALAN NUMBERS

CHAPTER 18

Historical Background

In 1751, in a letter to the Prussian mathematician Christian Goldbach (1670–1764), the prolific Swiss mathematician Leonard Euler (1707–1783) conjectured a method for counting the following geometric situations.

Start with a polygon of n (≥ 3) sides. Such a polygon is called *convex* if whenever P and Q are two points in the interior of the polygon, then all points on the segment PQ are also in the interior of the polygon—as shown, for example, for the convex hexagon in Fig. 18.1(a) on p. 148. In Fig. 18.1(b), the pentagon shown is not convex, for some of the points on the segment PQ are in the exterior of the pentagon.

Now consider the convex pentagon in Fig. 18.1(c). Here we have triangulated the interior of the pentagon by drawing the two diagonals AC and CE, which do not intersect within the interior of the pentagon. Part (d) of Fig. 18.1 provides a second such triangulation of the interior of the convex pentagon, this time by the diagonals AD and BD.

Euler was concerned with determining T_n, the total number of ways one can draw $n - 3$ diagonals within the interior of a convex polygon of n sides (for $n \geq 3$) so that no two of the diagonals intersect within the interior of the polygon, and the interior of the polygon is triangulated into $n - 2$ triangles. He calculated T_n for the first few small values of n and considered the ratio T_{n+1} / T_n.

In a later correspondence with the Hungarian mathematician, physicist, and physician Johann Andreas von Segner (1704–1777), Euler suggested this problem. Von Segner obtained a recurrence relation for T_n, calculating the results for $n \leq 20$. However, his solution did not prove, or disprove, the method suggested. Von Segner sent his results to Euler in 1756, not realizing that his answers were incorrect in the cases of $n = 15$ and $n = 20$—these errors due to mistakes in his arithmetic calculations. Euler, in turn, essentially solved the recurrence but without the details of a proof.

The French mathematician, physicist, and engineer Gabriel Lamé (1795–1870) was among the first to provide an elegant combinatorial proof for the Euler–von Segner result, in 1838, in a letter to the French mathematician Joseph Liouville (1809–1882). His proof was further developed and discussed in articles published in 1838

Fibonacci and Catalan Numbers: An Introduction, First Edition. Ralph P. Grimaldi.
© 2012 John Wiley & Sons, Inc. Published 2012 by John Wiley & Sons, Inc.

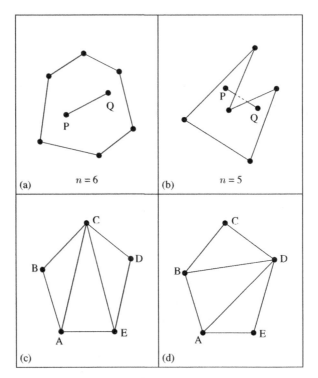

FIGURE 18.1

and 1839 by the Belgian mathematician Eugène Charles Catalan (1814–1894). (More about counting the triangulations of a convex polygon is covered in Example 23.1 and Example 30.1.)

Catalan wrote other articles dealing with this number sequence, including one in 1838 where he determined the number of ways a chain of $n + 1$ symbols can be parenthesized with n pairs of parentheses so that each pair surrounds two symbols, a parenthesized expression and a symbol, or two parenthesized expressions. (This is examined in Example 20.3.) In Reference [30], the German mathematician Eugen Otto Erwin Netto (1848–1919) attributed the solutions of the aforementioned parenthesization problem and the triangulation of a convex polygon to Catalan and the term "Catalan number" seems to have arisen from these citations.

Catalan also did research in such areas as continued fractions, infinite products, multiple integrals, differential equations, and geometry. In addition, he wrote the texts *Élements de Geometrié*, published in 1843, and *Notions d́ astronomie*, published in 1860. Another interesting historical discussion about the Catalan numbers is found in Reference [3].

A little known historical fact about the Catalan numbers is that they were discovered independently in China by the Mongolian mathematician Antu Ming (c. 1692–

1763). In the 1730's he derived several recurrences for the Catalan numbers and started a book titled *Efficient Methods for the Precise Values of Circular Functions*, wherein his knowledge of this sequence of numbers is clearly demonstrated. The book was finished by his student Chen Jixin in 1774, but it was not published until 1839.

CHAPTER 19

A First Example: A Formula for the Catalan Numbers

In this chapter, we shall find an example where the Catalan numbers arise. This example will deal with lattice paths in the Cartesian plane and it will provide us with the means to determine a formula for the Catalan numbers. In addition, we shall learn how the Catalan numbers are related to certain entries in Pascal's triangle.

Example 19.1:   Let us start at the point $(0, 0)$ in the xy-plane and consider two types of steps:

$$R: (x, y) \to (x + 1, y) \qquad U: (x, y) \uparrow (x, y + 1).$$

We want to count the number of ways we can travel from $(0, 0)$ to $(4, 4)$ using such steps—one unit to the right or one unit up. Consequently, any such path will be made up of four R's and four U's. Since we can arrange four R's and four U's in

$$\frac{8!}{4!4!} = \binom{8}{4} = 70$$

ways, this tells us that there are 70 such paths from $(0, 0)$ to $(4, 4)$. These paths are often referred to as *lattice* paths. In parts (a), (b), (c), and (e) of Fig. 19.1 we find four of these 70 possible lattice paths.

Now suppose that we once again start at the point $(0, 0)$ in the xy-plane and travel to the point $(4, 4)$ using the same types of steps—namely, four R's and four U's. But this time there is a catch! As we progress from $(0, 0)$ to $(4, 4)$, we can never travel above the line $y = x$. We will touch the line at $(0, 0)$ and $(4, 4)$, and perhaps at $(1, 1)$, $(2, 2)$, or $(3, 3)$, but we will not touch it at any other point on the line $y = x$, nor will we ever venture above the line. Consequently, the paths shown in parts (c) and (e) of Fig. 19.1 are *not* among the paths we are interested in counting here. The paths shown in Fig. 19.1(a), (b) are, however, two of the paths we want to consider.

Fibonacci and Catalan Numbers: An Introduction, First Edition. Ralph P. Grimaldi.
© 2012 John Wiley & Sons, Inc. Published 2012 by John Wiley & Sons, Inc.

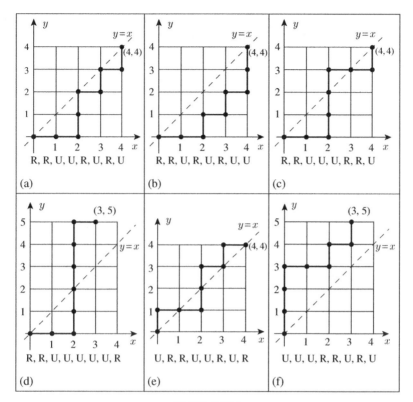

FIGURE 19.1

It is evident that each such arrangement of four R's and four U's must start with an R and end with a U. Then for each such arrangement of four R's and four U's, going from left to right, the number of R's at any point must always equal or exceed the number of U's. This happens in parts (a), (b) of Fig. 19.1, but not in parts (c) and (e). Consequently, if we can count the number of paths from $(0, 0)$ to $(4, 4)$ where the path rises above the line $y = x$ [as in Fig. 19.1(c), (e)], then we can solve the problem at hand. All we have to do is subtract the number of paths that rise above the line $y = x$ from the total number of 70 $\left[= \binom{8}{4} \right]$ paths.

Look again at the path in Fig. 19.1(c). Where does the situation there break down for the first time—that is, at what point do the number of U's first exceed the number of R's, as we examine the arrangement of four R's and four U's, going from left to right? After all, we start with the required R—then follow it by another R. So far, so good! Then we follow these first two R steps with two U's, and everything is still fine. But then there is a third U and, at this (first) time, the number of U's (namely, three) exceeds the number of R's (namely, two).

At this point, let us introduce the following transformation:

$$\text{R, R, U, U, U, } \vdots \text{ R, R, U} \leftrightarrow \text{R, R, U, U, U, } \vdots \text{ U, U, R.}$$

What have we done here? For the path on the left side of the transformation, we found the first step (the third U) where the path rose above the line $y = x$ for the first time. The moves up to and including this step (the third U) remain as is, but the steps that follow are interchanged—each R is replaced by a U and each U by an R. The result is the path on the right-hand side of the transformation—an arrangement of three R's and five U's, as shown in Fig. 19.1(d). Part (e) of Fig. 19.1 provides a second path we want to avoid; part (f) shows us what occurs when this path is transformed by the method given above.

Now suppose that we start with an arrangement of three R's and five U's, say

$$R, U, U, \vdots R, U, U, R, U.$$

These eight steps provide us with a path from $(0, 0)$ to $(3, 5)$. Focus on the first place where the number of U's exceeds the number of R's. This occurs in the third position, the location of the second U. This arrangement is now transformed as follows: The steps up to and including the second U remain as they appear; the last five moves are interchanged as before—each R is replaced by a U, each U by an R. This gives us the arrangement

$$R, U, U, \vdots U, R, R, U, R$$

—one of the *bad* arrangements (of four R's and four U's) that we want to avoid as we travel from $(0, 0)$ to $(4, 4)$. The correspondence provided by these transformations gives us a way to count all of the *bad* arrangements. We alternatively count the number of ways to arrange three R's and five U's. This number is $8!/(3!5!) = \binom{8}{3}$. Consequently, the number of paths from $(0, 0)$ to $(4, 4)$, which do not rise above the line $y = x$, is

$$\binom{8}{4} - \binom{8}{3} = \frac{8!}{4!4!} - \frac{8!}{3!5!} = \frac{5(8!) - 4(8!)}{5!4!} = \frac{8!}{5!4!}$$
$$= \left(\frac{1}{5}\right)\left(\frac{8!}{4!4!}\right) = \left(\frac{1}{4+1}\right)\binom{8}{4} = \left(\frac{1}{4+1}\right)\binom{2 \cdot 4}{4} = 14.$$

If, instead of going from $(0, 0)$ to $(4, 4)$, we would like to go from $(0, 0)$ to (n, n), for $n \geq 0$, then we find that the result for $n = 4$ generalizes as follows. For each integer $n \geq 0$, the number of lattice paths, made up of n R's and n U's, going from $(0, 0)$ to (n, n) and never rising above the line $y = x$, is the nth *Catalan number*, given by

$$C_n = \binom{2n}{n} - \binom{2n}{n-1}, \quad n \geq 1, \ C_0 = 1$$

or

$$C_n = \frac{1}{n+1}\binom{2n}{n}, \quad n \geq 0.$$

[The reader will be asked to verify that $\binom{2n}{n} - \binom{2n}{n-1} = (1/(n+1))\binom{2n}{n})$, for $n \geq 1$, in the exercises for this chapter.]

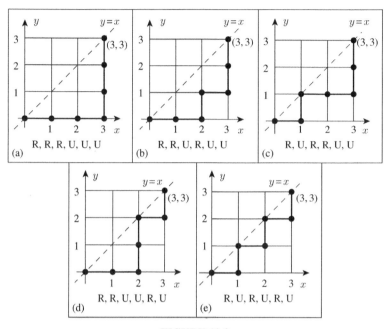

FIGURE 19.2

For $n = 3$, we have

$$C_3 = \frac{1}{3+1}\binom{2 \cdot 3}{3} = \frac{1}{4}\binom{6}{3} = \left(\frac{1}{4}\right)(20) = 5.$$

The five lattice paths from $(0, 0)$ to $(3, 3)$ that never rise above the line $y = x$ are exhibited in Fig. 19.2. As we shall see shortly, these paths will provide a source of comparison with other situations counted by the Catalan numbers.

Using either of the above formulas, we find that the first 20 Catalan numbers are as given in Table 19.1.

TABLE 19.1

$C_0 = 1$	$C_5 = 42$	$C_{10} = 16,796$	$C_{15} = 9,694,845$
$C_1 = 1$	$C_6 = 132$	$C_{11} = 58,786$	$C_{16} = 35,357,670$
$C_2 = 2$	$C_7 = 429$	$C_{12} = 208,012$	$C_{17} = 129,644,790$
$C_3 = 5$	$C_8 = 1430$	$C_{13} = 742,900$	$C_{18} = 477,638,700$
$C_4 = 14$	$C_9 = 4862$	$C_{14} = 2,674,440$	$C_{19} = 1,767,263,190.$

Since binomial coefficients appear in both of our formulas for the Catalan numbers, let us close this chapter by examining how one can obtain these numbers from Pascal's triangle. In Fig. 19.3 we have the rows for $n = 0$ to $n = 8$ in Pascal's triangle.

$$1$$
$$1 \quad 1$$
$$1 \quad 2 \quad 1$$
$$1 \quad 3 \quad 3 \quad 1$$
$$1 \quad 4 \quad 6 \quad 4 \quad 1$$
$$1 \quad 5 \quad 10 \quad 10 \quad 5 \quad 1$$
$$1 \quad 6 \quad 15 \quad 20 \quad 15 \quad 6 \quad 1$$
$$1 \quad 7 \quad 21 \quad 35 \quad 35 \quad 21 \quad 7 \quad 1$$
$$1 \quad 8 \quad 28 \quad 56 \quad 70 \quad 56 \quad 28 \quad 8 \quad 1$$

FIGURE 19.3

The central binomial coefficients—namely, $1 = \binom{2 \cdot 0}{0}$, $2 = \binom{2 \cdot 1}{1}$, $6 = \binom{2 \cdot 2}{2}$, $20 = \binom{2 \cdot 3}{3}$, and $70 = \binom{2 \cdot 4}{4}$—appear down the center of the array in Fig. 19.3. When each such term of the form $\binom{2n}{n}$ is divided by $n + 1$, we find that

$$(n = 0) \quad \frac{1}{0+1}\binom{2 \cdot 0}{0} = \frac{1}{1}\binom{0}{0} = \frac{1}{1}(1) = 1 = C_0$$

$$(n = 1) \quad \frac{1}{1+1}\binom{2 \cdot 1}{1} = \frac{1}{2}\binom{2}{1} = \frac{1}{2}(2) = 1 = C_1$$

$$(n = 2) \quad \frac{1}{2+1}\binom{2 \cdot 2}{2} = \frac{1}{3}\binom{4}{2} = \frac{1}{3}(6) = 2 = C_2$$

$$(n = 3) \quad \frac{1}{3+1}\binom{2 \cdot 3}{3} = \frac{1}{4}\binom{6}{3} = \frac{1}{4}(20) = 5 = C_3$$

$$(n = 4) \quad \frac{1}{4+1}\binom{2 \cdot 4}{4} = \frac{1}{5}\binom{8}{4} = \frac{1}{5}(70) = 14 = C_4.$$

Alternatively, for each $n \geq 1$, consider the entry in the center of the row for $2n$ (in Pascal's triangle). This is the term $\binom{2n}{n}$. From this, subtract the entry to its left in the triangle—namely, the term $\binom{2n}{n-1}$. When this is done, we find that

$(n = 1)$ $\binom{2 \cdot 1}{1} - \binom{2 \cdot 1}{1 - 1} = \binom{2}{1} - \binom{2}{0} = 2 - 1 = 1 = C_1$

$(n = 2)$ $\binom{2 \cdot 2}{2} - \binom{2 \cdot 2}{2 - 1} = \binom{4}{2} - \binom{4}{1} = 6 - 4 = 2 = C_2$

$(n = 3)$ $\binom{2 \cdot 3}{3} - \binom{2 \cdot 3}{3 - 1} = \binom{6}{3} - \binom{6}{2} = 20 - 15 = 5 = C_3$

$(n = 4)$ $\binom{2 \cdot 4}{4} - \binom{2 \cdot 4}{4 - 1} = \binom{8}{4} - \binom{8}{3} = 70 - 56 = 14 = C_4.$

EXERCISES FOR CHAPTER 19

1. For $n \geq 1$, prove that

$$\binom{2n}{n} - \binom{2n}{n - 1} = \left(\frac{1}{n + 1}\right)\binom{2n}{n}.$$

2. For $n \geq 1$, prove that

$$C_n = \frac{1}{n}\binom{2n}{n + 1}.$$

3. For $n \geq 1$, prove that

$$C_n = \frac{1}{n}\binom{2n}{n - 1}.$$

4. For $n \geq 0$, prove that

$$C_n = 2\binom{2n}{n} - \binom{2n + 1}{n}.$$

5. For $n \geq 1$, prove that

$$C_n = 4\binom{2n - 1}{n} - \binom{2n + 1}{n}.$$

6. (a) For $n \geq 1$, prove that

$$C_n = \frac{4n - 2}{n + 1} C_{n-1}.$$

(b) Determine

$$\lim_{n \to \infty} \frac{C_{n+1}}{C_n}.$$

7. For $n \geq 0$, prove that

$$C_n = \frac{1}{2n + 1} \binom{2n + 1}{n}.$$

8. For $n \geq 1$, prove that

$$C_n = \binom{2n - 1}{n - 1} - \binom{2n - 1}{n - 2}.$$

9. For $n \geq 0$, prove that

$$C_n = \binom{2n + 1}{n + 1} - 2\binom{2n}{n + 1}.$$

10. For $n \geq 5$, prove that

$$\text{(a) } \binom{2n}{n} < 4^{n-1} \quad \text{and} \quad \text{(b) } C_n < \frac{4^{n-1}}{n + 1}.$$

11. For $n \geq 2$, prove that

$$C_{n+1} = 2C_n + \frac{2}{n}\binom{2n}{n - 2}.$$

12. For $n \geq 2$, prove that

$$C_{n+1} = \binom{2n}{n} - \binom{2n}{n - 2}.$$

13. For $n \geq 1$, prove that

$$C_{2n} = \binom{4n}{2n} - \binom{4n}{2n - 1}.$$

14. For $n \geq 1$, prove that

$$C_n = \frac{2 \cdot 6 \cdot 10 \cdot 14 \cdots (4n - 2)}{1 \cdot 2 \cdot 3 \cdot 4 \cdots (n + 1)}. \text{ (C. D. Olds, 1947) [31]}.$$

15. The *gamma function* Γ was developed by the French mathematician Adrien-Marie Legendre (1752–1833) as an extension of the factorial function. For a real number $r > 0$, we have

$$\Gamma(r) = \int_0^{+\infty} t^{r-1} e^{-t} dt.$$

For $r > 1$ this function satisfies the recurrence relation

$$\Gamma(r) = (r-1)\Gamma(r-1).$$

If r is an integer and $r \geq 1$, then

$$\Gamma(r) = (r-1)!.$$

Also, $\Gamma\left(\frac{1}{2}\right) = \sqrt{\pi}$.

(a) Prove that, for a positive integer n,

$$\Gamma\left(n + \frac{1}{2}\right) = \left(n - \frac{1}{2}\right)\left(n - \frac{3}{2}\right)\left(n - \frac{5}{2}\right) \cdots \left(n - \left(\frac{2n-1}{2}\right)\right)\Gamma\left(\frac{1}{2}\right)$$
$$= \frac{(2n)!\sqrt{\pi}}{4^n n!}.$$

(b) For n a positive integer, prove that

$$C_n = \left(\frac{4^n}{\sqrt{\pi}}\right) \frac{\Gamma(n + (1/2))}{\Gamma(n+2)}.$$

16. Consider the moves

$$R: (x, y) \to (x+1, y) \qquad U: (x, y) \uparrow (x, y+1),$$

as in Example 19.1. In how many ways can one travel in the xy-plane from $(0, 0)$ to $(4, 4)$ if the path taken may touch but *never* fall below the line $y = x$? In how many ways from $(0, 0)$ to $(7, 7)$?

17. For the same types of moves as in Example 19.1 and the previous exercise, in how many ways can one travel from

   (a) $(3, 2)$ to $(8, 7)$ and not rise above the line $y = x - 1$?

   (b) $(3, 8)$ to $(11, 16)$ and not rise above the line $y = x + 5$?

18. In how many ways can one travel from $(0, 0)$ to $(8, 8)$ using the steps $R: (x, y) \to (x+1, y)$ and $U: (x, y) \uparrow (x, y+1)$, so that the path taken never rises above the line $y = x$ but touches the line at the point $(3, 3)$?

19. (a) In how many ways can one go from $(0, 0)$ to $(12, 7)$ if the only moves permitted are R: $(x, y) \rightarrow (x + 1, y)$ and U: $(x, y) \uparrow (x, y + 1)$, and the number of U's may never exceed the number of R's along the path taken?

    (b) Let m and n be positive integers with $m > n$. Answer the question posed in part (a), upon replacing 12 by m and 7 by n.

20. Cori's theatre group is putting on a production of Shakespeare's *The Comedy of Errors*. To help with the project, Meg has volunteered to sell tickets at the entrance to the theatre on the night of the play's premier. When she arrives at the box office, 22 patrons, 11 each with a $10 bill and the other 11 each with a $20 bill, are the first to arrive at the theatre, where the price of admission is $10. In how many ways can these 22 individuals (all loners) line up so that the number with a $10 bill is never exceeded by the number with a $20 bill (and, as a result, Meg is always able to make any necessary change from the bills taken in from the first 21 of these 22 patrons)?

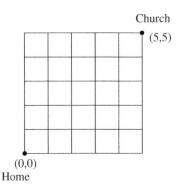

Church
(5,5)

(0,0)
Home

FIGURE 19.4

21. The church that Mary Lou attends is five blocks east and five blocks north of where she lives. (See Fig. 19.4.) Unless the weather is bad, every Sunday Mary Lou enjoys walking the ten blocks from her home to the church. How many different routes can Mary Lou travel if

    (a) she never crosses (but may touch) the diagonal from her home to the church?

    (b) she never crosses or touches the diagonal from her home to the church?

22. A particle starts from the origin and moves a total of seven unit steps to the right and seven unit steps to the left, along the x-axis one unit step at a time. Following these 14 moves, the particle has returned to the origin. In how many ways could this have happened so that the path taken by the particle was left of the origin at some point?

23. For $n \geq 0$, prove that

$$C_n = \frac{1}{n+1} \sum_{i=0}^{n} \binom{n}{i}^2.$$

■■■■■■ **CHAPTER 20**

Some Further Initial Examples

Although we now have a formula for C_n, the nth Catalan number, things here will prove to be somewhat different from what we did when we first encountered the Fibonacci numbers. Unlike what we saw in Part One, at this point we do not have any type of recurrence relation for the Catalan numbers.

When we questioned whether a new problem was another example of the Fibonacci numbers (sometimes shifted one or two subscripts), we always made an effort to count the structures in our new situation by means of the second-order linear recurrence relation satisfied by the Fibonacci numbers. However, for the Catalan numbers, at least at the start, we shall have to take a different approach. This will be accomplished by setting up a one-to-one correspondence (or bijection) between the elements in our new example and those of an example whose elements we already know are counted by the Catalan numbers. Although we presently know only one example where the Catalan numbers arise, we shall soon find several others.

Example 20.1: Let us start by counting all *balanced* strings made up of n left parentheses and n right parentheses, where the number of right parentheses never exceeds the number of left parentheses, as the string is read from left to right. For example, if $n = 3$, we are interested in counting strings like ()()() and (())() but not (()))(or ())((. The five balanced strings made up from three left parentheses and three right parentheses are shown in Fig. 20.1.

(a) ((()))   (b) (())()   (c) ()(())   (d) (())()   (e) ()()()

FIGURE 20.1

To show that such balanced strings of n left parentheses and n right parentheses are counted by the Catalan numbers, consider the paths in Fig. 19.2. For the path in Fig. 19.2 (a) replace each R with a "(" and each U with a ")". As a result, the path R, R, R, U, U, U is transformed into the balanced string ((())) in Fig. 20.1 (a). In the same way, the paths in parts (b) through (e) in Fig. 19.2 are transformed into the respective balanced strings in parts (b) through (e) in Fig. 20.1. Furthermore, this transformation

---

Fibonacci and Catalan Numbers: An Introduction, First Edition. Ralph P. Grimaldi.
© 2012 John Wiley & Sons, Inc. Published 2012 by John Wiley & Sons, Inc.

can be reversed so that each balanced string in Fig. 20.1 determines a corresponding path in Fig. 19.2. Consequently, for $n \geq 1$, this one-to-one correspondence tells us that there are C_n balanced strings made up of n left parentheses and n right parentheses. Furthermore, in the case of $n = 0$, we even have $1(= C_0)$ way where we can arrange no left parentheses and no right parentheses to obtain the unique *blank* balanced string.

Example 20.2:   Another somewhat immediate example deals with binary strings. Given three 0's and three 1's, there are 5 $(= C_3)$ ways to list these six symbols so that, for each list, the number of 0's never exceeds the number of 1's (as a list is read from left to right). These five binary strings are given in Fig. 20.2.

(a)   111000   (b)   110100   (c)   101100   (d)   110010   (e)   101010

FIGURE 20.2

Comparing the results in Figs. 20.1 and 20.2, we can readily set up a one-to-one correspondence. Consider, for example, the balanced string of three left parentheses followed by three right parentheses in Fig. 20.1 (a). Replace each "(" with a 1 and each ")" with a 0. The result is the binary string in Fig. 20.2 (a). Continuing in this way, each balanced string in parts (b) through (e) in Fig. 20.1 is transformed into the respective binary string in parts (b) through (e) in Fig. 20.2. This one-to-one correspondence indicates that for $n \geq 0$, there are C_n binary strings made up of n 0's and n 1's, where the number of 0's never exceeds the number of 1's, as the string is read from left to right. [For $n = 0$, we have the 1 $(= C_0)$ blank binary string.]

For the case of $n = 4$, the 14 possible binary strings are shown in Table 20.1.

TABLE 20.1

10101010	11001010	11100010
10101100	11001100	11100100
10110010	11010010	11101000
10110100	11010100	
10111000	11011000	11110000

Before we progress to our next example, we should realize that there is nothing really special about the symbols 0 and 1, as used in this example. Nor is there any significance about the fact that the number of 0's may not exceed the number of 1's. If, instead, we wanted the binary strings made up of n 0's and n 1's, where the number of 1's never exceeds the number of 0's, as the binary string is read from left to right, we would get the same number, C_n, of binary strings. Consequently, the number of ways we can list four x's and four y's as a string of eight symbols, where the number of x's never exceeds the number of y's, as the string is read from left to right, is 14 $(= C_4)$. Should we prefer to have the number of y's never exceed the number of x's, the resulting number of such strings is still 14 $(= C_4)$.

Example 20.3:   Here we shall modify Example 20.1 so that we can examine another situation where the Catalan numbers arise. This example is the one that was solved by Eugène Charles Catalan in 1838, when he was exploring the Towers of Hanoi puzzle. It seems to be one of the reasons his name is attached to the sequence.

This time we shall count the number of ways we can parenthesize the product $x_0 x_1 x_2 \cdots x_n$ of the $n + 1$ symbols x_0, x_1, x_2, ..., x_n using n pairs of parentheses under the following condition: Within each pair of parentheses, there are (i) two symbols; (ii) a symbol and a parenthesized expression; or (iii) two parenthesized expressions.

TABLE 20.2

$(((x_0 x_1) x_2) x_3)$	$(((x_0 x_1 x_2$	111000
$((x_0 (x_1 x_2)) x_3)$	$((x_0 (x_1 x_2$	110100
$(x_0 ((x_1 x_2) x_3))$	$(x_0 ((x_1 x_2$	101100
$((x_0 x_1)(x_2 x_3))$	$((x_0 x_1 (x_2$	110010
$(x_0 (x_1 (x_2 x_3)))$	$(x_0 (x_1 (x_2$	101010

In the first column of Table 20.2, we find the five ways that the product $x_0 x_1 x_2 x_3$ can be parenthesized under these conditions. The first of these is $(((x_0 x_1) x_2) x_3)$. Reading from left to right, we list the three occurrences of the left parenthesis "(" and the symbols x_0, x_1, x_2—maintaining the order in which these six symbols occur. This results in $(((x_0 x_1 x_2$, the first expression in the second column of Table 20.2. Likewise, the second expression $((x_0 (x_1 x_2)) x_3)$ in column 1 is transformed into the corresponding expression $((x_0 (x_1 x_2$ in column 2—and so on, for the other three pairs of corresponding expressions in columns 1 and 2.

Now we want to go in the opposite direction. So take any expression in column 2 and append "$x_3)$" to its right end. For example, the fourth expression in column 2—namely, $((x_0 x_1 (x_2$—becomes $((x_0 x_1 (x_2 x_3)$. As we read this new expression from left to right, we now insert a right parenthesis ")" whenever a product of two results arises. Consequently, the expression $((x_0 x_1 (x_2 x_3)$ is transformed into

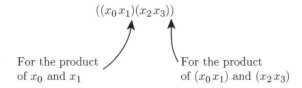

For the product
of x_0 and x_1

For the product
of $(x_0 x_1)$ and $(x_2 x_3)$

A one-to-one correspondence between the expressions in columns 2 and 3 readily follows. For an expression in column 2, simply replace each left parenthesis "(" with a "1" and each symbol with a "0." To go in the opposite direction, we now replace each "1" with a left parenthesis "(", the first 0 with x_0, the second 0 with x_1, and the third with x_2. In this way we transform the expressions in column 3 into the corresponding expressions in column 2.

Now that we have a one-to-one correspondence between the expressions in columns 1 and 2 and another between the expressions in columns 2 and 3, we can put these one-to-one correspondences together (using the composition of functions) to get a one-to-one correspondence between the expressions in columns 1 and 3. It now follows from Example 20.2 that for $n \geq 1$, there are C_n ways to parenthesize the product $x_0 x_1 x_2 \cdots x_n$, under the conditions specified at the beginning of this example. [For $n = 0$, we consider (x_0) as the $1 (= C_0)$ way we can parenthesize the one factor product made up of just x_0.]

Example 20.4: (a) Consider the following scenario. James and Colin are the final candidates running for the position of president of their high school's mathematics club. If they each receive 20 votes, then the number of ways the votes can be read, one at a time, so that James' count is never smaller than Colin's count is the number of ways we can arrange 20 J's and 20 C's in a row, so that as the letters are read from left to right, the number of C's never exceeds the number of J's. The idea presented in Example 20.2 tells us that tallying the votes in this way can be accomplished in $C_{20} (= 6, 564, 120, 420)$ ways.

(b) A comparable situation arises as follows. Suppose that Desiree flips a fair coin 20 times and gets 10 heads and 10 tails. In how many ways could she have done this so that, as the flips occurred, the number of heads was never exceeded by the number of tails? Once again the idea in Example 20.2 tells us that there are $C_{10} (= 16, 796)$ ways in which the resulting flips could have occurred in the manner prescribed.

Example 20.5: (a) Terri has three identical manila folders. In how many ways can she place these folders in her attaché case—upright, and perhaps nested within one another?

Referring to Fig. 20.3, consider the dotted line through each of the five configurations of three manila folders. For the configuration in Fig. 20.3 (a) write a 1 when a

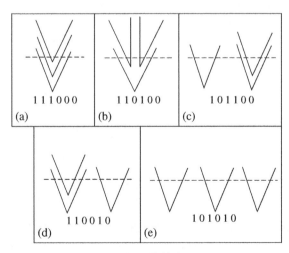

FIGURE 20.3

folder is encountered for the first time (on the left side) and a 0 for when that folder is encountered for the second time (on the right side). This results in the binary string 111000. In like manner, the other four configurations in Fig. 20.3 correspond, respectively, with the remaining binary strings in parts (b)–(e) of Fig. 20.2. This establishes a one-to-one correspondence between the binary strings of Example 20.2 and the configurations of identical manila folders given here. Consequently, Terri can place the three folders in her attaché case in C_3 ($= 5$) ways. If she had four such folders she could place them in her case in C_4 ($= 14$) ways.

(b) Now suppose that after chairing a meeting on the finances of three departments, Terri places a summary sheet for each department into a separate manila folder. Under these circumstances, she can place the (no longer identical) folders into her attaché case in $(3!)C_3 = (6)(5) = 30$ different ways.

Example 20.6:   A slight change in wording takes us from the 0's and 1's of Examples 20.2 and 20.3 to the following situation.

Suppose we want to arrange three 1's and three -1's so that all six resulting partial sums (starting with the first summand) are nonnegative. For instance, one possibility is

$$1, \ 1, \ -1, \ -1, \ 1, \ -1,$$

where the partial sums are

$$1 = 1 \geq 0$$
$$1 + 1 = 2 \geq 0$$
$$1 + 1 - 1 = 1 \geq 0$$
$$1 + 1 - 1 - 1 = 0 \geq 0$$
$$1 + 1 - 1 - 1 + 1 = 1 \geq 0$$
$$1 + 1 - 1 - 1 + 1 - 1 = 0 \geq 0.$$

We see that for such an arrangement to have all partial sums (starting with the first summand) nonnegative, the first summand must be 1, and as we examine the summands (from left to right) the number of -1's may never exceed the number of 1's. Consequently, if we do not change the 1's but replace the -1's with 0's, we can set up a one-to-one correspondence between the arrangements here and those in Example 20.2 for $n = 3$. The five arrangements in Fig. 20.2 now lead us to the five corresponding arrangements in Fig. 20.4, where all the partial sums (starting with the first summand) are nonnegative:

(a) $1,1,1,-1,-1,-1$ (b) $1,1,-1,1,-1,-1$ (c) $1,-1,1,1,-1,-1$

(d) $1,1,-1,-1,1,-1$ (e) $1,-1,1,-1,1,-1$

FIGURE 20.4

In general, for $n \geq 1$, there are C_n such arrangements of n 1's and n −1's where all the partial sums (starting with the first summand) are nonnegative. For $n = 0$, we accept 1 ($= C_0$) such arrangement (of nothing) made up of no 1's and no −1's and having the one partial nonnegative sum of 0.

Example 20.7: Consider arranging the positive integers 1, 2, 3, 4, 5, and 6, subject to the following conditions: (i) 1 appears before 3, which appears before 5; (ii) 2 appears before 4, which appears before 6; and, (iii) 1 appears before 2, 3 appears before 4, and 5 appears before 6. In Fig. 20.5, we find the five possible arrangements satisfying these three conditions.

(a)   135246   (b)   132546   (c)   123546   (d)   132456   (e)   123456

FIGURE 20.5

For each of these arrangements, replace each odd integer with a 1 and each even integer with a −1. Then, for instance, the arrangement in Fig. 20.5 (c)—namely, 123546—would correspond with the sequence 1, −1, 1, 1, −1, −1 in Fig 20.4 (c). In this way, a one-to-one correspondence is established between the arrangements given here in parts (a)–(e) of Fig. 20.5 and the sequences in parts (a)–(e), respectively, of Fig. 20.4 (in Example 20.6).

When this correspondence is generalized, we learn that, for $n \geq 1$, C_n counts the numbers of ways one can arrange the positive integers $1, 2, 3, \ldots, 2n - 1, 2n$ so that (i) the odd integers occur in increasing order; (ii) the even integers occur in increasing order; and, (iii) $2k - 1$ appears before $2k$, for all $1 \leq k \leq n$.

Example 20.8: In closing this chapter, we shall investigate lattice paths once again. This time, however, we shall consider them in pairs.

(a) For $n \geq 0$, each path will be made up of $n + 1$ steps, where each step is of type R: $(x, y) \rightarrow (x + 1, y)$ or U: $(x, y) \uparrow (x, y + 1)$. Both paths start at $(0, 0)$. The first path starts with an R, while the second path starts with a U. Each path then continues for n additional steps, and after a total of $n + 1$ steps, they arrive at the same point for the first time. Hence these paths have only their starting and finishing points in common. One such pair of paths, for the case of $n = 10$, is shown in Fig. 20.6. The region enclosed by this pair of paths is an example of a *parallelogram polyomino*. The paths both start at $(0, 0)$ and terminate at $(4, 7)$, and they do not meet at any other point. The interior of this figure is made up of 12 unit squares arranged in four columns and seven rows. We designate the columns as C_1, C_2, C_3, and C_4.

At this point, our objective is to associate a sequence of 1's and −1's with this configuration. To this end, we shall let a_i count the number of squares in column C_i, for each $1 \leq i \leq 4$. So

$$a_1 = 3, \quad a_2 = 4, \quad a_3 = 4, \quad a_4 = 1.$$

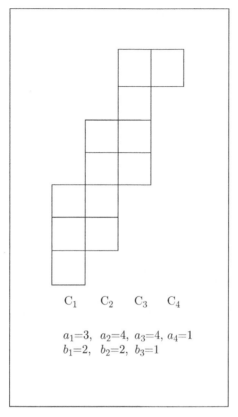

FIGURE 20.6

Then for $1 \leq i \leq 3 \, (= 4 - 1)$, we let b_i count the number of rows where squares appear in both columns C_i and C_{i+1}. (So $a_i \geq b_i$ for $1 \leq i \leq 3$.) Consequently, here

$$b_1 = 2, \quad b_2 = 2, \quad b_3 = 1.$$

Now we shall consider a sequence of 1's and -1's that we generate as follows:

We start with a_1 1's,

which are followed by $a_1 - b_1 + 1$ -1's,

which are followed by $a_2 - b_1 + 1$ 1's,

which are followed by $a_2 - b_2 + 1$ -1's,

which are followed by $a_3 - b_2 + 1$ 1's,

which are followed by $a_3 - b_3 + 1$ -1's,

which are followed by $a_4 - b_3 + 1$ 1's,

and we end with a_4 -1's.

Consequently, this sequence contains $a_1 + a_2 + a_3 + a_4 - (b_1 + b_2 + b_3) + 3$ 1's and the same number of -1's. But how do we know that $a_1 \geq a_1 - b_1 + 1$? If not, then $a_1 < a_1 - b_1 + 1$, from which it follows that $b_1 < 1$, and this means that $b_1 = 0$. But if this happens, then the two paths meet at the point $(1, 3)$, contradicting their first meeting point as $(4, 7)$. Likewise, if $a_1 \geq a_1 - b_1 + 1$ but $a_1 + (a_2 - b_1 + 1) < (a_1 - b_1 + 1) + (a_2 - b_2 + 1)$, then $b_2 < 1$ so $b_2 = 0$ and now the paths meet for the first time at $(2, 5)$, not $(4, 7)$. Continuing this type of argument, we see that the sequence of 1's and -1's we obtain is such that the number of 1's is never exceeded by the number of -1's, as the string is read from left to right.

[Also note that if we are given such a sequence, the value of a_1 is the number of 1's at the start of the sequence. The values of a_1 and $a_1 - b_1 + 1$ (the number of -1's in the first subsequence of -1's) provide the value of b_1. Continuing in this way,

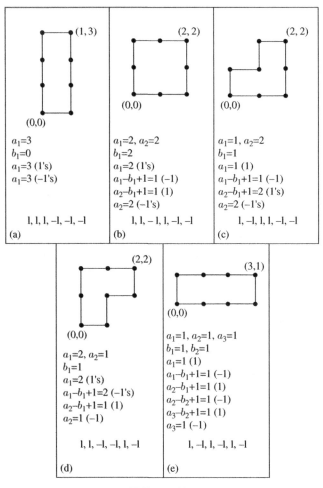

FIGURE 20.7

the values of a_2, b_2, a_3, b_3, and a_4 are determined and from the values of the seven terms a_i, $1 \le i \le 4$, and b_j, $1 \le j \le 3$, the given parallelogram polyomino can be reconstructed.]

The sequence of 1's and -1's for the parallelogram polyomino in Fig. 20.6 is

$$1, 1, 1, -1, -1, 1, 1, 1, -1, -1, -1, 1, 1, 1, -1, -1, -1, -1, 1, -1.$$

With this idea in hand, we now turn to the pairs of lattice paths in Fig. 20.7. These configurations are for the case where $n = 3$, so each of the two lattice paths (in a given configuration) is made up of 4 $(= 3 + 1)$ steps. In each of parts (a)–(e) of the figure, we determine the sequence of 1's and -1's that corresponds to the parallelogram polyomino in that part. This provides us with a one-to-one correspondence between the configurations given here in parts (a)–(e) and the respective arrangements in parts (a)–(e) of Fig. 20.4 in Example 20.6. Since this type of correspondence can be developed for each $n \ge 0$, it follows that the number of parallelogram polyominoes (for pairs of paths with $n + 1$ steps) is the nth Catalan number C_n.

(b) Finally, look back at the configurations in Fig. 20.7. In each case, remove the first and last steps from each path in a given pair, then coalesce the new starting and terminating points, so that the lower path never rises above the higher one. This variation of the result in part (a) is demonstrated in Fig. 20.8 and provides another situation that is counted by the Catalan numbers. (This variation is due to Louis W. Shapiro.)

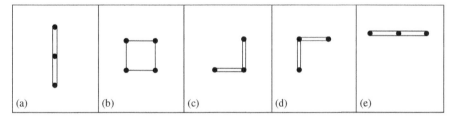

FIGURE 20.8

EXERCISES FOR CHAPTER 20

1. In how many ways can one parenthesize the product $abcdefgh$?
2. There are 42 ways in which the product $uvwxyz$ can be parenthesized.
   (a) Determine, as in Example 20.4, the list of five 1's and five 0's that corresponds to each of the following.
      (i) $(((uv)w)(x(yz)))$
      (ii) $((((uv)(wx))y)z)$
      (iii) $(u(v(w(x(yz)))))$

   (b) Find, as in Example 20.4, the way to parenthesize $uvwxyz$ that corresponds to each given list of five 1's and five 0's.

     (i) 1110010100

     (ii) 1011100100

     (iii) 1100110010

3. In the fall of next year, Samuel will start his college education. He plans to major in computer science but also wants to earn a minor in engineering management. To earn this minor, Samuel must take four courses s_1, s_2, s_3, s_4 in statistics and four courses e_1, e_2, e_3, e_4 in engineering. His plan is to take one of these courses each semester during his eight semesters of college. The statistics courses are such that s_i is a prerequisite for s_{i+1}, for $i = 1, 2, 3$, and the engineering courses likewise require e_i to be completed before e_{i+1}, for $i = 1, 2, 3$. In addition, for $i = 1, 2, 3, 4$, the statistics course s_i must be completed before the engineering course e_i. In how many ways can Samuel order these eight courses so that he can satisfy all of the prerequisites and complete the minor in engineering management in eight semesters?

CHAPTER 21

Dyck Paths, Peaks, and Valleys

This chapter will provide us with examples that modify and extend what we learned about lattice paths in Chapter 19.

Example 21.1: Let us start by returning to the lattice paths in Example 19.1. This time, however, we shall introduce the following diagonal steps. Replace each

$$R: (x, y) \to (x + 1, y) \quad \text{with} \quad D: (x, y) \nearrow (x + 1, y + 1)$$

and each

$$U: (x, y) \uparrow (x, y + 1) \quad \text{with} \quad D^*: (x, y) \searrow (x + 1, y - 1).$$

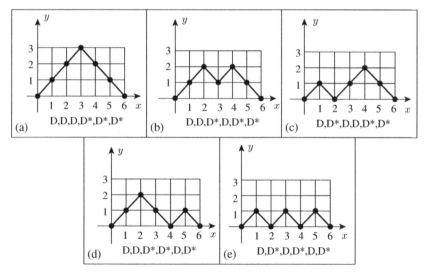

FIGURE 21.1

Fibonacci and Catalan Numbers: An Introduction, First Edition. Ralph P. Grimaldi.
© 2012 John Wiley & Sons, Inc. Published 2012 by John Wiley & Sons, Inc.

The resulting paths in Fig. 21.1 are called *Dyck* (pronounced "Dike") paths, after the German mathematician Walther Franz Anton von Dyck (1856–1934). In general, for $n \geq 0$, there are C_n Dyck paths of length $2n$, made up of n D's and n D*'s. When these paths start at $(0, 0)$ and end at $(2n, 0)$, they never fall below the x-axis. Considering their shape, one also finds the configurations in Fig. 21.1 referred to as *mountain ranges*.

Example 21.2: Figure 21.2 provides us with what are called the *left factors* of the Dyck paths in Fig. 21.1. As seen in the figure, the left factor of each Dyck path is made up of all the diagonal steps that precede the last D step. Consequently, each left factor is seen to have the same number of diagonal steps D—namely, $2 (= 3 - 1)$. We shall establish a one-to-one correspondence between these left factors with $n - 1$ diagonal steps D and the Dyck paths of length $2n$, for the case of $n = 3$. One can then generalize this construction for all $n \geq 1$.

Take, for example, the left factor in Fig. 21.2 (a). If we append an additional diagonal step D at the (right) end of the factor and then add on three additional diagonal steps D*, we arrive at the Dyck path in Fig. 21.1 (a). This is demonstrated in parts (a)–(c) of Fig. 21.3. For the left factor in Fig. 21.2 (b) we once again append an additional D step at the (right) end of the factor, but this time we only add two additional D* steps in order to reach the x-axis. Parts (d)–(f) of Fig. 21.3 show how this left factor evolves into the Dyck path in Fig. 21.3 (f) [and Fig. 21.1 (b)]. So, in general, given a left factor with $n - 1$ D steps, we append a D step to the (right) end of the factor and then add on the additional (from 1 to n) D* steps needed to return

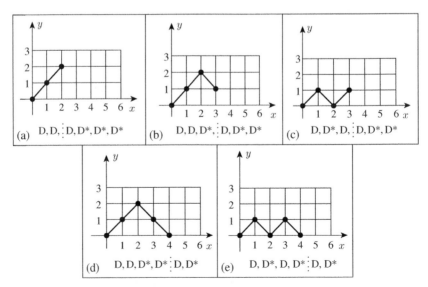

FIGURE 21.2

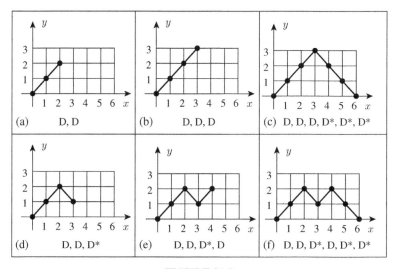

FIGURE 21.3

us to the *x*-axis. In this way, we establish a one-to-one correspondence between the structures in Fig. 21.1 and Fig. 21.2.

Example 21.3: (a) Now we shall investigate paths where the following four types of steps are available:

(1) D: $(x, y) \nearrow (x + 1, y + 1)$ (2) D*: $(x, y) \searrow (x + 1, y - 1)$

(3) R_1: $(x, y) \xrightarrow{\text{red}} (x + 1, y)$ (4) R_2: $(x, y) \xrightarrow{\text{blue}} (x + 1, y)$.

For $n \geq 2$, we want to count the number of paths from $(0, 0)$ to $(n - 1, 0)$ using these four types of steps—and, in addition, we never allow the paths to dip below the *x*-axis. These paths are known as the *two-colored Motzkin paths*, named for the (German-born) American mathematician Theodore Samuel Motzkin (1908–1970).

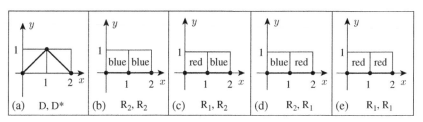

FIGURE 21.4

When $n = 3$, we find the five two-colored Motzkin paths given in Fig. 21.4. To set up a one-to-one correspondence between these paths and the ones shown in Fig. 21.1, we first observe that each path in Fig. 21.1 starts with a D from $(0, 0)$ to $(1, 1)$ and ends with a D* from $(5, 1)$ to $(6, 0)$. In between there are four steps—two of type D and two of type D*. Now we examine these remaining four steps in pairs and replace each (consecutive) pair in a Dyck path with a single step in a two-colored Motzkin path according to the correspondence shown in Fig. 21.5. In this way, we arrive at a one-to-one correspondence between the five Dyck paths in parts (a)–(e) of Fig. 21.1 and the five respective two-colored Motzkin paths in parts (a)–(e) of Fig. 21.4. [When we start with a two-colored Motzkin path we notice, for example, from part (a) of Fig. 21.4 and Fig. 21.5, that the low point of a D, D Dyck path is at the same level as the top point of a D Motzkin path. From the two-colored Motzkin path in part (c) of Fig. 21.4 and 21.5, we realize that the Motzkin R_2 step is at the same level as the endpoints of a Dyck D, D* path. These observations, along with similar ones for the Motzkin steps in parts (b) and (d) of Fig. 21.5, show us why the resulting Dyck paths never fall below the x-axis.]

For the general case, if $n \geq 2$, then for $1 \leq i \leq n - 1$, we replace steps $2i$ and $2i + 1$ of a given Dyck path [from $(0, 0)$ to $(2n, 0)$] with the ith step of the corresponding two-colored Motzkin path, according to the transformations shown in parts (a)–(d) of Fig. 21.5.

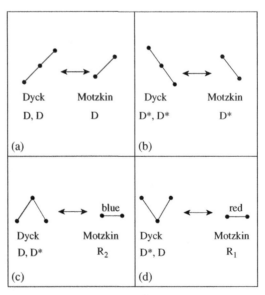

FIGURE 21.5

(b) Our next result is related to that given in part (a) and was discovered by David Callan. First, however, let us recall that for $k \geq 2$, a collection S_1, S_2,

..., S_k of subsets of a set S is called *pairwise disjoint* if $S_i \cap S_j = \varnothing$ for all $1 \leq i < j \leq k$.

Now for $n \geq 1$, we want to count the number of triples (A, B, C) of pairwise disjoint subsets of $\{1, 2, \ldots, n-1\}$ where (i) $A = \{a_1, a_2, \ldots, a_r\}$, $B = \{b_1, b_2, \ldots, b_r\}$, with $0 \leq 2r \leq n-1$; (ii) $a_i < a_{i+1}$ and $b_i < b_{i+1}$ for all $1 \leq i \leq r-1$; and, (iii) $a_i < b_i$ for all $1 \leq i \leq r$. Considering the two-colored Motzkin paths above in part (a), label the steps in each path with $1, 2, \ldots, n-1$, as the path is traversed from left to right. Place (i) the labels for all D steps in the path in set A; (ii) the labels for all D* steps in the path in set B; and, (iii) the labels for all the (red) R_1 steps in set C. For the case where $n = 3$, the following triples of pairwise disjoint subsets of $\{1, 2\}$ are in a one-to-one correspondence with parts (a)–(e), respectively, of Fig. 21.4: (a) $(\{1\}, \{2\}, \varnothing)$; (b) $(\varnothing, \varnothing, \varnothing)$; (c) $(\varnothing, \varnothing, \{1\})$; (d) $(\varnothing, \varnothing, \{2\})$; (e) $(\varnothing, \varnothing, \{1, 2\})$.

Example 21.4: This example is also due to David Callan and deals with Dyck paths of length $2n + 2$ where the first occurrence of a D* ($\searrow$) step is followed by another step of this type—or, alternatively, the first occurrence of a D ($\nearrow$) step cannot be followed by a D* ($\searrow$) step, since Dyck paths cannot dip below the x-axis.

For $n = 3$, the five possible paths of length 8 $[= 2(3) + 2]$ that satisfy the given condition are shown in Fig. 21.6. To set up a one-to-one correspondence between these Dyck paths and those of length 6 in Fig. 21.1, we proceed as follows. Consider the Dyck path of length 8 in Fig. 21.6 (a). Remove the first occurrence of the two consecutive steps D and D*—in this case, these are steps 4 and 5. Now coalesce

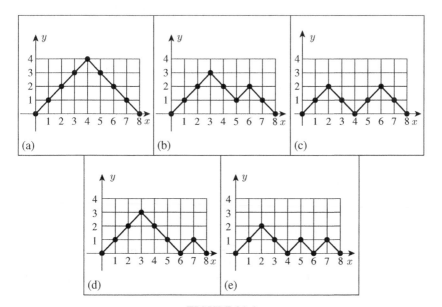

FIGURE 21.6

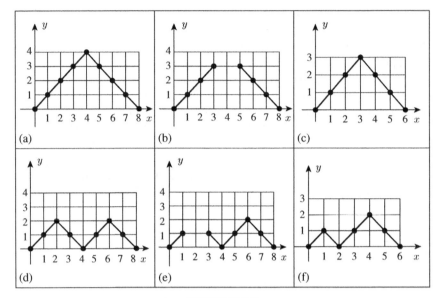

FIGURE 21.7

the right endpoint of the left part of the resulting structure with the left endpoint of the right part of the structure. This is demonstrated in parts (a)–(c) of Fig. 21.7. The result is the Dyck path of length 6 [$= 2(3)$] in Fig. 21.1 (a). Parts (d)–(f) of Fig. 21.7 demonstrate the correspondence between the Dyck path of length 8 in Fig. 21.6 (c) with the Dyck path of length 6 in Fig. 21.1 (c). When this procedure is applied to the five paths in parts (a)–(e) of Fig. 21.6 we obtain the corresponding respective paths in parts (a)–(e) of Fig. 21.1.

The preceding can be adapted for the general case where $n \geq 1$ and so we find that there are C_n Dyck paths of length $2n + 2$ satisfying the stated condition.

Example 21.5: In Fig. 21.8, we have the two ($= C_2$) Dyck paths from $(0, 0)$ to $(4, 0)$ (that do not dip below the x-axis). We see that these paths determine five ($= C_3$) points on the x-axis. There are two such points for the path in part (a) of the figure and three in part (b). If we examine the five ($= C_3$) Dyck paths in Fig. 21.1, we find a total of 14 ($= C_4$) points where these paths meet the x-axis.

These observations suggest our next result—due to Robert A. Sulanke.

For $n \geq 1$, the number of points where the C_{n-1} Dyck paths [of length $2(n-1)$] from $(0, 0)$ to $(2(n-1), 0)$ meet the x-axis equals C_n. To verify this claim, we shall set up a one-to-one correspondence between the points where the C_{n-1} Dyck paths of length $2(n-1)$ meet the x-axis and the C_n Dyck paths of length $2n$.

We demonstrate with the case for $n = 3$. Consider the two Dyck paths of length 4 in Fig. 21.8. If $(p, 0)$ is a point of contact with the x-axis, we place a D ($\nearrow$) step there and then add a D* ($\searrow$) step at the end of the path. This is demonstrated for the

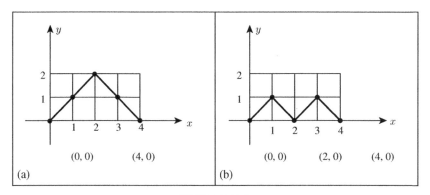

FIGURE 21.8

second contact point at (4, 0) for the Dyck path in part (a) of the figure. Parts (a)–(c) of Fig. 21.9 show us how we obtain the corresponding Dyck path from (0, 0) to (6, 0). In parts (d)–(f) of Fig. 21.9 we see how the Dyck path in part (f) corresponds with the first contact point at (0, 0) for the Dyck path in Fig. 21.8 (b). Table 21.1 on p. 176 provides the complete correspondence between the five contact points on the x-axis in Fig. 21.8 and the five Dyck paths in Fig. 21.1.

This technique can be used for any $n \geq 1$ and so it follows that the number of points where the C_{n-1} Dyck paths of length $2(n - 1)$ meet the x-axis is C_n.

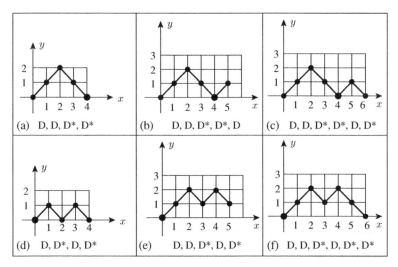

FIGURE 21.9

Example 21.6: When dealing with Dyck paths, a *peak* occurs as part of a path whenever we have a D ($\nearrow$) step followed by a D* ($\searrow$) step. For the five Dyck paths

TABLE 21.1

Points on the x-Axis in Fig. 21.8	Corresponding Dyck Path in Fig. 21.1
(1) (0,0) in part (a)	(1) Dyck path in part (a)
(2) (4,0) in part (a)	(2) Dyck path in part (d)
(3) (0,0) in part (b)	(3) Dyck path in part (b)
(4) (2,0) in part (b)	(4) Dyck path in part (c)
(5) (4,0) in part (b)	(5) Dyck path in part (e)

in Fig. 21.1, we find a total of ten peaks. In Fig. 21.10, we introduce an additional type of step, denoted R*, where

$$R^*: (x, y) \to (x + 2, y).$$

The paths given here are examples of *Schröder* paths [from $(0, 0)$ to $(6, 0)$] with no peaks. They are named after the German mathematician Friedrich Wilhelm Karl Ernst Schröder (1841–1902). As observed by Louis W. Shapiro, these paths are counted by the Catalan numbers.

For the case where $n = 3$, a one-to-one correspondence is readily obtained by taking the Dyck paths of Fig. 21.1 and replacing each peak D ($\nearrow$), D* ($\searrow$) with an R* ($\to$) step. As a result, the five Dyck paths in parts (a)–(e) of Fig. 21.1 then

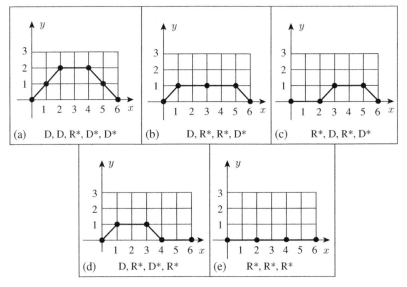

FIGURE 21.10

correspond respectively with the five Schröder paths, with no peaks, in parts (a)–(e) of Fig. 21.10.

Since this type of one-to-one correspondence can be defined for each $n \geq 1$, we learn that the number of Schröder paths from $(0, 0)$ to $(2n, 0)$, with no peaks, is C_n.

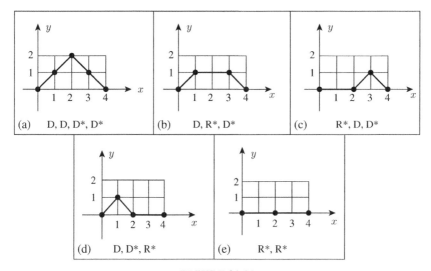

FIGURE 21.11

Example 21.7: Comparable to the notion of a peak (as given in the previous example), a *valley* in a Dyck path occurs when a D* ($\searrow$) step is followed by a D ($\nearrow$) step. Once again we use the D, D*, and R* steps of the previous example. This time we learn that there are C_n Schröder paths from $(0, 0)$ to $(2(n - 1), 0)$ with no valleys. For $n = 3$, the five such Schröder paths are shown in Fig. 21.11.

A one-to-one correspondence between the Schröder paths of Fig. 21.11 and the Dyck paths of Fig. 21.1 can be developed as follows. Starting with a Dyck path from $(0, 0)$ to $(2 \cdot 3, 0)$, remove the initial D ($\nearrow$) step and the final D* ($\searrow$) step. For the remaining four steps, replace each occurrence of a D* ($\searrow$) step followed by a D ($\nearrow$) step with an R* ($\rightarrow$) step. Now that all valleys have been deleted, start the first step of the new path at $(0, 0)$ and terminate the last step at $(2 \cdot 2, 0)$. The result is a Schröder path, with no valleys, from $(0, 0)$ to $(2 \cdot 2, 0)$—as demonstrated in Fig. 21.12 on p. 178 for the Dyck path in Fig. 21.1 (b). Using the same constructive idea, the five Dyck paths in parts (a)–(e) of Fig. 21.1 then correspond respectively with the five Schröder paths, with no valleys, in parts (a)–(e) of Fig. 21.11. Furthermore, this one-to-one correspondence is valid for each $n \geq 1$, thereby establishing that the number of Schröder paths from $(0, 0)$ to $(2(n - 1), 0)$, with no valleys, is C_n.

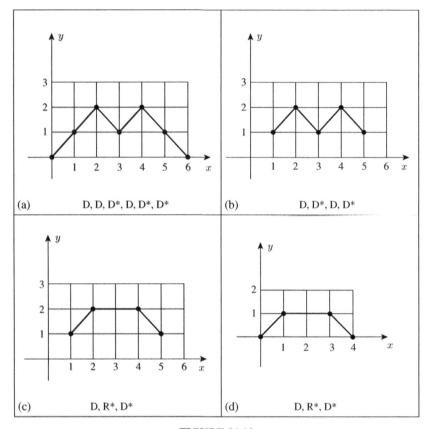

FIGURE 21.12

EXERCISES FOR CHAPTER 21

1. Determine the number of Dyck paths from $(0, 0)$ to $(14, 0)$ that include the point $(8, 0)$.

2. (a) Determine the number of Dyck paths from $(0, 0)$ to $(20, 0)$ that never touch the x-axis—except at $(0, 0)$ and $(20, 0)$.

   (b) Generalize the result in part (a).

3. (a) Find the left factor for each of the following Dyck paths:

   (i) D, D, D*, D, D, D, D*, D*, D*, D*

   (ii) D, D*, D, D, D*, D*, D, D, D, D*, D*, D*

   (b) Find the Dyck path for each of the following left factors:

   (i) D, D, D, D*, D*

   (ii) D, D*, D, D, D*, D*

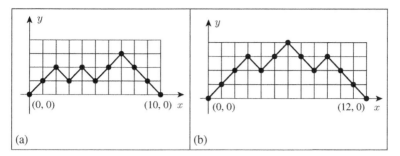

FIGURE 21.13

4. Find the two-colored Motzkin paths that correspond to each of the Dyck paths in Fig. 21.13.
5. Find the Dyck paths that correspond to each of the two-colored Motzkin paths in Fig. 21.14.
6. (a) A two-colored Motzkin path starts at $(0, 0)$ and ends at $(11, 0)$. If the corresponding Dyck path starts at $(0, 0)$, where does it end?
   (b) A Dyck path starts at $(0, 0)$ and ends at $(34, 0)$. If the corresponding two-colored Motzkin path starts at $(0, 0)$, where does it end?
7. Mona and Mary Ellen are developing software to draw and count the Dyck paths from $(0, 0)$ to $(16, 0)$ and also determine the number of times these paths meet the x-axis. Since these two young women are excellent computer scientists, they realize the need to validate their results. To do so, they need to know how many Dyck paths there are from $(0, 0)$ to $(16, 0)$ and how many times these paths meet the x-axis. What are these quantities?
8. Following a successful evening at the poker tables of "Hopping Henry's" Casino, a somewhat inebriated Scott takes his leave. Standing outside, with his back to the casino entrance (which faces south), Scott decides to turn left and walk east to his hotel. In so doing, he walks along a straight walkway made up of 4×4 pavement squares. At the one square directly in front of the casino and each square to his left, Scott decides to do one of the following: (i)

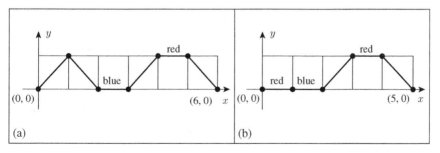

FIGURE 21.14

take one step forward, to the east (E); (ii) take one step backward, to the west (W); (iii) stay on the square where he is and turn around clockwise (C); or, (iv) stay on the square where he is and turn around counterclockwise (C*).

(a) In how many ways can Scott take two steps to find himself back at the entrance to "Hopping Henry's", if at no time is he to find himself on a square behind his starting point?

(b) Answer part (a) upon replacing 2 by n, where $n \geq 0$.

9. A *Motzkin path* from $(0, 0)$ to $(m, 0)$ is made up of three types of steps: D: $(x, y) \nearrow (x + 1, y + 1)$, D*: $(x, y) \searrow (x + 1, y - 1)$, and R: $(x, y) \longrightarrow (x + 1, y)$. There are m steps in total and the number of D steps equals the number of D* steps. Such a path is called *peakless* if there is no occurrence of a D step followed by a D* step.

Consider the peakless Motzkin paths where n counts the total number of all R steps and all D steps (and the number of D* steps equals the number of D steps). The case for $n = 3$ is shown in Fig. 21.15. Show that these peakless Motzkin paths are counted by the Catalan numbers. (This result is due to Emeric Deutsch.)

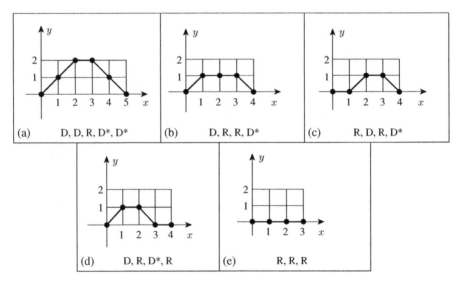

FIGURE 21.15

10. For $n \geq 1$, consider the lattice paths from $(0, 0)$ to $(n - 1, n - 1)$ made up of three types of steps: U: $(x, y) \uparrow (x, y + 1)$, R: $(x, y) \rightarrow (x + 1, y)$, and D: $(x, y) \nearrow (x + 1, y + 1)$, where such a path can never fall below the line $y = x$ and each D step only occurs along the line $y = x$. The case for $n = 3$ is exhibited in Fig. 21.16. Show that these lattice paths are counted by the Catalan numbers.

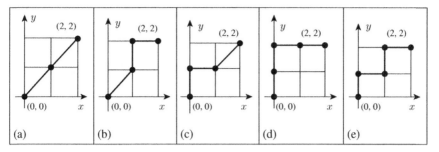

FIGURE 21.16

11. Consider the two-colored Motzkin paths of Example 21.3 (a). Now, however, for $n \geq 1$, focus on those paths from $(0, 0)$ to $(n, 0)$ where no R_1 step can occur on the x-axis. The case for $n = 3$ is shown in Fig. 21.17. Show that these two-colored Motzkin paths are counted by the Catalan numbers. (This result is due to Emeric Deutsch.)

12. For $n \geq 1$, consider the Schröder paths (of Example 21.6) from $(0, 0)$ to $(2(n - 1), 0)$. In this case, however, peaks, resulting from a D step followed by a D* step, only occur on the x-axis. The case for $n = 3$ is exhibited in Fig. 21.18 on p. 182. Show that these Schröder paths are counted by the Catalan numbers.

13. (a) Consider Example 21.3 (b) for the case where $n = 11$. In this case, how many triples (A, B, C) are there with $A = B = \varnothing$?

    (b) Generalize the result in part (a).

14. (a) For $n \geq 1$, show that the Catalan number C_n counts the number of quadruples (A, B, C, D) of pairwise disjoint subsets of $\{1, 2, \ldots, n - 1\}$ where (i) $A = \{a_1, a_2, \ldots, a_r\}$, $B = \{b_1, b_2, \ldots, b_r\}$, with $0 \leq 2r \leq n - 1$; (ii) $a_i < a_{i+1}$ and $b_i < b_{i+1}$ for all $1 \leq i \leq r - 1$; (iii) $a_i < b_i$ for all $1 \leq i \leq r$; and, (iv) $A \cup B \cup C \cup D = \{1, 2, \ldots, n - 1\}$. (Here the nonempty subsets among A, B, C, D provide a partition of $\{1, 2, \ldots, n - 1\}$.)

    (b) For a given $n \geq 1$, how many of the quadruples in part (a) have $A = B = \varnothing$?

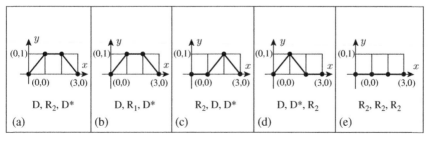

FIGURE 21.17

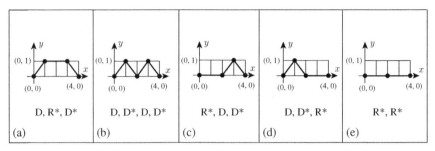

D, R*, D*	D, D*, D, D*	R*, D, D*	D, D*, R*	R*, R*
(a)	(b)	(c)	(d)	(e)

FIGURE 21.18

15. Margaret and Artie each take five turns tossing a fair coin. First Margaret tosses the coin, then Artie, then Margaret, and so on, with Artie's fifth toss the final toss.

   (a) What is the probability that there are equal numbers of heads and tails in the ten tosses?

   (b) What is the probability that there are equal numbers of heads and tails in the ten tosses and that the number of heads never exceeds the number of tails during the ten tosses?

16. Nancy and Paul each take six turns tossing a fair coin. First Nancy tosses the coin, then Paul, then Nancy, and so on, with Paul's sixth toss the final toss. What is the probability that there are equal numbers of heads and tails in the 12 tosses, that the number of heads never exceeds the number of tails during the 12 tosses, and that the numbers of heads and tails equal each other for the first time after the 12 tosses?

CHAPTER 22

Young Tableaux, Compositions, and Vertices and Arcs

New structures and familiar ones arise as we continue to investigate some of the many places where the Catalan numbers surface.

Example 22.1: This example introduces a combinatorial structure called a 2 by n *Young Tableaux*. These structures are named for the English clergyman and mathematician Alfred Young (1873–1940) and are made up of $2n$ cells divided into 2 rows and n columns. In using such a tableau, we want to count the number of ways we can place the integers from 1 to $2n$ into the $2n$ cells so that the entries in each row are in ascending order and, for each of the n columns, the entry in row 1 is smaller than the entry in row 2. For $n = 3$, we find that there are five such 2 by 3 Young Tableaux:

$$
\text{(a)} \begin{bmatrix} 1 & 2 & 3 \\ 4 & 5 & 6 \end{bmatrix} \quad
\text{(b)} \begin{bmatrix} 1 & 2 & 4 \\ 3 & 5 & 6 \end{bmatrix} \quad
\text{(c)} \begin{bmatrix} 1 & 3 & 4 \\ 2 & 5 & 6 \end{bmatrix}
$$

$$
\text{(d)} \begin{bmatrix} 1 & 2 & 5 \\ 3 & 4 & 6 \end{bmatrix} \quad
\text{(e)} \begin{bmatrix} 1 & 3 & 5 \\ 2 & 4 & 6 \end{bmatrix}
$$

FIGURE 22.1

To set up a one-to-one correspondence between the Dyck paths in Fig. 21.1 and these 2 by 3 Young Tableaux, consider the Dyck path in Fig. 21.1 (a). Where are the D ($\nearrow$) steps in this path? They are the first, second, and third steps in the path—and so we place 1, 2, and 3 (in order) in the first row of the tableau. Likewise, we now consider the locations of the D* ($\searrow$) steps in this path. Their locations, listed in order

Fibonacci and Catalan Numbers: An Introduction, First Edition. Ralph P. Grimaldi.
© 2012 John Wiley & Sons, Inc. Published 2012 by John Wiley & Sons, Inc.

of occurrence, give us the results in the second row of the tableau—namely, 4, 5, 6. Applying this idea to the other Dyck paths in Fig. 21.1, we find that the Dyck paths in parts (a)–(e) of the figure correspond respectively to the five Young Tableaux given in parts (a)–(e) of Fig. 22.1. Consequently, since this correspondence can be set up in a similar way for any $n \geq 1$, it follows that the number of 2 by n Young Tableaux is C_n.

Example 22.2:   In Chapter 6, we were introduced to the idea of compositions of an integer. First and foremost, we learned that the positive integer n has 2^{n-1} compositions. In this example, due to Astrid Reifegerste, we will be interested in pairs of compositions of a given positive integer n. The compositions in each pair will be of the form

$$a_1 + a_2 + \cdots + a_k \quad \text{and} \quad b_1 + b_2 + \cdots + b_k.$$

Consequently, each of these compositions of n has the same number of summands (or parts). In addition, we require that

$$a_1 \geq b_1, \quad a_1 + a_2 \geq b_1 + b_2, \quad a_1 + a_2 + a_3 \geq b_1 + b_2 + b_3, \ldots,$$

$$a_1 + a_2 + a_3 + \cdots + a_{k-1} \geq b_1 + b_2 + b_3 + \cdots + b_{k-1},$$

$$a_1 + a_2 + \cdots + a_k = b_1 + b_2 + \cdots + b_k.$$

For $n = 3$, we find five such pairs in Fig. 22.2.

Unfortunately, four of these five pairs have the same composition in both components, and the components of the pair $(2 + 1, 1 + 2)$ involve the same exact summands—except for their order. A little more variety arises for the pairs of compositions of $n = 4$. Here we find the 14 ordered pairs in Fig. 22.3.

In the case of $n = 3$, consider the pair $(2 + 1, 1 + 2)$. Here we have $a_1 = 2, a_2 = 1$, $b_1 = 1$, and $b_2 = 2$. To set up a one-to-one correspondence with the Dyck paths in Fig. 21.1, we construct the Dyck path for this pair of compositions by drawing a_1 D's, then b_1 D*'s, then a_2 D's, and finally b_2 D*'s. This is the Dyck path in Fig. 21.1 (b). Using this method, we find that the pairs of compositions in parts (a)–(e) of Fig. 22.2 correspond respectively with the Dyck paths in parts (a)–(e) of Fig. 21.1.

In Fig. 22.4 (a) we have the Dyck path that corresponds with pair (10) in Fig. 22.3—namely, $(3 + 1, 2 + 2)$. Part (b) of Fig. 22.4 provides the Dyck path for pair (14) in Fig. 22.3—this time, $(1 + 2 + 1, 1 + 1 + 2)$.

(a)	$(3, 3)$	(b)	$(2 + 1, 1 + 2)$	(c)	$(1 + 2, 1 + 2)$
(d)	$(2 + 1, 2 + 1)$	(e)	$(1 + 1 + 1, 1 + 1 + 1)$.		

FIGURE 22.2

(1)   (4, 4)   (2)   (3 + 1, 3 + 1)   (3)   (2 + 2, 2 + 2)

(4)   (2 + 1 + 1, 2 + 1 + 1)   (5)   (1 + 3, 1 + 3)

(6)   (1 + 2 + 1, 1 + 2 + 1)   (7)   (1 + 1 + 2, 1 + 1 + 2)

(8)   (1 + 1 + 1 + 1, 1 + 1 + 1 + 1)

(9)   (3 + 1, 1 + 3)   (10)   (3 + 1, 2 + 2)   (11)   (2 + 2, 1 + 3)

(12)   (2 + 1 + 1, 1 + 2 + 1)   (13)   (2 + 1 + 1, 1 + 1 + 2)

(14)   (1 + 2 + 1, 1 + 1 + 2).

FIGURE 22.3

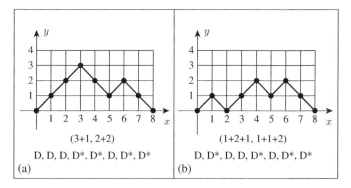

FIGURE 22.4

This idea can be used for each $n \geq 1$ and shows that the number of pairs of compositions of n, which satisfy the given conditions, is C_n.

Example 22.3: In each part of Fig. 22.5 on p. 186, there are six vertices lying along a horizontal line. Three arcs, each connecting two vertices, are drawn for each part, and these arcs satisfy the following: (i) each vertex is part of just one arc; (ii) the arcs lie above the vertices; and, (iii) no two of the arcs intersect. For each configuration, we shall travel along the horizontal (base) line—going from left to right. As we do so, we write a 1 for the first time we encounter an arc. The second time we encounter that arc, we write a −1. Each resulting sequence of three 1's and three −1's (where the number of 1's is never exceeded by the number of −1's or, equivalently, the partial sums are all nonnegative) is provided below the corresponding configuration. In this way, a one-to-one correspondence is established between the five configurations in parts (a)–(e) of Fig. 22.5 and the corresponding sequences in parts (a)–(e), respectively, of

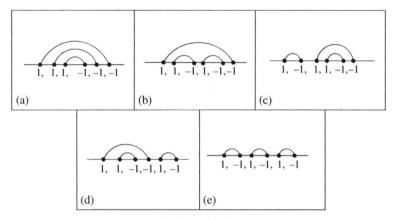

FIGURE 22.5

Fig. 20.4 (in Example 20.6). Consequently, there are 5 (= C_3) ways in which we can draw the six vertices and three arcs under the given conditions.

In general, for $n \geq 1$, we place $2n$ vertices along a horizontal (base) line and draw n arcs—each arc connecting two of the vertices. The number of ways in which these n arcs can be drawn, so that conditions (i), (ii), and (iii), as given above, are satisfied, is C_n.

Looking back at Fig. 20.1, we see that if we join the upper part of each pair of corresponding parentheses, the resulting configurations in parts (a)–(e) of that figure bear a marked resemblance to parts (a)–(e), respectively, of Fig. 22.5. Likewise, if we turn Terri's manila folders in Fig. 20.3 (in Example 20.5) upside down, then we find an immediate correspondence between these configurations and those in Fig. 22.5.

Another immediate example, dealing with the Catalan numbers, arises if we make the arcs of Fig. 22.5 into semicircles. For now we see that C_n counts the number of ways we can draw n semicircles, whose centers are on a horizontal (base) line and which satisfy the conditions that (i) all semicircles are above the horizontal (base) line; (ii) no two semicircles are tangent; and (iii) no two semicircles intersect. However, one or more semicircles may be enclosed within another semicircle. Finally, suppose these semicircles are reflected in the given horizontal (base) line. Upon considering each semicircle and its reflection, we now find that C_n counts the number of ways n circles can be drawn so that (i) all circles are centered on a given horizontal (base) line; (ii) no two circles are tangent; and (iii) no two circles intersect. Once again, however, one or more circles may be enclosed within another circle. These configurations are demonstrated in Fig. 22.6.

Example 22.4: Let A be a nonempty set.

A function $f: A \rightarrow A$ is called an *involution* if the composition of f with itself is the identity function on A—that is, $f^2 = f \circ f = 1_A$.

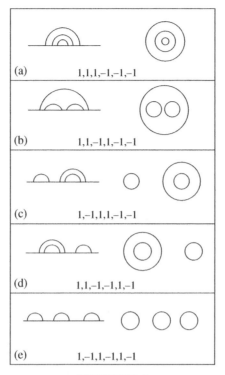

FIGURE 22.6

A function $f: A \rightarrow A$ is called a *permutation* if f is bijective (that is, one-to-one and onto).

Finally, if $f: A \rightarrow A$ and $x \in A$ with $f(x) = x$, we say that x is a *fixed point* for f.

For $n \geq 1$, we want to determine the number of permutations f for the set $\{1, 2, 3, \ldots, 2n - 1, 2n\}$, where f is an involution that has no fixed points and that satisfies the following *noncrossing* condition:

$$\text{If } a, b, c, d \text{ are distinct elements from } \{1, 2, 3, \ldots, 2n - 1, 2n\},$$

with $a < b < c < d$, then we cannot have $f(a) = c$ and $f(b) = d$.

Once again we focus our attention on the case for $n = 3$. In Fig. 22.7 on p. 188, we find the five possible permutations $f: \{1, 2, 3, 4, 5, 6\} \rightarrow \{1, 2, 3, 4, 5, 6\}$ that are involutions with no fixed points and that satisfy the noncrossing condition.

The reader who has studied permutation groups will recognize these functions as the elements of the symmetric group S_6 that can be represented as shown in Fig. 22.8 on p. 188. Here each permutation is written as a product of disjoint transpositions.

[Note, for example, that neither the permutation $(13)(24)(56)$ nor the permutation $(12)(35)(46)$ appears in Fig. 22.8. Both are involutions with no fixed points, but neither satisfies the noncrossing condition.]

f_1	f_2	f_3	f_4	f_5
$1 \to 6$	$1 \to 6$	$1 \to 2$	$1 \to 4$	$1 \to 2$
$2 \to 5$	$2 \to 3$	$2 \to 1$	$2 \to 3$	$2 \to 1$
$3 \to 4$	$3 \to 2$	$3 \to 6$	$3 \to 2$	$3 \to 4$
$4 \to 3$	$4 \to 5$	$4 \to 5$	$4 \to 1$	$4 \to 3$
$5 \to 2$	$5 \to 4$	$5 \to 4$	$5 \to 6$	$5 \to 6$
$6 \to 1$	$6 \to 1$	$6 \to 3$	$6 \to 5$	$6 \to 5$

FIGURE 22.7

$f_1 :$ $(16)(25)(34)$ $f_2 :$ $(16)(23)(45)$ $f_3 :$ $(12)(36)(45)$

$f_4 :$ $(14)(23)(56)$ $f_5 :$ $(12)(34)(56)$

FIGURE 22.8

Figure 22.9 provides the same arcs as in Fig. 22.5, but now we have labeled the points on the horizontal (base) axis. No matter if we use Fig. 22.7 or Fig. 22.8, we can determine these five functions from the arcs in Fig. 22.9. For example, part (c) of Fig. 22.9 shows that we pair 2 with 1, 6 with 3, and 5 with 4, just like f_3 does. This correspondence between the configurations in Fig. 22.9 and the functions in Fig. 22.7 (or Fig. 22.8) shows us that there are C_3 such functions.

For the general case, when $n \geq 1$, we have C_n permutations of the set $\{1, 2, 3, \ldots, 2n - 1, 2n\}$, which are involutions with no fixed points and which satisfy the noncrossing condition.

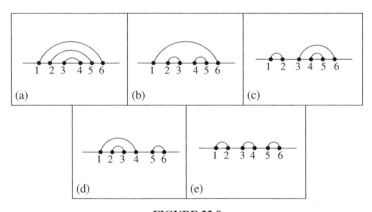

FIGURE 22.9

Example 22.5: For $n \geq 2$, here we start with $n - 1$ 1's and an arbitrary number k (between 0 and $n - 1$) of -1's. These $(n - 1) + k$ integers are then listed so that all partial sums (read from left to right) are nonnegative. When $n = 3$, for example, five such lists can be found—as in Fig. 22.10.

Look back at the left factors in Fig. 21.2 (of Example 21.2). A one-to-one correspondence between the structures given there and the lists in Fig. 22.10 can be obtained if each D ($\nearrow$) step is replaced by a 1 while each D* ($\searrow$) step is replaced by a -1. In this way the five left factors in parts (a)–(e) of Fig. 21.2 correspond, respectively, with the five lists in parts (a)–(e) of Fig. 22.10.

Since this construction can be defined for each $n \geq 2$, these lists of 1's and -1's are counted by C_n.

Before continuing to the next example, the reader may feel that the result here is very similar to that in Example 20.6. In the exercises, the reader will be asked to determine a one-to-one correspondence between the lists given here and the lists in Example 20.6.

(a) 1,1 (b) 1,1,-1 (c) 1,-1,1 (d) 1,1,-1,-1 (e) 1,-1,1,-1.

FIGURE 22.10

Example 22.6: Once again we have $n \geq 2$, but now we consider arcs and isolated vertices whose total number is $n - 1$. The endpoints of each arc and the isolated vertices lie along a horizontal (base) line, the arcs do not intersect, and the arcs lie above the vertices. Furthermore, no isolated vertex may be located below an arc, although one arc may be located below another.

There are five such possible configurations for $n = 3$—as exhibited in Fig. 22.11 on p. 190. A one-to-one correspondence between these configurations and the lists in Example 22.5 can be established as follows: Examining the vertices (from left to right) in each configuration in Fig. 22.11, write a 1 for each isolated vertex and for the left endpoint of each arc, and write a -1 for the right endpoint of each arc. This provides a one-to-one correspondence between the five configurations of isolated vertices and arcs in parts (a)–(e) of Fig. 22.11 and the respective lists in parts (a)–(e) of Fig. 22.10. Further, this type of correspondence can be defined for each $n \geq 2$ and exhibits another situation where the Catalan numbers arise.

Example 22.7: This example is due to David Callan. Like Example 22.6, it also deals with isolated vertices and arcs that lie above them. For $n \geq 2$, we now place $2n - 2$ vertices along a horizontal (base) line and label them consecutively from 1 to $2n - 2$. As before, these arcs lie above the horizontal (base) line and cannot intersect. Furthermore, for each arc that appears, the left endpoint is at an odd-numbered vertex, while the right endpoint is at an even-numbered vertex.

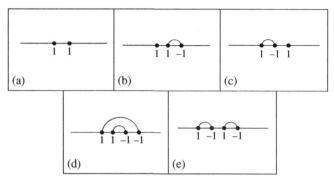

FIGURE 22.11

When $n = 3$, for example, we obtain the five configurations shown in Fig. 22.12. To set up a one-to-one correspondence between the results here and the Dyck paths in Fig. 21.1 (of Example 21.1), proceed as follows: (i) Start with a D ($\nearrow$) step. (ii) The left endpoint of an arc and an isolated vertex with an even label are each replaced with a D ($\nearrow$) step. (iii) The right endpoint of an arc and an isolated vertex with an odd label are each replaced with a D* ($\searrow$) step. (iv) Finally, add a D* ($\searrow$) step at the end. It then follows that the five configurations in parts (a)–(e) of Fig. 22.12 correspond, respectively, with the five Dyck paths in parts (a)–(e) of Fig. 21.1.

The preceding correspondence for the case of $n = 3$ can just as readily be developed for each $n \geq 2$. Therefore, these configurations of arcs and isolated vertices (with the stated conditions) on $2n - 2$ vertices are counted by the Catalan number C_n.

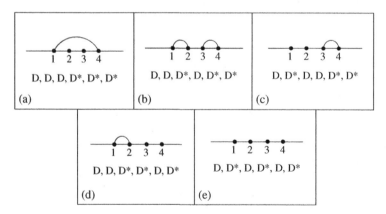

FIGURE 22.12

Example 22.8:   This last result for Chapter 22 is due to Emeric Deutsch. For $n \geq 1$, we shall consider configurations made up from arcs and labeled isolated vertices. Each vertex, either isolated or as the endpoint of an arc, will appear along a horizontal (base) axis. The total number of vertices (in a configuration) will vary—from n to $2n - 1$.

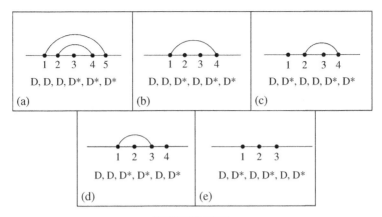

FIGURE 22.13

However, the total number of arcs and isolated vertices for each configuration will be n. Furthermore, the arcs are drawn above the vertices, do not intersect, and do not have endpoints labeled with consecutive integers. The case for $n = 3$ is presented in Fig. 22.13. Note that each configuration is made up of three pieces—each piece an isolated vertex or an arc on nonconsecutively labeled vertices.

To set up a one-to-one correspondence between these configurations and the Dyck paths in Fig. 21.1 (of Example 21.1), we replace (i) each isolated vertex with the pair of steps D ($\nearrow$), D* ($\searrow$); (ii) the left endpoint of an arc with the step D ($\nearrow$); and, (iii) the right endpoint of an arc with the step D* ($\searrow$). In this way, we find that the configurations in parts (a)–(e) of Fig. 22.13 correspond, respectively, with the Dyck paths in parts (a)–(e) of Fig. 21.1.

Since this type of correspondence can be defined for each $n \geq 1$, it follows that the number of these configurations, for a fixed n, is C_n.

EXERCISES FOR CHAPTER 22

1. Emma and Christopher are the sponsors for the women's fencing club at their university. There are 12 students involved in the club and it just so happens that no two of them have the same height. For a yearbook photo, Emma and Christopher must arrange these 12 students in two rows of six students each, so that, as all the students are facing forward, the heights of the students in each row are increasing from left to right. Furthermore, each student in the front row must be shorter than the student standing behind her. In how many ways can such arrangements be made?

2. Determine a one-to-one correspondence between the lists in Example 20.6 and those in Example 22.5.

CHAPTER 23

Triangulating the Interior of a Convex Polygon

At this point we return to the problem mentioned at the start, in Chapter 18.

Example 23.1: For $n \geq 1$, we start with a convex polygon P of $n + 2$ sides. We want to count the number of ways $n - 1$ diagonals can be drawn within the interior of P so that

(i) no two diagonals intersect within the interior of P; and,

(ii) the diagonals partition the interior of P into n triangles—hence, the verb, *triangulate*.

To show that C_n is the number of ways the interior of a convex $(n + 2)$-gon can be triangulated, we shall develop a one-to-one correspondence, for when $n = 3$, with the results in Table 20.2 (in Example 20.3). The method given here was developed in 1961 by Henry George Forder of the University of Auckland in New Zealand. His method is developed in Reference [13].

We replace x_0, x_1, x_2, and x_3 in Table 20.2 with a, b, c, and d, respectively. In Fig. 23.1, we find the five ways one can triangulate the interior of a convex pentagon with no intersecting diagonals. In each part, we have labeled four of the sides—with the letters a, b, c, d—as well as all five of the vertices. In part (a), for example, the labels on sides a and b are used to provide the label ab on the diagonal connecting vertices 2 and 4. This is due to the fact that the diagonal ab, together with the sides a and b, provides the leftmost interior triangle in the triangulation of the given convex pentagon. Then the diagonal ab and the side c provide the label $(ab)c$ for the diagonal connecting vertices 2 and 5. The sides labeled ab, c, and $(ab)c$ then determine the center interior triangle in this triangulation. Continuing in this way, we finally label the base edge connecting vertices 1 and 2 with the label $((ab)c)d$. Placing this label within a third set of parentheses, we arrive at the expression $(((ab)c)d)$ at the bottom of Fig. 23.1 (a). This is one of the five ways we can introduce parentheses in order to obtain the three products (of two numbers at a time) needed to compute $abcd$. In this

Fibonacci and Catalan Numbers: An Introduction, First Edition. Ralph P. Grimaldi.
© 2012 John Wiley & Sons, Inc. Published 2012 by John Wiley & Sons, Inc.

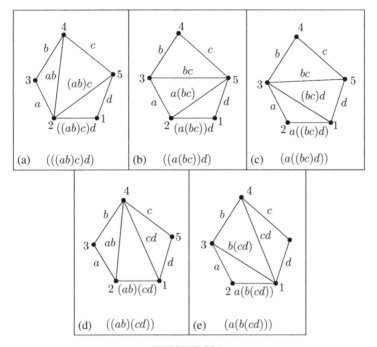

FIGURE 23.1

way, we correspond the triangulation in Fig. 23.1 (a) with the entry in the first row and first column of Table 20.2 (in Example 20.3). Applying this idea to the triangulations in parts (b)–(e) of Fig. 23.1 completes the one-to-one correspondence with the entries in the second through fifth rows (and first column), respectively, of Table 20.2.

Since this type of correspondence can be developed for each $n \geq 1$, it follows that T_n, the number of ways one can triangulate the interior of a convex $(n + 2)$-gon, is C_n.

Related to the above situation, let us now consider taking a convex n-gon and drawing 0 or more nonintersecting (within the polygon) diagonals within the interior of the n-gon. If the dissection results in an even number of regions (within the n-gon), the dissection is called *even* and $e(n)$ counts the number of all such even dissections. Likewise, a dissection that results in an odd number of regions is called *odd*, and $o(n)$ counts their number. If we have a convex 3-gon—that is, a triangle—then we can only draw 0 diagonals, resulting in one region. So here $e(3) = 0$ and $o(3) = 1$. For a convex 4-gon—that is, a convex quadrilateral—we get one region (with four sides), when we draw no diagonals, and two regions (with three sides) for each of the two (even) dissections where we draw one diagonal. Consequently, $e(4) = 2$ and $o(4) = 1$. Considering a convex pentagon, we get one odd dissection when we draw no diagonals, five even dissections when we draw one diagonal, and five odd dissections when we draw two diagonals. As a result, $e(5) = 5$ and $o(5) = 6$.

Now note that

$$e(3) - o(3) = 0 - 1 = -1 = (-1)^3$$
$$e(4) - o(4) = 2 - 1 = 1 = (-1)^4$$
$$e(5) - o(5) = 5 - 6 = -1 = (-1)^5.$$

This pattern continues and one finds that for all $n \geq 3$, $e(n) - o(n) = (-1)^n$. This was discovered by Emeric Deutsch in 2005. Two different proofs of this result are provided in Reference [7].

EXERCISES FOR CHAPTER 23

1. Find the parenthesized product associated with the triangulated convex hexagon [in part (a)] and the triangulated convex octagon [in part (b)] of Fig. 23.2.

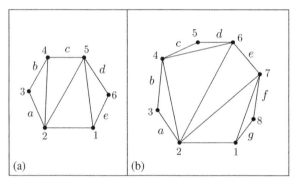

FIGURE 23.2

2. In how many ways can Yolanda triangulate the interior of the convex 12-sided polygon $ABC \ldots JKL$ if the triangulations must include the diagonal AH?

3. For $n \geq 4$, prove that the number of ways one can partition the interior of a convex polygon of n sides into $n - 4$ triangles and one quadrilateral, using $n - 4$ diagonals, with no two diagonals intersecting within the polygon, is $\binom{2n-5}{n-4}$. (This result is due to Christian Jones.)

CHAPTER 24

Some Examples from Graph Theory

As the title suggests, this chapter will introduce some examples where the Catalan numbers arise in graph theory.

In Chapter 12, we were introduced to the structure called an undirected graph. In our first example, we shall investigate a special collection of undirected graphs, which have no loops and are defined using closed intervals of unit length.

Example 24.1 (*Unit-Interval Graphs*):   For $n \geq 1$, we start with n closed intervals of unit length and draw the corresponding unit-interval graphs on n vertices, as shown in Fig. 24.1 (for the cases where $n = 1$ and $n = 2$). In Fig 24.1 (a) we find one unit interval. This corresponds to the single (isolated) vertex u_1. Here we can represent both the closed interval and the unit-interval graph by the binary string 01. When we consider two closed unit intervals, we can draw them so that they do not overlap, as in Fig. 24.1 (b), or overlapping, as in part (c) of the figure. When two closed unit intervals overlap, we draw an edge in the graph joining the vertices that correspond to the two closed unit intervals. Consequently, the unit-interval graph in Fig. 24.1 (b) consists of the two isolated vertices w_1 and w_2, which correspond respectively with the first and second closed unit intervals that do not overlap. The closed unit intervals in Fig. 24.1 (c) do overlap, so the resulting graph consists of the vertices w_1, w_2 and the edge $w_1 w_2$. The vertices w_1 and w_2 correspond, respectively, with the lower and

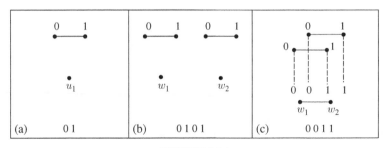

FIGURE 24.1

Fibonacci and Catalan Numbers: An Introduction, First Edition. Ralph P. Grimaldi.
© 2012 John Wiley & Sons, Inc. Published 2012 by John Wiley & Sons, Inc.

upper closed unit intervals in Fig. 24.1 (c). In both parts (b) and (c), we can represent the positioning of the two closed intervals and the corresponding unit-interval graphs by the binary string 0101 for part (b) and 0011 for part (c).

In Fig. 24.2, we have the five unit-interval graphs for the case of $n = 3$. At the bottom of each graph, we find its corresponding binary string of three 0's and three 1's. Consequently, we have a one-to-one correspondence between the five unit-interval graphs for three unit intervals and the five binary strings of three 0's and three 1's, where the number of 1's never exceeds the number of 0's.

For an arbitrary positive integer n, the preceding result for $n = 3$ generalizes and provides a one-to-one correspondence between the unit-interval graphs for n unit intervals and the binary strings of n 0's and n 1's, where the number of 1's never exceeds the number of 0's. From the discussion at the end of Example 20.2, it then follows that the number of such unit-interval graphs for n unit intervals is C_n.

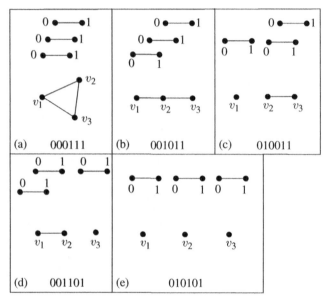

FIGURE 24.2

Before we can deal with our next two examples, we need to introduce some further material about trees. In Example 12.4, we learned about rooted binary trees. Now we shall extend the notion to rooted trees in general.

Once again there will be a unique vertex with no edges coming into it. This will be the *root r* of the tree and we say that the *in degree* of r is 0. All the other vertices in the tree have at least one edge coming into it, so the in degree of each vertex, other than the root, is greater than 0. The (directed) tree in part (1) of Fig. 24.3 is rooted with root r. In part (2) of Fig. 24.3 we have redrawn the (directed) tree in part (1) and here the directions are understood to be going down. So, for instance, we understand that there is a directed edge ae from a to e. (This is sometimes denoted by $\overrightarrow{ae}$.) Also, there is no edge from e to a. In such (directed) trees, the number of edges coming out

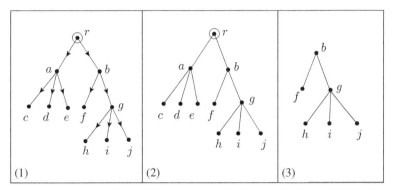

FIGURE 24.3

from a vertex is called the *out degree* of the vertex. In part (2) of Fig. 24.3 the vertex *g*, for example, has in degree 1 and out degree 3. A vertex with out degree 0 is called a *leaf*. Vertices *c*, *d*, *e*, *f*, *h*, *i*, and *j* are the leaves of the tree in [part (1) and] part (2) of Fig. 24.3. All the other vertices, including the root, are called *internal vertices*.

Vertex *b* is called the *parent* of the vertex *f*, while *f* is called a *child* of vertex *b*. Each of *f* and *g* is called a *sibling* of the other. Vertices *h*, *i*, and *j* are considered *descendants* of *g*, *b*, and *r*, while *g*, *b*, and *r* are called *ancestors* of *h*, *i*, and *j*. We consider the root *r* to be at *level* 0, while the vertices *a* and *b* are said to be at *level* 1. Similarly, the vertices *c*, *d*, *e*, *f*, and *g* are at *level* 2, while the vertices *h*, *i*, and *j* are at *level* 3. Finally, in part (3) of Fig. 24.3 we have the *subtree* [of the given tree in part (2)] *rooted at b*. Note that this tree is made up from the vertex *b* and all of its descendents (and the edges they determine). A single vertex, such as *d*, with no descendants, as in part (2) of Fig. 24.3, is still considered as the subtree rooted at *d*.

When presented with a rooted tree, we want to develop methods whereby we can systematically visit all of the vertices of such a tree. In the study of data structures, two of the most prevalent such methods are the preorder and postorder. These are defined recursively in the following.

Definition 24.1:   Let *T* be a rooted tree with root *r*, vertex set *V*, and edge set *E*. If *T* has no other vertices, then the root *r* by itself constitutes the *preorder* and *postorder traversals* of *T*. If $|V| > 1$, let $T_1, T_2, T_3, \ldots, T_k$ denote the subtrees of *T* as we go from left to right (as in Fig. 24.4 on p. 198).

(a) The *preorder traversal* of *T* first visits the root *r* and then traverses the vertices of T_1 in preorder, then the vertices of T_2 in preorder, and so on until the vertices of T_k are traversed in preorder.

(b) The *postorder traversal* of *T* traverses in postorder the vertices of the subtrees $T_1, T_2, \ldots, T_k$ and then visits the root *r*.

We shall demonstrate these traversals in the following example.

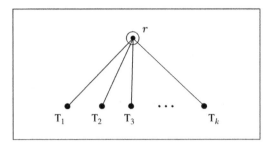

FIGURE 24.4

Example 24.2: Consider the rooted tree in Fig. 24.5.

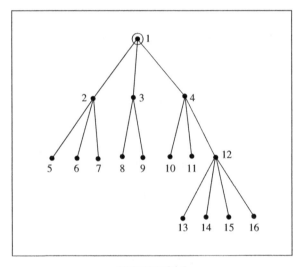

FIGURE 24.5

(a) ***Preorder***: After visiting vertex 1, we visit the subtree T_1 rooted at vertex 2. After we visit vertex 2, we proceed to the subtree rooted at vertex 5. This subtree is simply the vertex 5, so we visit vertex 5 and then return to vertex 2 from which we visit, in succession, vertices 6 and 7. Following this, we *backtrack* (7 to 2 to 1) to the root 1 and then proceed to visit the vertices in the subtree T_2 (rooted at vertex 3) in the preorder 3, 8, 9. Now we return to the root 1 for the final time and then visit the vertices in the subtree T_3 (rooted at vertex 4) in the preorder 4, 10, 11, 12, 13, 14, 15, 16. Consequently, the preorder listing of the vertices in this rooted tree is

$$1, 2, 5, 6, 7, 3, 8, 9, 4, 10, 11, 12, 13, 14, 15, 16.$$

In this ordering, we start at the root r and build a path as far as possible. At each level, we go to the leftmost vertex (not previously visited) at the next level, until we reach a leaf l. Then we backtrack to the parent p of l and visit, if it exists, the sibling s (and the subtree for which it is the root) of l that is directly to the right of l. If no such sibling exists, we backtrack one step further to the parent g of p—hence, the grandparent of l. Now we visit, if it exists, a vertex v that is a sibling of p and is directly to the right of p. As we continue, eventually we visit (the first time each vertex is encountered) all the vertices in the rooted tree.

(b) ***Postorder***: For this type of traversal, we start at the root r and build the longest path, going to the leftmost child of each internal vertex, whenever possible. Upon arriving at a leaf l, we visit l and then backtrack to its parent p. However, we do *not* visit p until after we have visited all of the descendants of p. The next vertex we visit is found by applying the same procedure, at the vertex p, that we originally applied at r in order to get to the leaf l. However, now we first go from p to the sibling of l that is directly to the right of l. Throughout this procedure, no vertex is visited more than once or before any of its descendants.

For the rooted tree in Fig. 24.5, the postorder traversal starts with the postorder traversal of the subtree T_1 rooted at vertex 2. This results in the initial listing 5, 6, 7, 2. Proceeding to the subtree rooted at vertex 3, the postorder listing continues with 8, 9, 3. Then, for the subtree rooted at vertex 4, we have the postorder listing 10, 11, 13, 14, 15, 16, 12, 4. Finally, the root 1 is visited. Consequently, for this rooted tree, the postorder listing of its vertices is

$$5, 6, 7, 2, 8, 9, 3, 10, 11, 13, 14, 15, 16, 12, 4, 1.$$

Having developed the concepts of the preorder and postorder traversal of a rooted tree, it is time to examine two examples where these ideas come into play.

Example 24.3: A rooted tree T is called *binary* if each internal vertex has no more than two children. If each internal vertex has exactly two children—distinguished as the left child and the right child—then we call such a rooted tree a *complete binary* tree.

For $n \geq 0$, we consider the complete binary trees which have $2n + 1$ vertices. The cases for $0 \leq n \leq 3$ are covered in Fig. 24.6 on p. 200.

Below each tree in the figure, we list the vertices as they are visited in a preorder traversal. We see that for the cases where $1 \leq n \leq 3$, there is a list of n L's and n R's below each preorder traversal. These lists are determined as follows. When $n = 2$, for example, the list for the second complete binary tree is L, R, L, R because, after we visit the root r, we go to the left (L) subtree rooted at a and visit vertex a. Backtracking then to r, we go to the right (R) subtree rooted at b. Now we visit vertex b and then go to the left (L) subtree of b rooted at c and visit vertex c. Lastly, we backtrack to vertex b and go to the right (R) subtree of b rooted at d. Here we visit vertex d and this completes the preorder traversal and accounts for the list L, R, L, R that appears beneath the preorder listing r, a, b, c, d. The other seven lists of L's and R's that appear in the figure, for $1 \leq n \leq 3$, are developed in the same way.

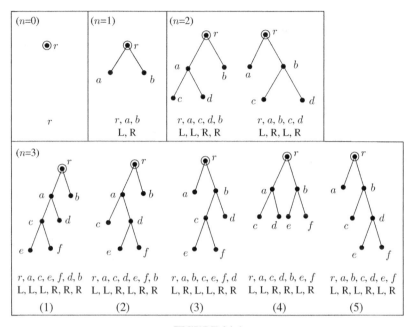

FIGURE 24.6

As we traverse these trees in preorder, each of the lists (for the case of $n = 3$) starts with an L. There is an equal number of L's and R's in each list because these trees are complete binary trees. Also, the number of R's never exceeds the number of L's as a specific list is read from left to right—again, because we are traversing these trees in preorder. Comparing the lists in this example with the strings in Fig. 20.2 in Example 20.2, a one-to-one correspondence between these complete binary trees and the strings of that example can readily be defined—just correspond L with 1 and R with 0. In this way, the five complete binary trees in parts (1)–(5) [for the case of $n = 3$ in Fig. 24.6] correspond, respectively, with the binary strings in parts (a)–(e) of Fig. 20.2 in Example 20.2. Since this type of correspondence can be given for each $n \geq 0$, it follows that there are C_n complete binary trees on $2n + 1$ vertices.

Example 24.4: For $n \geq 0$, we now want to count the number of ordered rooted trees on $n + 1$ vertices. The case for $n = 3$ is covered by the five trees in parts (1)–(5) of Fig. 24.7. [Note that the two trees in parts (6) and (7) of the figure are distinct as binary rooted trees, where we distinguish left from right. The tree in part (6) has a left subtree rooted at a, while the tree in part (7) has a right subtree rooted at a. However, when considered as ordered rooted trees, where the subtrees are not considered as left or right, but as first, second, and third, these two trees are considered the same tree and each is accounted for by the tree in part (2) of Fig. 24.7.]

Here we shall perform a postorder traversal of the vertices of these trees. For example, for the tree in part (1), we traverse each edge twice—once going down and once coming back up. When we traverse an edge going down, we shall write "1"

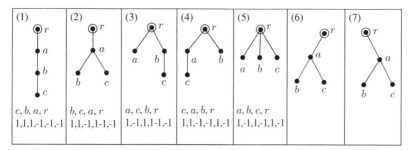

FIGURE 24.7

and when we traverse this edge coming back up, we shall write "−1". So for this particular ordered rooted tree on 4 (= 3 + 1) vertices, we have the postorder traversal

$$c, b, a, r$$

and the list

$$1, 1, 1, -1, -1, -1.$$

When the tree in part (2) of Fig. 24.7 is traversed in postorder, the vertices are visited in the order

$$b, c, a, r$$

and the list this time is

$$1, 1, -1, 1, -1, -1.$$

Parts (3)–(5) of Fig. 24.7 provide the postorder traversals and corresponding lists for the other three ordered rooted trees on four vertices. These lists provide us with a one-to-one correspondence between the ordered rooted trees on four vertices in parts (1)–(5) of Fig. 24.7 and the arrangements in parts (a)–(e), respectively, of Fig. 20.4 (in Example 20.6). This correspondence can just as readily be given for any $n \geq 0$, so we may conclude that the number of ordered rooted trees on $n + 1$ vertices is C_n.

[Looking back at the peakless Motzkin paths in Figure 21.15 in Exercise 9, we see that the number of R steps in each of parts (a)–(e) in that figure equals the number of leaves in each of parts (1)–(5), respectively, in Fig. 24.7. Also, if an R step is k units above the x-axis, for $k \geq 0$, then that R step corresponds with a leave at level $k + 1$.]

EXERCISES FOR CHAPTER 24

1. List the vertices in the rooted tree shown in Fig. 24.8 on p. 202 when they are visited in a preorder traversal and in a postorder traversal.
2. For $n \geq 0$, let T be a rooted tree with vertex set $\{1, 2, 3, \ldots, n, n + 1\}$, where 1 is the root. The vertices are arranged in a clockwise manner, in increasing order, on the circumference of a circle, and no two edges of the tree intersect

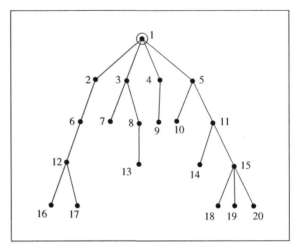

FIGURE 24.8

within the interior of the circle. Finally, the vertices increase along each path that starts at vertex 1 (the root). Such a tree T is called a *noncrossing increasing tree* on the vertex set $\{1, 2, 3, \ldots, n, n + 1\}$. The trees in Fig. 24.9 exemplify the case for $n = 3$.

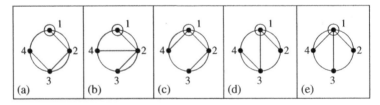

FIGURE 24.9

Show that for $n \geq 0$, the number of noncrossing increasing trees on the vertex set $\{1, 2, 3, \ldots, n, n + 1\}$ is the Catalan number C_n.

3. In Example 12.4, we examined connected graphs and introduced the idea of components for graphs that are disconnected (that is, not connected). In Fig. 24.2 (of Example 24.1), we see that the graphs in parts (a) and (b) are connected—the graph in part (a) being a cycle on three vertices, while that in part (b) is a path on three vertices. The graphs in parts (c) and (d) are disconnected, each having two components. Finally, the graph in part (e) is also disconnected and consists of three components, each an isolated vertex.

(a) The following binary strings correspond to unit-interval graphs like those shown in Example 24.1. For each string, determine whether the corresponding unit-interval graph is connected. If disconnected, determine the number of components for the graph.

(i) 01010101	(iii) 00101011	(v) 0011001101
(ii) 00001111	(iv) 0000111011	(vi) 0100001111

(b) Can we tell when a unit-interval graph is connected without drawing the graph? Can we determine the number of components of the graph by just looking at the binary string?

4. Let G be a unit-interval graph with at least two vertices. Prove that if G is connected, then the last two bits in the binary string associated with G are 11. Is the converse true—that is, if G is a unit-interval graph whose associated binary string ends in 11, then G is connected?

5. (a) Set up a one-to-one correspondence between the unit-interval graphs on n vertices and the Dyck paths from $(0, 0)$ to $(2n, 0)$ in Example 21.1.

   (b) Of the C_n unit-interval graphs on n vertices, how many are connected?

6. For $n \geq 0$, a *rooted trivalent tree* on $2n + 2$ vertices is a rooted tree where each vertex of the tree has degree 1 or degree 3. Here the degree of a vertex is the number of edges incident with the vertex. This is also the sum of the in degree of a vertex and the out degree of that vertex. Furthermore, in such a tree, a left child of a vertex is distinguished from a right child of that vertex.

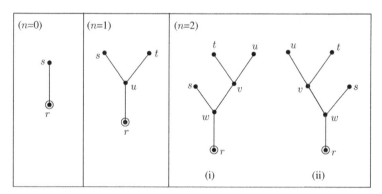

FIGURE 24.10

The cases for $n = 0$, 1, and 2 are shown in Fig. 24.10. When $n = 1$, there is only one such tree. Here r is the root and vertex u is the only vertex of degree 3, as there are three edges incident with vertex u—namely, us, ut, and ru. The other three vertices in this tree all have degree 1. For the two rooted trivalent trees that arise when $n = 2$, we find that for either tree, there are two vertices of degree 3 and four of degree 1. Also, in tree (i), vertex t is the left child of vertex v, while it is the right child of vertex v in tree (ii).

For the case where $n = 3$, one such tree is shown in part (3) of Fig. 24.11 on p. 204. In part (1) of Fig. 24.11 we have the triangularization of the convex pentagon shown in part (a) of Fig. 23.1. Here we have placed a vertex at the center of each of the five sides and at the center of each of the two diagonals of the pentagon, and added a vertex r outside the pentagon, below the vertex w.

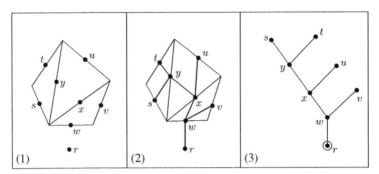

FIGURE 24.11

The vertex r will serve as the root for our trivalent tree. To obtain the result in part (2) of Fig. 24.11, we start by drawing the edge rw. Then from w we draw edges to the other two vertices in the perimeter of the triangle in which w appears. This gives us the edges wx and wv. At this point we see that the vertex v is not in the perimeter of any other triangle, but vertex x is. So from vertex x, edges are now drawn to each of the vertices y and u. At this point vertex u is no longer in the perimeter of any other triangle, but vertex y is. Finally, we finish by drawing the edges from y to each of the vertices s and t. The vertices r, s, t, u, v, w, x, y and the edges rw, wv, wx, xu, xy, ys, yt provide the rooted trivalent tree shown in part (3) of Fig. 24.11.

Set up a one-to-one correspondence between the other four triangularizations of the convex pentagon and the other four rooted trivalent trees on $2(3) + 2 = 8$ vertices.

CHAPTER 25

Partial Orders, Total Orders, and Topological Sorting

In Example 13.2, we were introduced to a special type of binary relation on a set A—namely, the notion of the partial order. Furthermore, when the set A is finite, we found that a partial order on A could be studied by means of its Hasse diagram.

Let us recall these ideas for the set A of all (positive integer) divisors of 12. Hence $A = \{1, 2, 3, 4, 6, 12\}$ and here the relation $\mathcal{R}$ is defined on A by $x\mathcal{R}y$ (that is, x is related to y) when x divides y. (Recall that we may also write $(x, y) \in \mathcal{R}$ in place of $x\mathcal{R}y$.) The Hasse diagram for this partial order is shown in Fig. 25.1.

Continuing, the following ideas will prove useful.

Definition 25.1: For a partial order $\mathcal{R}$ on a set A [often denoted by the pair $(A, \mathcal{R})$], an element $x \in A$ is called *maximal* if for each $a \in A$, if $x\mathcal{R}a$, then $x = a$.

The partial order shown in Fig. 25.1 has 12 as its unique maximal element.

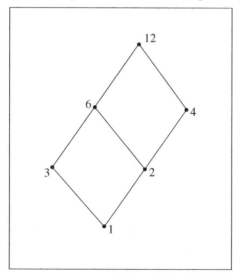

FIGURE 25.1

Fibonacci and Catalan Numbers: An Introduction, First Edition. Ralph P. Grimaldi.
© 2012 John Wiley & Sons, Inc. Published 2012 by John Wiley & Sons, Inc.

For another example, let B be the following set of subsets of $\{1, 2, 3\}$: $B = \{\{1\}, \{2\}, \{1, 2\}, \{2, 3\}\}$. Here the relation on B is the subset relation—that is, for X, $Y \in B$, X is related to Y when $X \subseteq Y$. This provides a partial order for B and its Hasse diagram is shown in Fig. 25.2. For this partial order we find two maximal elements—namely, $\{1, 2\}$ and $\{2, 3\}$.

A partial order need not have a maximal element. For example, consider the set $\mathbf{N} = \{1, 2, 3, \ldots\}$ with the relation $\mathcal{R}$, where for x, $y \in \mathbf{N}$, $x\mathcal{R}y$ when $x \leq y$ (the "less than or equal to" relation).

If we have a partial order $(A, \mathcal{R})$ and A ($\neq \varnothing$) is finite, then the partial order has a maximal element. Since $A \neq \varnothing$, let $a_1 \in A$. Should a_1 be maximal, we are done. If not, there exists an element $a_2 \in A$ such that $a_2 \neq a_1$ and $a_1\mathcal{R}a_2$. If a_2 is maximal, we are done. Otherwise, there exists an element $a_3 \in A$ such that $a_3 \neq a_2$, $a_3 \neq a_1$, and $a_2\mathcal{R}a_3$. Fortunately, this process cannot continue indefinitely because A is finite, so at some point we find distinct elements $a_1, a_2, a_3, \ldots, a_k$ with

$$a_1\mathcal{R}a_2\mathcal{R}a_3\mathcal{R}\ldots\mathcal{R}a_k$$

and a_k maximal.

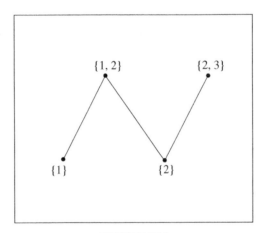

FIGURE 25.2

Dual to the concept of a maximal element for a partial order is the concept of a minimal element.

Definition 25.2:   For a partial order $(A, \mathcal{R})$, an element $y \in A$ is called a *minimal element* if for each $a \in A$, if $a\mathcal{R}y$, then $a = y$.

The element 1 is a minimal element for the Hasse diagram of the partial order shown in Fig. 25.1. The Hasse diagram in Fig. 25.2 has two minimal elements—namely, $\{1\}$ and $\{2\}$. Although the partial order $(\mathbf{N}, \leq)$ has no maximal element, the integer 1 is the unique minimal element. If we let $M = \{\ldots, -3, -2, -1\}$

with the relation $\leq$ (the ordinary "less than or equal to" relation), then for this partial order the integer -1 is the unique maximal element but there is no minimal element. However, if we have a partial order $(A, \mathcal{R})$ where $A \neq \varnothing$ and A is finite, then there is a minimal element for this partial order. (A proof for this result is requested in the exercises for this chapter.)

Now let us note something about two of the previous partial orders we have discussed above. Examining the partial order $(\mathbf{N}, \leq)$ a little closer, we see that for any $a, b \in \mathbf{N}$, either $a \leq b$ or $b \leq a$—that is, a is related to b or b is related to a. For our initial example, where $A = \{1, 2, 3, 4, 6, 12\}$ and the relation is given by $x\mathcal{R}y$ for $x, y \in A$ when x divides y, if we consider $3, 6 \in A$ we see that $3\mathcal{R}6$. However, if instead, we consider $4, 6 \in A$, then we find that 4 is *not* related to 6 *and* 6 is *not* related to 4. So 4 and 6 are labeled *incomparable*. These observations lead to the following idea.

Definition 25.3: Let $(A, \mathcal{R})$ be a partial order. We call $(A, \mathcal{R})$ a *total order* if for all $x, y \in A$, either $x\mathcal{R}y$ or $y\mathcal{R}x$.

At this point we know that a partial order $(A, \mathcal{R})$, where A is a finite set, need not be a total order. However, we shall now provide a method whereby such a partial order $(A, \mathcal{R})$ can be enlarged to a larger partial order $(A, \mathcal{T})$, where $\mathcal{R} \subseteq \mathcal{T}$ and $(A, \mathcal{T})$ is a total order. The way to accomplish this is via the Topological Sorting algorithm.

Topological Sorting Algorithm

(for a partial order $\mathcal{R}$ on a set A with $|A| = n$)

Step 1: Set the counter $i = 1$ and let H_1 be the Hasse diagram for the partial order $(A, \mathcal{R})$.

Step 2: Select a maximal element v_i in H_i—that is, an element for which there is no (implicitly directed) edge in H_i starting at the vertex v_i.

Step 3: If $i = n$, the process terminates and we arrive at the total order $(A, \mathcal{T})$, where

$$v_n \, \mathcal{T} v_{n-1} \, \mathcal{T} v_{n-2} \, \mathcal{T}... \, \mathcal{T} v_2 \, \mathcal{T} v_1$$

and $(A, \mathcal{T})$ contains $(A, \mathcal{R})$.

If $i < n$, then remove from H_i the vertex v_i and all (implicitly directed) edges of H_i that end at v_i. Call the resulting (implicitly directed) subgraph H_{i+1}. Then increase the counter i by 1 and return to step (2).

Before we apply this algorithm to a particular partial order, note that the selection in step (2) calls for "a" maximal element , not "the" maximal element. Consequently, it is possible to arrive at several different total orders (A, T) that contain the given partial order $(A, \mathcal{R})$.

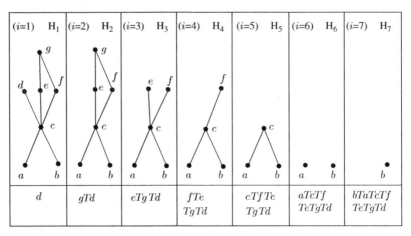

(i=1) H_1	(i=2) H_2	(i=3) H_3	(i=4) H_4	(i=5) H_5	(i=6) H_6	(i=7) H_7
d	gTd	$eTgTd$	fTe $TgTd$	$cTfTe$ $TgTd$	$aTcTf$ $TeTgTd$	$bTaTcTf$ $TeTgTd$

FIGURE 25.3

In Fig. 25.3 we find the Hasse diagrams that may come about when the Topological Sorting algorithm is applied to the partial order shown as H_1 (for the initial case of $i = 1$). The Hasse diagrams for this given partial order and the total order resulting from this particular application of the Topological Sorting algorithm are shown in parts (1) and (2), respectively, of Fig. 25.4. We see from part (1), for instance, that for $d, f \in A$, it happens that $(d,\ f) \notin \mathcal{R}$ and $(f,\ d) \notin \mathcal{R}$—that is, d is not related to f, nor is f related to d. So the partial order $(A,\ \mathcal{R})$ is not a total order. However, this is not the case in part (2) of Fig. 25.4. Here we see that $f \, Td$. Also, we realize that every ordered pair in the partial order $(A,\ \mathcal{R})$ is contained in the total order $(A,\ T)$. Part

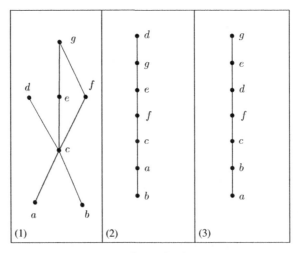

FIGURE 25.4

(3) of Fig. 25.4 provides another possible total order that contains the given partial order $(A, \mathcal{R})$.

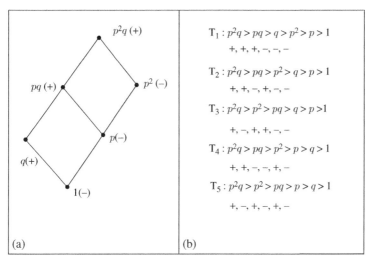

$T_1 : p^2q > pq > q > p^2 > p > 1$
$+, +, +, -, -, -$

$T_2 : p^2q > pq > p^2 > q > p > 1$
$+, +, -, +, -, -$

$T_3 : p^2q > p^2 > pq > q > p > 1$
$+, -, +, +, -, -$

$T_4 : p^2q > pq > p^2 > p > q > 1$
$+, +, -, -, +, -$

$T_5 : p^2q > p^2 > pq > p > q > 1$
$+, -, +, -, +, -$

(a)   (b)

FIGURE 25.5

At this point the reader may be wondering what the preceding material on total orders has to do with the Catalan numbers. The next example will provide a situation where the number of different total orders for the partial order in part (a) of Fig. 25.5 is determined. (Note that the partial order in Fig. 25.1 is the special case of Fig. 25.5, where $p = 2$ and $q = 3$.)

Example 25.1:   Let p, q be two distinct primes. Part (a) of Fig. 25.5 provides the Hasse diagram for the partial order $\mathcal{R}$ of all positive integer divisors of p^2q. (So $A = \{1, p, q, p^2, pq, p^2q\}$ and, for $x, y \in A$, $x\mathcal{R}y$ when x divides y.) Applying the Topological Sorting algorithm to this Hasse diagram, in part (b) of Fig. 25.5 we list the five possible total orders $\mathcal{T}_k$, where $\mathcal{R} \subseteq \mathcal{T}_k$, for $1 \leq k \leq 5$.

Now let us examine the Hasse diagram in part (a) of Fig. 25.5 a little closer. Here we have placed a plus sign at each vertex that has a q in the label for that vertex, and a minus sign at each vertex where the label has no occurrence of q. This time we shall focus on the three plus signs and three minus signs in the figure as well as the list below each total order in part (b) of Fig. 25.5. When the Topological Sorting algorithm is applied to the Hasse diagram for $(A, \mathcal{R})$, step (2) of the algorithm implies that the first maximal element selected will be unique—in this case, the divisor p^2q. This accounts for the first plus sign that appears for each total order $\mathcal{T}_k$, $1 \leq k \leq 5$. Continuing to apply the algorithm, we get two additional plus signs and three minus signs.

As the Topological Sorting algorithm is applied, is it possible that there could be more minus signs than plus signs at some point in any of the corresponding lists that arise? For instance, could a list start with $+, -, -$? If so, then we have failed to apply step (2) of the Topological Sorting algorithm correctly—for pq (not p) should have been recognized as the unique maximal element after p^2q and p^2. More to the point, for $0 \leq j \leq 2$, p^jq must be selected before p^j can be selected. Consequently, for each list of three plus signs and three minus signs, the number of minus signs will never exceed the number of plus signs, as the list is read from left to right. Replacing each plus sign with a 1 and each minus sign with a -1, we arrive at a one-to-one correspondence between the total orders T_k, $1 \leq k \leq 5$, in part (b) of Fig. 25.5 and the respective sequences of 1's and -1's in Fig. 20.4 (in Example 20.6).

Furthermore, for $n \geq 1$, the Topological Sorting algorithm can be applied to the partial order of all positive integer divisors of $p^{n-1}q$ to yield $(1/(n+1))\binom{2n}{n}$ total orders, thus providing one more instance where the Catalan numbers come to the forefront.

EXERCISES FOR CHAPTER 25

1. Let $(A, \mathcal{R})$ be a partial order where A ($\neq \varnothing$) is finite. Prove that there is a minimal element for this partial order.

2. How many different total orders are possible for the partial order in part (1) of Fig. 25.4?

3. Find the number of ways to totally order the partial order of all positive integer divisors of (a) 45; (b) 54; and (c) 160.

4. Let p and q be distinct primes and k a positive integer. If there are 1430 ways to totally order the partial order of positive integer divisors of p^kq, how many positive integer divisors are there for this partial order?

CHAPTER 26

Sequences and a Generating Tree

In this chapter we shall investigate a collection of sequences that are counted by the Catalan numbers. Examples related to these sequences are also examined. We start with the following.

Example 26.1: For $n \geq 1$, let $a_1, a_2, \ldots, a_n$ be a nondecreasing sequence of positive integers where $a_i \leq i$ for all $1 \leq i \leq n$. When $n = 1$, there is only one such sequence—namely, 1. There are two such sequences for $n = 2$: they are 1, 1 and 1, 2.

For $n = 3$, we shall set up a one-to-one correspondence between the lists of three 1's and three -1's in Fig. 20.4 (in Example 20.6) and the sequences of positive integers a_1, a_2, a_3 where $a_1 \leq 1$, $a_2 \leq 2$, and $a_3 \leq 3$. Start with the list 1, 1, 1, -1, -1, -1. For each 1 in this list, count the number of -1's in the list to the left of this 1. We find the following:

$$\text{List:} \quad 1, \ 1, \ 1, \ -1, \ -1, \ -1$$
$$\text{Number of } -1\text{'s to the left:} \quad 0, \ 0, \ 0$$
$$\text{Add 1 to each entry:} \quad 1, \ 1, \ 1.$$

We realize that 1, 1, 1 is one of the sequences (of length 3) that we are counting for the case of $n = 3$.

Next consider the second list in Fig. 20.4—that is, the list 1, 1, -1, 1, -1, -1. Counting the -1's to the left of each of the three 1's, as we did above, we have

$$\text{List:} \quad 1, \ 1, \ -1, \ 1, \ -1, \ -1$$
$$\text{Number of } -1\text{'s to the left:} \quad 0, \ 0, \quad 1$$
$$\text{Add 1 to each entry:} \quad 1, \ 1, \quad 2.$$

For the other three lists, we obtain the corresponding three sequences:

$$\text{List:} \quad 1, -1, 1, 1, -1, -1 \qquad 1, 1, -1, -1, 1, -1$$
$$\text{Number of } -1\text{'s to the left:} \quad 0, \quad 1, 1 \qquad\quad 0, 0, \qquad 2$$
$$\text{Add 1 to each entry:} \quad 1, \quad 2, 2 \qquad\quad 1, 1, \qquad 3$$

Fibonacci and Catalan Numbers: An Introduction, First Edition. Ralph P. Grimaldi.
© 2012 John Wiley & Sons, Inc. Published 2012 by John Wiley & Sons, Inc.

List: $1, -1, 1, -1, 1, -1$

Number of -1's to the left: 0,   1,   2

Add 1 to each entry: 1,   2,   3

Consequently, in Fig. 26.1, we now find the five sequences a_1, a_2, a_3, where $1 \leq a_i \leq i$, for all $1 \leq i \leq 3$.

(a)   1,1,1   (b)   1,1,2   (c)   1,2,2   (d)   1,1,3   (e)   1,2,3

FIGURE 26.1

To go in the reverse direction (for $n = 3$), let a_1, a_2, a_3 be a nondecreasing sequence of three positive integers where $a_i \leq i$ for $i = 1, 2, 3$. The corresponding list of three 1's and three -1's, whose partial sums are all nonnegative, is constructed as follows: (1) Start with a 1. (2) Then write $a_2 - a_1$ -1's (where $0 \leq a_2 - a_1 \leq 1$). (3) Write the second 1. (4) Now write $a_3 - a_2$ -1's (where $0 \leq a_3 - a_2 \leq 2$). (5) Write the third 1. (6) Finally, write $3 - (a_3 - a_1)$ -1's. As a result, we now have a one-to-one correspondence between the lists in parts (a)–(e) of Fig. 20.4 (in Example 20.6) and the respective sequences in parts (a)–(e) of Fig. 26.1.

In so much as this correspondence can be developed for each $n \geq 1$, it follows that the number of nondecreasing sequences $a_1, a_2, \ldots, a_n$, where, for $1 \leq i \leq n$, a_i is a positive integer and $a_i \leq i$, is equal to the nth Catalan number C_n.

Alternatively, we can show that these nondecreasing sequences are counted by the Catalan numbers, by setting up a one-to-one correspondence with the lattice paths of Example 19.1. For the case of $n = 3$, consider the lattice path shown in Fig. 19.2 (a). Examine the ith horizontal step in this path, for $i = 1, 2, 3$. This step takes the path from $(i - 1, \ 0)$ to $(i, \ 0)$ and is 0 units above the x-axis. So we assign a_i the value $0 + 1$. In this way, the lattice path in Fig. 19.2 (a) determines the sequence $1, 1, 1$. Examining the horizontal steps for the lattice path in Fig. 19.2 (b), we now see that the steps from $(0, 0)$ to $(1, 0)$ and from $(1, 0)$ to $(2, 0)$ are both 0 units above the x-axis and, once again, we have $a_1 = 0 + 1 = 1$ and $a_2 = 0 + 1 = 1$. The third horizontal step takes the path from $(2, 1)$ to $(3, 1)$ and is 1 unit above the x-axis, so a_3 is assigned the value $1 + 1 = 2$, in this case. Consequently, we develop the sequence $1, 1, 2$ from the lattice path in Fig. 19.2 (b). In like manner the lattice paths in parts (c)–(e) of Fig. 19.2 determine the respective sequences in parts (c)–(e) of Fig. 26.1.

To go in the reverse direction (for $n = 3$), start with a nondecreasing sequence a_1, a_2, a_3, with $1 \leq a_i \leq i$, for $1 \leq i \leq 3$. Draw the ith horizontal step in the corresponding lattice path [from $(0, 0)$ to $(3, 3)$, which never rises above the line $y = x$] from $(i - 1, a_i - 1)$ to $(i, a_i - 1)$. For $i = 1, 2$, if $a_i \neq a_{i+1}$, draw the $a_{i+1} - a_i$ vertical step(s) from $(i, a_i - 1)$ to $(i, a_{i+1} - 1)$. Finally, draw the $3 - (a_3 - 1)$ vertical step(s) from $(3, a_3 - 1)$ to $(3, 3)$. In this way we correspond the five sequences in parts (a)–(e) of Fig. 26.1 with the respective lattice paths shown in parts (a)–(e) of Fig. 19.2.

This correspondence can be given for any $n \geq 1$. Consequently, the number of nondecreasing sequences $a_1, a_2, \ldots, a_n$, where, for $1 \leq i \leq n$, a_i is a positive integer and $a_i \leq i$, is equal to the nth Catalan number C_n.

Example 26.2:   Related to the sequences of Example 26.1, we consider the following. For $n \geq 1$, let $b_1, b_2, \ldots, b_n$ be a sequence of n nonnegative integers where $\sum_{i=1}^{k} b_i \geq k$, for $k = 1, 2, \ldots, n - 1$ (when $n \geq 2$) and $\sum_{i=1}^{n} b_i = n$.

There is only one such sequence for $n = 1$—namely, $a_1 = 1$. For $n = 2$, there are two such sequences:

> (1) $b_1 = 2$, $\quad b_2 = 0$, $\quad$ where $\sum_{i=1}^{1} b_i = b_1 = 2 \geq 1$ and
> $\sum_{i=1}^{2} b_i = b_1 + b_2 = 2 + 0 = 2$
>
> (2) $b_1 = 1$, $\quad b_2 = 1$, $\quad$ where $\sum_{i=1}^{1} b_i = b_1 = 1 \geq 1$ and
> $\sum_{i=1}^{2} b_i = b_1 + b_2 = 1 + 1 = 2$.

There are five such sequences for $n = 3$, as shown in Fig. 26.2.

> (a) $\quad$ 3,0,0 $\quad$ (b) $\quad$ 2,1,0 $\quad$ (c) $\quad$ 1,2,0 $\quad$ (d) $\quad$ 2,0,1 $\quad$ (e) $\quad$ 1,1,1

FIGURE 26.2

To set up a one-to-one correspondence between the sequences in Fig. 26.1 and Fig. 26.2, let a_1, a_2, a_3 be one of the sequences of Fig. 26.1. Define the sequence b_1, b_2, b_3, where b_i is the number of i's that appear among a_1, a_2, a_3. Consequently, for the sequence 1, 1, 1 in Fig. 26.1 (a) we obtain the sequence 3, 0, 0 in Fig. 26.2 (a), because the sequence 1, 1, 1 contains three 1's, zero 2's, and zero 3's. Likewise, the sequences in parts (b)–(e) of Fig. 26.1 determine, respectively, the sequences in parts (b)–(e) of Fig. 26.2.

To go in the reverse direction, start this time with a sequence b_1, b_2, b_3 from Fig. 26.2. Here $b_1 \geq 1$, $b_1 + b_2 \geq 2$, and $b_1 + b_2 + b_3 = 3$. Now we write down the sequence that starts with b_1 1's, followed by b_2 2's, and then followed by b_3 3's. For example, the sequence 3, 0, 0 in Fig. 26.2 (a) has $b_1 = 3$ and $b_2 = b_3 = 0$. This then determines the sequence of three 1's, followed by zero 2's, and then zero 3's, resulting in the sequence 1 $(= a_1)$, 1 $(= a_2)$, 1 $(= a_3)$, which we find in Fig. 26.1 (a). In the same way, the sequences in parts (b)–(e) of Fig. 26.2, determine the sequences in parts (b)–(e), respectively, of Fig. 26.1.

This one-to-one correspondence that we have established (for the case of $n = 3$) can be developed in a similar way for each $n \geq 1$. This verifies that these sequences $b_1, b_2, \ldots, b_n$ (as defined at the start of this example) are counted by the nth Catalan number C_n.

Example 26.3:   The result in Example 26.2 will now appear in the following, which is due to Emeric Deutsch. As in all of our previous examples, we consider the case for $n = 3$.

Expanding the product $w(w + x)(w + x + y)$, we obtain

$$w^3 + 2w^2 x + wx^2 + w^2 y + wxy$$
$$= w^3 x^0 y^0 + 2w^2 x^1 y^0 + w^1 x^2 y^0 + w^2 x^0 y^1 + w^1 x^1 y^1.$$

Examine the first term in the expansion—namely, the monomial $w^3 x^0 y^0 (= w^3)$. If we list the exponents on w, x, and y for this term, we obtain the sequence $3, 0, 0$, which appears in Fig. 26.2 (a). Doing likewise for the other four terms in the above expansion, we then obtain the respective sequences in parts (b)–(e) of Fig. 26.2. This one-to-one correspondence indicates that there are 5 ($= C_3$) monomials in the expansion of $w(w + x)(w + x + y)$.

To check this idea for the case of $n = 4$, consider the expansion of the following product:

$$\begin{aligned}
w(w + x)(w + x + y)(w + x + y + z) = {} & w^4 + 3w^3 x + 2w^3 y + w^3 z + 3w^2 z^2 \\
& + 4w^2 xy + 2w^2 xz + w^2 y^2 + w^2 yz + wx^3 \\
& + 2wx^2 y + wx^2 z + wxy^2 + wxyz
\end{aligned}$$

Note that here we have a total of 14 ($= C_4$) monomials in the expansion.

In general, for $n \geq 1$, the total number of monomials that appear in the expansion of

$$\prod_{i=1}^{n}(x_1 + x_2 + \cdots + x_i)$$

is C_n.

Example 26.4: The following sequences arose on p. 555 of the article by Eugene Paul Wigner in Reference [46]. [Eugene Paul Wigner (1902–1995), of Princeton University, received the Nobel Prize for Physics in 1963.] For $n \geq 1$, we are interested in counting the sequences consisting of $2n$ nonnegative integers $a_1, a_2, \ldots, a_{2n}$, where

$$a_1 = 1, \quad a_{2n} = 0, \quad \text{and} \quad |a_{i+1} - a_i| = 1, \quad \text{for } 1 \leq i \leq 2n - 1.$$

For $n = 1$ the sole example is the sequence 1, 0. The sequences 1, 2, 1, 0 and 1, 0, 1, 0 are the only examples for when $n = 2$. In the case of $n = 3$, there are five such sequences as listed in Fig. 26.3.

| (a) | 1,2,3,2,1,0 | (b) | 1,2,1,2,1,0 | (c) | 1,0,1,2,1,0 |

| (d) | 1,2,1,0,1,0 | (e) | 1,0,1,0,1,0 |

FIGURE 26.3

Starting with the sequence in Fig. 26.3 (a), upon taking successive differences we obtain

$$2 - 1 = 1, \quad 3 - 2 = 1, \quad 2 - 3 = -1, \quad 1 - 2 = -1, \quad 0 - 1 = -1$$

which we rewrite as the list 1, 1, −1, −1, −1. This list and the corresponding lists for parts (b)–(e) of Fig. 26.3 are given in Fig. 26.4.

(a)   1,1,−1,−1,−1   (b)   1,−1,1,−1,−1   (c)   − 1,1,1,−1,−1

(d)   1,−1,−1,1,−1   (e)   − 1,1,−1,1,−1

FIGURE 26.4

Appending "1," at the start of each list in Fig. 26.4 leads to the following five lists:

(a)   1, 1, 1, −1, −1, −1   (b)   1, 1, −1, 1, −1, −1   (c)   1, −1, 1, 1, −1, −1
(d)   1, 1, −1, −1, 1, −1   (e)   1, −1, 1, −1, 1, −1,

which are precisely the lists that appear earlier in Fig. 20.4 (in Example 20.6).

To go in the reverse direction from the lists in Fig. 20.4 to the sequences in Fig. 26.3, we start with a "1" in each case and then continue with the successive partial sums. So, for example, starting with the list $1, 1, 1, -1, -1, -1$, we obtain the sequence

$$1, \ 1+1 = 2, \quad 1+1+1 = 3, \quad 1+1+1-1 = 2,$$
$$1+1+1-1-1 = 1, \quad 1+1+1-1-1-1 = 0$$

— that is, the sequence 1, 2, 3, 2, 1, 0, which appears in Fig. 26.3 (a). In like manner, the lists in parts (b)–(e) in Fig. 20.4 (in Example 20.6) determine the sequences found in parts (b)–(e), respectively, of Fig. 26.3.

As this type of one-to-one correspondence can be established for each $n \geq 1$, we find that the number of sequences of $2n$ integers $a_1, a_2, \ldots, a_{2n}$, where $a_1 = 1$, $a_{2n} = 0$, and $|a_{i+1} - a_i| = 1$, for $1 \leq i \leq 2n - 1$, is counted by the nth Catalan number, C_n.

One final note before leaving Example 26.4. Referring back to Example 21.1, we find that the sequences in parts (a)–(e) of Fig. 26.3 provide the heights (above the x-axis) of the vertices where $x = 1, 2, 3, 4, 5, 6$ on the Dyck paths in parts (a)–(e), respectively, of Fig. 21.1.

In our next example the result from Example 26.2 reappears.

Example 26.5:   Here we want to count the number of integer sequences $c_1, c_2, \ldots, c_n$, where $c_i \geq -1$ for each $1 \leq i \leq n$, and

$$c_1 \geq 0, \quad c_1 + c_2 \geq 0, \quad \ldots, \quad c_1 + c_2 + \cdots + c_{n-1} \geq 0,$$

and

$$c_1 + c_2 + \cdots + c_{n-1} + c_n = 0.$$

The sequence $c_1 = 0$ is the only such sequence for $n = 1$. For the case where $n = 2$, there are two such sequences—namely, $c_1 = c_2 = 0$ and $c_1 = 1$, $c_2 = -1$. We provide the five sequences for $n = 3$ in Fig. 26.5.

(a) $2, -1, -1$ (b) $1, 0, -1$ (c) $0, 1, -1$ (d) $1, -1, 0$ (e) $0, 0, 0$

FIGURE 26.5

A one-to-one correspondence between the sequences given here and those in Fig. 26.2 is readily given as follows: Consider the sequence $3, 0, 0$ in Fig. 26.2 (a). Subtracting 1 from each term of this sequence, the result is the sequence $2, -1, -1$—namely, the sequence in Fig. 26.5 (a). Likewise, the sequences in parts (b)–(e) of Fig. 26.5 are obtained in the same way from the respective sequences in parts (b)–(e) of Fig. 26.2. To go in the reverse direction, we add 1 to each term in a sequence in Fig. 26.5 to obtain the corresponding sequence in Fig. 26.2.

Example 26.6:   Our last example for this chapter deals with a special tree. For $n \geq 2$, we now want to count the number of vertices at level $n - 1$ in the following rooted tree T. This tree T is such that (i) the root r has two children and (ii) if a vertex has m children $c_1, c_2, \ldots, c_m$ (from left to right), then c_k has $k + 1$ children, for each $1 \leq k \leq m$.

Part of this tree T is shown in Fig. 26.6. In part (a) of the figure, we list the number of children for each of the vertices shown. Part (b) of the figure has the root labeled as r, the children of the root labeled as a and b, and the five vertices at level 2 of T labeled as v, w, x, y, z. Next to vertex v, we find the triple $(2, 2, 2)$, since the (unique) path from the root r to v uses the (directed) edges $\overrightarrow{ra}$ and $\overrightarrow{av}$, and the labels on r, a, and v in Fig. 26.6 (a) are all 2. In a similar way, we associate the triple $(2, 3, 4)$ with the vertex z because of the (unique) path from r to z. This path uses the (directed) edges $\overrightarrow{rb}$ and $\overrightarrow{bz}$.

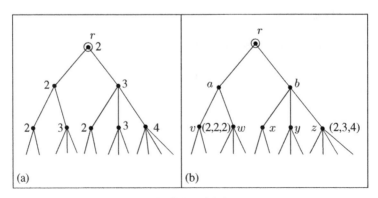

FIGURE 26.6

The triples that arise for the five vertices at level 2 are given as follows: (The reader may find the order given for the vertices a bit strange, but the reason for this order will become apparent shortly.)

(a) z: $(2, 3, 4)$ (b) y: $(2, 3, 3)$ (c) w: $(2, 2, 3)$ (d) x: $(2, 3, 2)$ (e) v: $(2, 2, 2)$

Consider the label (l_1, l_2, l_3) for each vertex on level 2. We associate a sequence s_1, s_2, s_3 with the label (l_1, l_2, l_3) as follows:

$$s_i = i + 2 - l_i, \quad 1 \leq i \leq 3.$$

Consequently, the sequence associated with the label $(2, 3, 4)$ for z is determined as

$$s_1 = 1 + 2 - 2 = 1$$
$$s_2 = 2 + 2 - 3 = 1$$
$$s_3 = 3 + 2 - 4 = 1.$$

In Table 26.1 we find the labels and sequences for each of the five vertices at level 2 of T.

TABLE 26.1

	Vertex	Label	Sequence
(a)	z	(2,3,4)	1,1,1
(b)	y	(2,3,3)	1,1,2
(c)	w	(2,2,3)	1,2,2
(d)	x	(2,3,2)	1,1,3
(e)	v	(2,2,2)	1,2,3

Now it is time to look back at the sequences in Fig. 26.1. The sequences given there in parts (a)–(e) are precisely those given in parts (a)–(e), respectively, of Table 26.1. Also, we can go (in the reverse direction) from the sequences to the labels by solving the equations $s_i = i + 2 - l_i$, $1 \leq i \leq 3$, for l_i—obtaining $l_i = i + 2 - s_i$, $1 \leq i \leq 3$. As a result, we find that the number of vertices at level 2 ($= 3 - 1$) is 5 ($= C_3$).

Using the same type of constructions (as above) for each $n \geq 2$, we now learn that the number of vertices in T at level $n - 1$ is the nth Catalan number C_n.

Furthermore, if we consider the vertices at level $n - 2$ of T, for $n \geq 3$, we see that the sum of the numbers of children for the vertices at this level of T is the same as the number of vertices at the next level—namely, the level $n - 1$. Consequently, the sum of the numbers (of children for the vertices) at level $n - 2$ is the nth Catalan number C_n. Because of this property of the tree T, we say that T generates the Catalan numbers, and refer to it as a *generating tree* for this sequence of numbers.

(Reference [17] provides another way to establish this result.)

EXERCISES FOR CHAPTER 26

1. Determine the number of ways to place n identical objects into n distinct boxes—numbered 1, 2, 3, ..., n—so that there is at most one object in box 1, at most two objects in boxes 1 and 2, at most three objects in boxes 1, 2, 3, ..., but exactly n objects in boxes 1, 2, 3, ..., n.

2. Determine the sequences b_1, b_2, b_3, b_4 of nonnegative integers where $\sum_{i=1}^{k} b_i \geq k$, for $k = 1, 2, 3$ and $\sum_{i=1}^{4} b_i = 4$.

3. (a) Determine the number of monomials in the expansion of

$$\prod_{i=1}^{8} (x_1 + x_2 + \cdots + x_i).$$

   (b) Determine the sum of the coefficients of all the monomials in the expansion in part (a).

4. (a) Determine the number of monomials in the expansion of

$$u(2u + v)(u + 3v + w)(3u + v + w + 3x)(u + v + w + 2x + y) \cdot$$
$$(6u + 5v + 4w + 3x + 2y + z).$$

   (b) Determine the sum of the coefficients of all the monomials in the expansion in part (a).

5. (i) Determine the sequences b_1, b_2 where b_1, b_2 are positive integers with $b_1 \leq 2$, $b_2 \leq 4$, and $b_1 < b_2$.

   (ii) For $n \geq 2$, determine the number of sequences of positive integers b_1, b_2, b_3, ..., b_{n-1} where $b_i \leq 2i$, for $1 \leq i \leq n - 1$, and $b_1 < b_2 < b_3 < \cdots < b_{n-1}$.

CHAPTER 27

Maximal Cliques, a Computer Science Example, and the Tennis Ball Problem

Our first example deals with graph theory and uses the result of Example 26.6. As we progress, we shall learn how the three topics for this chapter are related.

Example 27.1: If $G = (V, E)$ is an undirected graph with vertex set V and edge set E, we say that the nonempty subset W of V induces a *clique* in G, if for all vertices $a, b \in W$, the edge $ab (= ba) \in E$. Hence the subgraph of G made up of the vertices in W together with all the possible edges determined by the vertices from W is a *complete graph*.

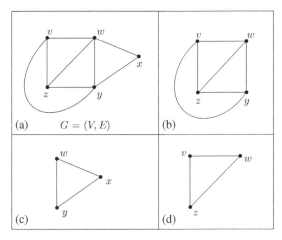

FIGURE 27.1

In Fig. 27.1 (a) we have an undirected graph $G = (V, E)$, with $V = \{v, w, x, y, z\}$ and $E = \{vw, vy, vz, wx, wy, wz, xy, yz\}$. The undirected graph in part (b) of the figure is a clique for the undirected graph G. This is the clique induced by the vertex set $\{v, w, y, z\}$. In Fig. 27.1 (c) we find another clique from G—this one induced by the

Fibonacci and Catalan Numbers: An Introduction, First Edition. Ralph P. Grimaldi.
© 2012 John Wiley & Sons, Inc. Published 2012 by John Wiley & Sons, Inc.

vertex set $\{w, x, y\}$. Note that Fig. 27.1 (d) provides the clique induced by $\{v, w, z\}$. This clique is a subgraph (actually, a subclique) of the clique in Fig. 27.1 (b).

We consider the clique in Fig. 27.1 (b) to be a *maximal clique* of G, since it is not contained (as a subgraph) in any larger clique of G. The clique induced by $\{w, x, y\}$ is also a maximal clique for G. However, the clique in Fig. 27.1 (d) is *not* a maximal clique for G because it is contained in the larger clique shown in Fig 27.1 (b).

Having introduced the notions of cliques and maximal cliques, we now turn our attention to counting the number of graphs with vertex set $V = \{1, 2, \ldots, n\}$, where the vertex set for each maximal clique is made up of a set of *consecutive* integers. This example was introduced to the author by Charles Anderson.

As we see in Fig. 27.2, there is one such graph for $n = 1$. The graph consisting of just the vertex 1 is a maximal clique (of size 1). We code this graph as (**1**). For $n = 2$, there are two such graphs—as shown in Fig. 27.2. These are coded as follows:

(1) The graph coded as (1, **1**) is obtained from the one vertex graph for $n = 1$ by performing **one** new task—namely, adding the vertex 2. This graph has two

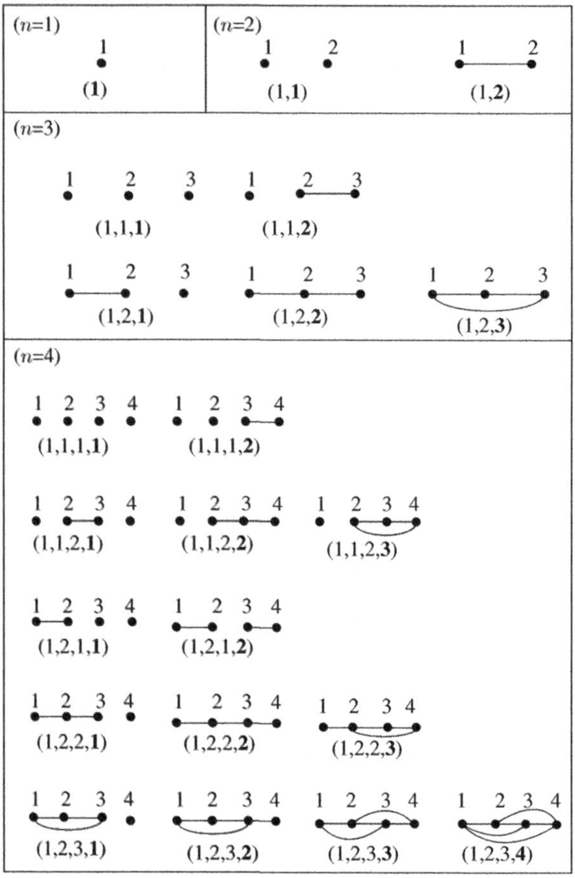

FIGURE 27.2

maximal cliques: one consists of just the vertex 1, while the other is made up of just the vertex 2.

(2) The graph coded as $(1, 2)$ is obtained from the one vertex graph for $n = 1$ by performing **two** new tasks: (i) adding the vertex 2; and, (ii) adding an (undirected) edge from 2 (back) to 1. This graph has only one maximal clique, made up of the vertices 1 and 2 and the edge 12 (connecting the vertices 1 and 2).

To obtain the five graphs for $n = 3$, we start with each of the two graphs for $n = 2$.

Starting with the graph coded as $(1, 1)$, how do we obtain the graphs coded as $(1, 1, 1)$ and $(1, 1, 2)$? For $(1, 1, 1)$ we simply perform the **one** new task where we add the vertex 3. The resulting graph has three maximal cliques, each consisting of a single vertex. To obtain the graph coded as $(1, 1, 2)$, we perform **two** new tasks: add the vertex 3 and then add the edge from 3 back to 2. This graph has two maximal cliques: one made up of the single vertex 1 and the other consisting of the vertices 2 and 3 and the edge 23.

Starting now with the graph coded as $(1, 2)$, the other three graphs for $n = 3$ are generated as follows:

(1) The graph coded as $(1, 2, 1)$ is obtained by performing the **one** new task: add the vertex 3 to the graph consisting of the vertices 1 and 2 and the edge 12. This graph has two maximal cliques: one consists of the vertices 1 and 2 and the edge 12, the other consists of just the vertex 3.

(2) The graph coded as $(1, 2, 2)$ is obtained by performing the **two** new tasks: add the vertex 3 to the graph consisting of the vertices 1 and 2 and the edge 12, and then the edge from 3 back to 2. This graph also has two maximal cliques: one consists of the vertices 1 and 2 and the edge 12, the other consists of the vertices 2 and 3 and the edge 23.

(3) Finally, the third graph is the one coded as $(1, 2, 3)$. It is obtained from the graph coded as $(1, 2)$ by performing **three** new tasks: add the vertex 3 to the graph consisting of the vertices 1 and 2 and the edge 12, then add the edge from 3 back to 2, and finally add the edge from 3 back to 1. This graph is a maximal clique (also referred to as the complete graph on three vertices).

In the bottom part of Fig. 27.2, we find the 14 graphs on the vertex set $\{1, 2, 3, 4\}$, where the vertex set for each maximal clique is made up of a set of *consecutive* integers. We shall describe how two of these graphs are generated.

(1) For instance, the graph coded as $(1, 2, 2, 3)$ is obtained from the graph coded as $(1, 2, 2)$ by performing **three** new tasks: add the vertex 4 to the graph consisting of the vertices 1, 2 and 3 and the edges 12 and 23, then add the edge from 4 back to 3, and finally add the edge from 4 back to 2. This graph has two maximal cliques: one consists of the vertices 1 and 2 and the edge 12; the other is made up of the vertices 2, 3 and 4 and the edges 23, 34, and 24.

(2) The graph coded as $(1, 2, 3, 4)$ is obtained from the graph coded as $(1, 2, 3)$ by performing **four** new tasks: start by adding the vertex 4 to the graph consisting

of the vertices 1, 2 and 3, and the edges 12, 23, and 13, then add the edge from 4 back to 3, then the edge from 4 back to 2, and finally the edge from 4 back to 1. This graph is a maximal clique (also referred to as the complete graph on four vertices).

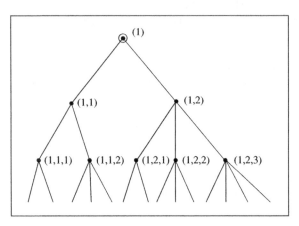

FIGURE 27.3

The rooted tree in Fig. 27.3 shows how these graphs with $n + 1$ vertices are obtained from those with n vertices, for $n = 1, 2$. Comparing this tree with the one in Fig. 26.6, we see that the number of graphs on the vertex set $\{1, 2, 3, \ldots, n\}$, where the vertex sets for all maximal cliques contain only consecutive integers, is the nth Catalan number C_n.

Now, perhaps, is a good time to switch gears and leave the trees and cliques of graph theory behind. Our next example surely seems to do so and simply asks us to determine the value of the (integer) variable *counter* following the execution of a given (pseudocode) procedure, made up of n nested **for** loops.

Example 27.2:   We start with the following (pseudocode) procedure:

```
counter := 0
for i₁ := 1 to 1
  for i₂ := 1 to i₁ + 1
    for i₃ := 1 to i₂ + 1
      ⋮
      for iₙ := 1 to iₙ₋₁ + 1
        counter := counter +1
```

We would like to know the value of the variable *counter* after this procedure is executed.

When $n = 1$, there is only one **for** loop (as we consider $i_0 = 0$) and when we exit this procedure, the value of the variable *counter* has been incremented from 0 to 1. For

$n = 2$, the two nested **for** loops result in incrementing the variable *counter* twice—once for when $i_1 = 1$ and $i_2 = 1$ and then for when $i_1 = 1$ and $i_2 = 2 \, (= i_1 + 1)$. Consequently, when we exit this procedure for this case, the value of *counter* has been incremented from 0 to 2.

In the case of $n = 3$, there are three nested **for** loops and the variable *counter* is incremented five times—once for each of the assignments in Table 27.1.

TABLE 27.1

	i_1	i_2	i_3
(1)	1	1	1
(2)	1	1	2
(3)	1	2	1
(4)	1	2	2
(5)	1	2	3

These results for $n = 1, 2$, and 3 are precisely the labels for the vertices in the rooted tree of Fig. 27.3 for Example 27.1. Therefore, we have another instance where the Catalan numbers arise. So we now know that, for a given positive integer n, after the execution of the above procedure, the value of the variable *counter* is the nth Catalan number C_n.

Example 27.3 (The Tennis Ball Problem):   This last example is due to Joseph Moser. Its roots are found in an infinite situation developed in Reference [44] and proceeds as follows. Initially, suppose that you are in a box and that tennis balls labeled 1 and 2 are tossed into the box. You are allowed to throw one of these tennis balls out of the box onto the lawn beside the box. Say you throw the tennis ball labeled 1 out onto the lawn. Then the tennis balls labeled 3 and 4 are tossed into the box. Now you are allowed to throw one of the tennis balls labeled 2, 3, and 4 out onto the lawn. Continuing in this way, for a positive integer i, the two tennis balls labeled $2i - 1$ and $2i$ are tossed into the box (during the ith toss), and you are now allowed to throw out onto the lawn any of the $2i - (i - 1) = i + 1$ labeled tennis balls in the box with you.

If this procedure is continued infinitely many times, how many labeled tennis balls are left in the box with you? One possible answer is none—for you can first throw out (onto the lawn) the tennis ball labeled 1, followed by the ball labeled 2, then that labeled 3, and so on. A second possibility leaves you with a finite number of labeled tennis balls in the box with you. For any positive integer n, throw out, in succession, the tennis balls with the labels $2, 4, 6, \ldots, 2n$—keeping those with the labels $1, 3, 5, \ldots, 2n - 1$ in the box with you. After that, throw out, in succession, the tennis balls with the labels $2n + 1, 2n + 2, 2n + 3, \ldots$. Finally, another possibility results in an infinite number of labeled tennis balls remaining in the box with you. This happens, for instance, if you first throw out the tennis ball labeled 1, then follow with the ball labeled 3, then that labeled 5, and so on. In this case there are infinitely many tennis balls on the lawn—all those with odd labels—and infinitely many remaining with you in the box—this time, all those tennis balls with even labels.

At this point the reader is probably asking "What has this got to do with the Catalan numbers?." Well, we are still going to throw those tennis balls, two at a time, into the box with you, and the balls will still be labeled 1, 2 (for the first toss), then 3, 4 (for the second toss), then 5, 6 (for the third toss), ..., and $2n - 1, 2n$ (for the nth and *final* toss). Yes, this time we shall stop after n tosses into the box, each such toss followed by a labeled ball being thrown from the box onto the lawn.

If we perform this procedure only once, then there is only one tennis ball on the lawn—the possibilities for the labels are 2 and 1. Should we do this twice, the possible sets of two labels (on the tennis balls on the lawn) are

(1) $\{2, 4\}$ (2) $\{2, 3\}$ (3) $\{1, 4\}$ (4) $\{1, 3\}$ (5) $\{1, 2\}$.

We should realize that if, for instance, we find the balls with labels 1 and 2 on the lawn we do not know if the ball labeled 1 was thrown out first and then the ball labeled 2 was thrown out second, or if the ball labeled 2 was thrown out first, and this was then followed by throwing out the ball labeled 1. There is no concern about the order in which this happened! We only care that 1, 2 is one of the five possible sets of two labels that can result in this process. We also note that 3, 4 is a set of two labels that cannot be obtained in this way.

We use the rooted tree in Fig. 27.4 to see how the possible subsets of labels of tennis balls on the lawn can be generated for $n = 1, 2, 3$. Comparing the results

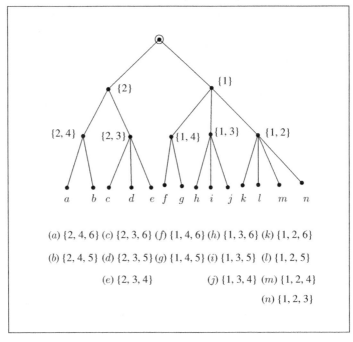

(a) $\{2, 4, 6\}$ (c) $\{2, 3, 6\}$ (f) $\{1, 4, 6\}$ (h) $\{1, 3, 6\}$ (k) $\{1, 2, 6\}$

(b) $\{2, 4, 5\}$ (d) $\{2, 3, 5\}$ (g) $\{1, 4, 5\}$ (i) $\{1, 3, 5\}$ (l) $\{1, 2, 5\}$

(e) $\{2, 3, 4\}$ (j) $\{1, 3, 4\}$ (m) $\{1, 2, 4\}$

(n) $\{1, 2, 3\}$

FIGURE 27.4

for this rooted tree with those for the rooted trees in Fig. 26.6 and Fig. 27.3, we see how the results here can be placed in a one-to-one correspondence with these previous examples, where the Catalan numbers arise. Consequently, when we count the number of possible sets of n labels that can be on the n tennis balls on the lawn, we find their number to be the $(n + 1)$st Catalan number, C_{n+1}.

This result is derived in another way in Reference [18]. Generalizations of the problem are investigated in References [6, 26]. If, for each $n \geq 1$, one sums all of the n labels for all the possible sets (on the lawn) taken from the $2n$ labeled tennis balls tossed into the box, the resulting sequence starts with 3, 23, 131, 664. This sequence is examined in detail in Reference [25].

EXERCISE FOR CHAPTER 27

1. Consider the Tennis Ball Problem in Example 27.3.

   (a) Suppose you are in the box and the tennis balls labeled 1 and 2 are tossed into the box, and you throw one of them out onto the lawn next to the box. Then suppose the balls labeled 3 and 4 are tossed into the box and you throw out one of the three balls onto the lawn next to the box. (i) How many of the possible sets of two labeled balls (on the lawn) have ball 2 as the ball with the smaller label? (ii) How many of these sets of two labeled balls contain the ball with the label 1?

   (b) Now suppose that we continue from part (a) and the two balls labeled 5 and 6 are tossed into the box and you throw out one of the four labeled balls in the box onto the lawn beside the box. (i) How many of the possible sets of three labeled balls (on the lawn) have ball 2 as the ball with the smallest label? (ii) How many of these sets of three labeled balls contain the ball with the label 1?

   (c) For $n \geq 4$, continue from part (b) until the balls labeled $2n - 1$ and $2n$ are tossed into the box and you throw out one of the $n + 1$ balls onto the lawn next to the box. (i) How many of the possible sets of n labeled balls (on the lawn) have ball 2 as the ball with the smallest label? (ii) How many of these sets of n labeled balls contain the ball with the label 1?

CHAPTER 28

The Catalan Numbers at Sporting Events

This chapter is novel in that it takes us from our typical mathematical applications to one involving probability concepts in conjunction with well-known sporting events.

Example 28.1: The following example is due to Louis Shapiro and Wallace Hamilton. The discussion given here follows that presented in Reference [35].

Given a positive integer n, consider a series of at least n games but no more than $2n - 1$ games, where the winner is the first to win n of the games played. For instance, at a grand slam (such as Wimbledon or the U. S. Open), the women's final is won by the first woman to win two of the three (possible) sets. So in this case, $n = 2$. When the men play the final at such a grand slam, the champion is the first to win three of the five (possible) sets. This time $n = 3$.

Each October, the American and National League pennant winners square off for the world series—or October classic. For this series, $n = 4$ and the first team to win four of the seven possible games is the world champion (for that year).

Let A and B denote the two opponents for the three preceding situations. Assume the probability opponent A wins a given game (or set) is p; for opponent B, the probability is then $1 - p = q$. Now we shall let E_n denote the expected number of games played, if the first opponent to win n of at most $2n - 1$ games (or sets) wins the championship. It can be shown that

$$E_1 = 1$$
$$E_2 = 2(1 + pq)$$
$$E_3 = 3(1 + pq + 2p^2q^2)$$
$$E_4 = 4(1 + pq + 2p^2q^2 + 5p^3q^3)$$
$$E_5 = 5(1 + pq + 2p^2q^2 + 5p^3q^3 + 14p^4q^4).$$

Fibonacci and Catalan Numbers: An Introduction, First Edition. Ralph P. Grimaldi.
© 2012 John Wiley & Sons, Inc. Published 2012 by John Wiley & Sons, Inc.

We confirm the result for E_3 as follows. In this case, $n = 3$, so the opponents are involved in at least three but no more than five games (or sets). Consequently,

$$E_3 = 3[p^3 + q^3] + 4[3(p^2q)p + 3(pq^2)q] + 5[6(p^2q^2)p + 6(p^2q^2)q].$$

For instance, the summand $4[3(p^2q)p + 3(pq^2)q]$ takes into account the case where the championship is decided after four games. To compute the probability for this, suppose that A is the winner. In this case, A wins two of the first three games and follows this by then winning the fourth game. This can be accomplished by opponent A as follows:

Game 1	Game 2	Game 3	Game 4	Probability
Win	Win	Lose	Win	$ppqp = (p^2q)p$
Win	Lose	Win	Win	$pqpp = (p^2q)p$
Lose	Win	Win	Win	$qppp = (p^2q)p$

Consequently, the probability that opponent A wins after the fourth game is $3(p^2q)p$. In a similar way, it can be shown that the probability that opponent B wins after the fourth game is $3(pq^2)q$.

Consequently, we now find that

$$\begin{aligned}
E_3 &= 3[p^3 + q^3] + 4[3(p^2q)p + 3(pq^2)q] + 5[6(p^2q^2)p + 6(p^2q^2)q] \\
&= 3(p + q)(p^2 - pq + q^2) + 12\ pq(p^2 + q^2) + 30(p^2q^2)(p + q) \\
&= 3[p^2 - pq + q^2 + 4\ pq(p^2 + q^2) + 10\ p^2q^2], \quad \text{because } p + q = 1.
\end{aligned}$$

Since $p^2 + q^2 = p^2 + 2\ pq + q^2 - 2\ pq = (p + q)^2 - 2\ pq = 1^2 - 2\ pq = 1 - 2\ pq$, we continue with

$$\begin{aligned}
E_3 &= 3[p^2 - pq + q^2 + 4\ pq(p^2 + q^2) + 10\ p^2q^2] \\
&= 3[1 - 3\ pq + 4\ pq(1 - 2\ pq) + 10\ p^2q^2] \\
&= 3[1 - 3\ pq + 4\ pq - 8\ p^2q^2 + 10\ p^2q^2] \\
&= 3[1 + pq + 2\ p^2q^2].
\end{aligned}$$

The expressions for E_1, E_2, E_3, E_4, and E_5 suggest that for $n \geq 1$,

$$\frac{E_{n+1}}{n + 1} - \frac{E_n}{n} = C_n\ p^n q^n.$$

To verify this for all $n \geq 1$, we consider a series where the opponents play at most $2n - 1$ games and the winner is the first to win n of these games. For example, this type of series ends after $n + 1$ games if one opponent, say B, wins one of the first n games and opponent A wins the other n games. We can select one game from the first n in $\binom{n}{1}$ ways, so the probability that only $n + 1$ games are played with

A as the winner is $\binom{n}{1}(p^{n-1}q)p = \binom{n}{1}(p^n q)$. Interchanging the roles of A and B, the probability is $\binom{n}{1}(q^n p)$. Consequently, the probability that the series is over after $n + 1$ games is $\binom{n}{1}(p^n q + q^n p)$. Similar reasoning for the cases where the opponents play a total of $n + 2$, $n + 3$, ..., $2n - 1$ games leads us to

$$E_n = n(p^n + q^n) + (n + 1)\binom{n}{1}(p^n q + q^n p) + (n + 2)\binom{n+1}{2}(p^n q^2 + q^n p^2)$$

$$+ \cdots + (2n - 1)\binom{2n-2}{n-1}(p^n q^{n-1} + q^n p^{n-1})$$

$$= \sum_{k=0}^{n-1}(n + k)\binom{n-1+k}{k}(p^n q^k + q^n p^k)$$

$$= n\sum_{k=0}^{n-1}\binom{n+k}{k}(p^n q^k + q^n p^k),$$

because

$$(n + k)\binom{n-1+k}{k} = (n + k)\left(\frac{(n-1+k)!}{k!(n-1)!}\right) = \frac{(n+k)!}{k!(n-1)!}$$

$$= n\frac{(n+k)!}{k!n!} = n\binom{n+k}{k}.$$

From this form for E_n, we see that

$$\frac{E_n}{n} = \sum_{k=0}^{n-1}\binom{n+k}{k}(p^n q^k + q^n p^k).$$

Consequently,

$$\frac{E_{n+1}}{n+1} = \sum_{k=0}^{n}\binom{n+k+1}{k}(p^{n+1}q^k + q^{n+1}p^k)$$

$$= \sum_{k=0}^{n}\binom{n+k+1}{k}[p^n(1 - q)q^k + q^n(1 - p)p^k], \quad \text{since } p + q = 1$$

$$= \sum_{k=0}^{n}\binom{n+k+1}{k}(p^n q^k + q^n p^k)$$

$$- \sum_{k=0}^{n}\binom{n+k+1}{k}(p^n q^{k+1} + q^n p^{k+1})$$

$$= \sum_{k=0}^{n}\binom{n+k}{k}(p^n q^k + q^n p^k) + \sum_{k=1}^{n}\binom{n+k}{k-1}(p^n q^k + q^n p^k)$$

$$- \sum_{k=1}^{n+1}\binom{n+k}{k-1}(p^n q^k + q^n p^k),$$

because (i) for $n \geq 0$, $k \geq 0$, $\binom{n+k+1}{k} = \binom{n+k}{k} + \binom{n+k}{k-1}$, with $\binom{n}{-1} = 0$, and (ii) by replacing $k+1$ by k and reindexing, it follows that $\sum_{k=0}^{n} \binom{n+k+1}{k}(p^n q^{k+1} + q^n p^{k+1}) = \sum_{k=1}^{n+1} \binom{n+k}{k-1}(p^n q^k + q^n p^k)$. Continuing, we find that

$$\frac{E_{n+1}}{n+1} = \sum_{k=0}^{n} \binom{n+k}{k}(p^n q^k + q^n p^k) - \binom{n+(n+1)}{(n+1)-1}(p^n q^{n+1} + q^n p^{n+1})$$

$$= \sum_{k=0}^{n} \binom{n+k}{k}(p^n q^k + q^n p^k) - \binom{2n+1}{n}(p^n q^{n+1} + q^n p^{n+1})$$

$$= \left[\sum_{k=0}^{n-1} \binom{n+k}{k}(p^n q^k + q^n p^k) + \binom{2n}{n}(2p^n q^n) \right]$$

$$\quad - \binom{2n+1}{n}(p^n q^n)(q+p)$$

$$= \sum_{k=0}^{n-1} \binom{n+k}{k}(p^n q^k + q^n p^k) + \left[2\binom{2n}{n} - \binom{2n+1}{n} \right] p^n q^n$$

$$= \sum_{k=0}^{n-1} \binom{n+k}{k}(p^n q^k + q^n p^k) + C_n \, p^n q^n,$$

because $\left[2\binom{2n}{n} - \binom{2n+1}{n} \right] = 2\left(\frac{(2n)!}{n!n!}\right) - \frac{(2n+1)!}{n!(n+1)!} = \frac{(2n)!}{n!n!}\left[2 - \frac{2n+1}{n+1} \right] = \frac{(2n)!}{n!n!}\left[\frac{2(n+1)-(2n+1)}{n+1} \right] = \frac{1}{n+1}\frac{(2n)!}{n!n!} = C_n$.

Therefore, the suggested pattern is true for all $n \geq 1$ and it follows that

$$\frac{E_{n+1}}{n+1} - \frac{E_n}{n} = C_n \, p^n q^n$$

or

$$\frac{E_{n+1}}{n+1} = \frac{E_n}{n} + C_n \, p^n q^n.$$

As a result, we have

$$\frac{E_2}{2} = \frac{E_1}{1} + C_1 p^1 q^1 = 1 + C_1 p^1 q^1 = C_0 p^0 q^0 + C_1 p^1 q^1$$

$$\frac{E_3}{3} = \frac{E_2}{2} + C_2 p^2 q^2 = C_0 p^0 q^0 + C_1 p^1 q^1 + C_2 p^2 q^2$$

$$\frac{E_4}{4} = \frac{E_3}{3} + C_3 p^3 q^3 = C_0 p^0 q^0 + C_1 p^1 q^1 + C_2 p^2 q^2 + C_3 p^3 q^3,$$

and, in general, for $n \geq 1$,

$$\frac{E_n}{n} = \sum_{k=0}^{n-1} C_k \, p^k q^k$$

or

$$E_n = n \sum\nolimits_{k=0}^{n-1} C_k \, p^k q^k.$$

EXERCISES FOR CHAPTER 28

1. Carol and Patricia are the best of friends. However, when it comes to the summer softball teams they each coach, they display a bitter rivalry! Carol puts her team through a very rigorous spring training program. As a result, her team has a slight edge in beating Patricia's team. In fact, the probability her team will beat Patricia's in any given game is 0.55, no matter what has happened earlier in the season. The two teams are scheduled to play once a week for up to 11 weeks, the first team to win six games being declared the summer champ for that season. What is the expected number of softball games these teams will play?

2. Kati and Justin are playing against Leah and Peter in a mixed doubles tennis tournament. To accommodate the wishes of the sponsors, the winner will be the first team to win five matches (of the nine matches that may have to be scheduled). The two teams are very evenly matched and each wins one of the first two matches. What is the expected number of additional matches they will play?

CHAPTER 29

A Recurrence Relation for the Catalan Numbers

In dealing with the Fibonacci numbers in Part One, many of our examples were established using a second-order linear recurrence relation. However, in our dealings with the Catalan numbers so far, we have established new examples (where we believe that the Catalan numbers arise) by placing the results in each new example in a one-to-one correspondence with results from an example that was previously shown to be counted by the Catalan numbers. Now we shall introduce a recurrence relation that is satisfied by the Catalan numbers. Then in Chapters 31 and 32, we shall use this recurrence relation to introduce some further examples where this number sequence arises.

Before we get started, the reader should not expect the recurrence relation here to be as simple as the second-order linear recurrence relation we found for the Fibonacci and Lucas numbers. The relation here will be nonlinear and of a special form.

Once again, for n a nonnegative integer, we shall count the number of lattice paths from $(0, 0)$ to (n, n). The only steps allowed are still just R: $(x, y) \rightarrow (x + 1, y)$ and U: $(x, y) \uparrow (x, y + 1)$ and the path may never rise above the line $y = x$, although it may touch the line at some point (k, k) where k is an integer and $0 \leq k \leq n$. We know from Example 19.1 that the number of these lattice paths is the nth Catalan number $C_n = \frac{1}{n+1}\binom{2n}{n}$.

This time our approach will examine those lattice paths that start at $(0, 0)$, finish at (n, n), never rise above the line $y = x$, and touch (but not cross) the line $y = x$ for the *first* time at (k, k), with $1 \leq k \leq n$. We know that such a path starts with one step to the right, going from $(0, 0)$ to $(1, 0)$, and reaches the point (k, k) by going one step up from $(k, k - 1)$ to (k, k). As we see in Fig. 29.1 on p. 232, the number of lattice paths that take us from $(0, 0)$ to (k, k), in this way, equals the number of lattice paths from $(1, 0)$ to $(k, k - 1)$ that never rise above the line $y = x - 1$. This is the same as the number of lattice paths from $(0, 0)$ to $(k - 1, k - 1)$ that never rise above the line $y = x$—and this number we know to be C_{k-1}. So now we need to determine the number of ways in which we can complete these lattice paths from $(0, 0)$ to (n, n). For the remainder of such a lattice path—that is, the part from (k, k) to (n, n), that never

Fibonacci and Catalan Numbers: An Introduction, First Edition. Ralph P. Grimaldi.
© 2012 John Wiley & Sons, Inc. Published 2012 by John Wiley & Sons, Inc.

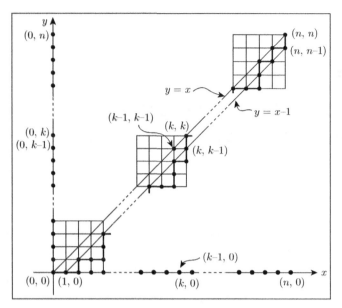

FIGURE 29.1

rises above the line $y = x$—the number of possible ways we can travel from (k, k) to (n, n) (and never rise above the line $y = x$) is the same as the number of lattice paths from $(0, 0)$ to $(n - k, n - k)$ that never rise above the line $y = x$. This number is C_{n-k}. Therefore, for $1 \le k \le n$, the number of lattice paths from $(0, 0)$ to (n, n), using only the steps R and U, and never rising above the line $y = x$, but touching the line $y = x$ for the first time at (k, k), is the product $C_{k-1}C_{n-k}$. This now leads us to the recurrence relation

$$C_n = C_0C_{n-1} + C_1C_{n-2} + C_2C_{n-3} + \cdots + C_{n-2}C_1 + C_{n-1}C_0$$
$$= \sum_{k=1}^{n} C_{k-1}C_{n-k}, \quad C_0 = 1. \tag{29.1}$$

Note that for $k = 1$, the path starts with the step R: $(0, 0) \to (1, 0)$, followed by the step U: $(1, 0) \uparrow (1, 1)$. Thus, for $n > 1$, the path has touched the line $y = x$ for the first time at $(1, 1)$. This can take place in $1 = C_0$ way, which is also the number of ways one can get from $(1, 0)$ to $(1, 0)$ along the line $y = x - 1$. Then C_{n-1} counts the number of ways the path can proceed from $(1, 1)$ to (n, n) without rising above the line $y = x$, and this is the same as the number of ways one can go from $(1, 0)$ to $(n, n - 1)$ without rising above the line $y = x - 1$. The paths for $k = n$ start with R: $(0, 0) \to (1, 0)$ and end with U: $(n, n - 1) \uparrow (n, n)$—and this happens in $1 = C_0$ way. Furthermore, these paths cannot rise above the line $y = x$ or touch the line $y = x$ at any point (i, i), where $1 \le i \le n - 1$. The number of possibilities here is the number of paths from $(1, 0)$ to $(n, n - 1)$ that never rise above the line

$y = x - 1$. This in turn equals the number of paths from $(0, 0)$ to $(n - 1, n - 1)$ that never rise above the line $y = x$, and this number is C_{n-1}. Hence the number of paths from $(0, 0)$ to (n, n) that only touch the line $y = x$ at $(0, 0)$ and (n, n) is $C_{n-1}C_0$.

Before we examine this result more closely, let us recall the following values, given earlier in Table 19.1.

n	C_n	n	C_n	n	C_n
0	1	3	5	6	132
1	1	4	14	7	429
2	2	5	42	8	1430

Checking the recurrence relation for C_n in Eq. (29.1) for when $n = 7$, we find that

$$C_0C_6 + C_1C_5 + C_2C_4 + C_3C_3 + C_4C_2 + C_5C_1 + C_6C_0$$
$$= 1 \cdot 132 + 1 \cdot 42 + 2 \cdot 14 + 5 \cdot 5 + 14 \cdot 2 + 42 \cdot 1 + 132 \cdot 1$$
$$= 132 + 42 + 28 + 25 + 28 + 42 + 132 = 429 = C_7.$$

Also, we see that

$$C_7 = 2[C_0C_6 + C_1C_5 + C_2C_4] + C_3C_3.$$

In the case of $n = 8$, we have

$$C_0C_7 + C_1C_6 + C_2C_5 + C_3C_4 + C_4C_3 + C_5C_2 + C_6C_1 + C_7C_0$$
$$= 1 \cdot 429 + 1 \cdot 132 + 2 \cdot 42 + 5 \cdot 14 + 14 \cdot 5 + 42 \cdot 2 + 132 \cdot 1 + 429 \cdot 1$$
$$= 429 + 132 + 84 + 70 + 70 + 84 + 132 + 429 = 1430 = C_8,$$

and

$$C_8 = 2[C_0C_7 + C_1C_6 + C_2C_5 + C_3C_4].$$

The preceding provide instances of the following:

$$C_n = \begin{cases} 2\left[C_0C_{n-1} + C_1C_{n-2} + \cdots + C_{((n-2)/2)}C_{(n/2)}\right], & n \text{ even} \\ 2\left[C_0C_{n-1} + C_1C_{n-2} + \cdots + C_{((n-3)/2)}C_{((n+1)/2)}\right] + \left(C_{((n-1)/2)}\right)^2, & n \text{ odd.} \end{cases}$$

From this we learn that if n is even and positive, then C_n is even.

Could C_n ever be odd? Following the presentation in Reference [23], for $n > 0$, we find that C_n is odd if and only if n is odd and $C_{((n-1)/2)}$ is odd. But then $C_{((n-1)/2)}$ is odd if and only if (i) $((n - 1)/2) = 0$ or (ii) $((n - 1)/2)$ is odd and $C_{(((n-1)/2)-1)/2} = C_{(n-3)/4} = C_{(n-(2^2-1))/2^2}$ is odd. Continuing, $C_{(n-(2^2-1))/2^2}$ is odd if and only if (i) $(n - (2^2 - 1))/2^2 = 0$ or (ii) $(n - (2^2 - 1))/2^2$ is odd and

$C_{(((n-(2^2-1))/2^2)-1)/2} = C_{(n-7)/8} = C_{(n-(2^3-1))/2^3}$ is odd. Descending further, we arrive at the situation where C_n is odd if and only if $C_{(n-(2^k-1))/2^k}$ is odd, for some positive integer k, and $(n - (2^k - 1))/2^k = 0$. Consequently, this occurs if and only if $n = 2^k - 1$, for some $k \geq 1$. These values of n—namely, 1, 3, 7, 15, 31, ...—are called *Mersenne numbers*, in honor of Marin Mersenne (1588-1648), the French theologian, philosopher, mathematician, and music theorist, often referred to as the "Father of Acoustics." Referring back to the Catalan numbers listed in Table 19.1, we find the first four cases that demonstrate this result:

$$C_1 = 1, \quad C_3 = 5, \quad C_7 = 429, \quad C_{15} = 9694845.$$

Let us now take this discussion one step further and ask which Catalan numbers are prime. We know that $C_2 = 2$ and $C_3 = 5$, but are there any others? Once again we refer to the presentation in Reference [23]. Before starting, however, we state the following two results which will prove useful:

(i) For $n \geq 4$, $C_n > n + 2$. (This can be established using the Principle of Mathematical Induction. We leave this for the exercises for this chapter.)

(ii) If a, b, and p are positive integers with p prime and p divides ab, then p divides a or p divides b. (This is sometimes referred to as Euclid's Lemma. A proof for this result can be found on P. 7 of Reference [14], P. 237 of Reference [19], or P. 109 of Reference [34]).

Starting with $C_n = \frac{1}{n+1}\binom{2n}{n}$ we have

$$\left[C_n = \frac{1}{n+1}\binom{2n}{n}\right] \Rightarrow \left[C_{n+1} = \frac{1}{(n+1)+1}\binom{2(n+1)}{n+1}\right] \Rightarrow$$

$$(n+2)\,C_{n+1} = \frac{(2n+2)!}{(n+1)!(n+1)!} = \frac{(2n+2)(2n+1)(2n)!}{(n+1)(n+1)n!n!}$$

$$= \frac{2(2n+1)}{n+1}\frac{(2n)!}{n!n!} = (4n+2)\left[\frac{1}{n+1}\binom{2n}{n}\right]$$

$$= (4n+2)\,C_n.$$

If C_n is prime, then from $(4n+2)\,C_n = (n+2)\,C_{n+1}$, it follows from (ii) above that C_n divides $(n+2)$ or C_n divides C_{n+1}. Since $n \geq 4$, from (i) we know that $C_n > n+2$, so C_n divides C_{n+1}. Say that $C_{n+1} = mC_n$, where m is a positive integer. Then $(4n+2)\,C_n = (n+2)\,C_{n+1} = m\,(n+2)\,C_n$. Since $C_n \neq 0$, it follows that $(4n+2) = m(n+2)$. If $m \geq 4$, then $4n+2 = mn+2m \Rightarrow 0 > (2-2m) = (m-4)n \geq 0$. This impossibility implies that $m = 1$, 2, or 3.

($m = 1$): If $m = 1$, then $4n + 2 = n + 2$, so $3n = 0$ and $n = 0$. This contradicts $n \geq 4$.

($m = 2$): If $m = 2$, then $4n + 2 = 2(n + 2) \Rightarrow 2n = 2 \Rightarrow n = 1$. This also contradicts $n \geq 4$.

($m = 3$): If $m = 3$, then $4n + 2 = 3(n + 2) \Rightarrow n = 4$. But $C_4 = 14$ is not prime.

Consequently, the only Catalan numbers that are primes are $C_2 = 2$ and $C_3 = 5$. Let us close this chapter by summarizing what we have learned in the following.

Theorem 29.1:

(a) For $n > 0$, if n is even, then C_n is even.

(b) For n odd, C_n is odd if and only if n is a Mersenne number—that is, an integer of the form $2^k - 1$, for k a positive integer. (In all other cases, C_n is even.)

(c) The only Catalan numbers that are primes are $C_2 = 2$ and $C_3 = 5$.

EXERCISES FOR CHAPTER 29

1. For $n \geq 4$, prove that $C_n > n + 2$.

2. Prove that Euclid's Lemma is not true for arbitrary integers—that is, find three positive integers a, b, and c, such that c divides ab, but c does not divide a or b.

3. For the C_{10} ($= 16,796$) lattice paths from $(0, 0)$ to $(10, 10)$ that never rise above the line $y = x$, how many contain (a) the point $(3, 3)$; (b) the point $(7, 7)$; (c) both the point $(3, 3)$ and the point $(7, 7)$; and, (d) all three of the points $(3, 3)$, $(7, 7)$, and $(9, 9)$?

4. Determine the value of the positive integer n if the number of lattice paths from $(0, 0)$ to (n, n) that never rise above the line $y = x$ and contain the point $(9, 9)$ is 641,784.

5. Determine the value of the positive integer n if the number of lattice paths from $(0, 0)$ to (n, n) that never rise above the line $y = x$ and contain the points $(3, 3)$ and $(7, 7)$ is 2940.

CHAPTER 30

Triangulating the Interior of a Convex Polygon for the Second Time

Now it is time to revisit the situation we examined in Example 23.1.

Example 30.1: In this example, we shall take a second look at the problem of determining T_n, the number of ways the interior of a convex polygon of n sides can be triangulated into $n - 2$ triangles—by drawing $n - 3$ diagonals, no two of which intersect within the interior.

In Fig. 30.1 (a), we have a convex polygon with $n + 1$ sides, where we have drawn the diagonals $v_1 v_3$ and $v_3 v_{n+1}$. These diagonals break up the interior of this convex polygon into three regions: (i) the interior of triangle $v_1 v_2 v_3$; (ii) the interior of triangle $v_1 v_3 v_{n+1}$; and (iii) the interior of the $(n - 1)$-sided convex polygon $v_3 v_4 \cdots v_n v_{n+1}$. The interior of the first region can be triangulated in T_3 ways, while there are $T_{n-1} = T_{(n+1)-2}$ ways to triangulate the interior of the third region. Therefore, there are $T_3 T_{(n+1)-2}$ ways to triangulate the interior of the given convex polygon with $n + 1$ sides—where the triangle $v_1 v_3 v_{n+1}$ is part of the triangulation.

Now let us look at Fig. 30.1 (b). This time we have drawn the diagonals $v_1 v_4$ and $v_4 v_{n+1}$. They decompose the interior of the given convex polygon into the three regions: (i) the interior of the convex quadrilateral $v_1 v_2 v_3 v_4$; (ii) the interior of triangle

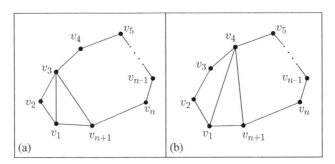

FIGURE 30.1

Fibonacci and Catalan Numbers: An Introduction, First Edition. Ralph P. Grimaldi.
© 2012 John Wiley & Sons, Inc. Published 2012 by John Wiley & Sons, Inc.

$v_1 v_4 v_{n+1}$; and (iii) the interior of the $(n-2)$-sided convex polygon $v_4 v_5 \cdots v_n v_{n+1}$. The interior of the first region can be triangulated in T_4 ways; the interior of the third region in $T_{(n+1)-3}$ ways. Consequently, there are $T_4 T_{(n+1)-3}$ ways to triangulate the interior of this $(n+1)$-sided polygon—where, this time, the triangulation includes the triangle $v_1 v_4 v_{n+1}$.

For $2 \le i \le n$, there are $T_i T_{(n+1)-(i-1)}$ ways to triangulate the interior of the $(n+1)$-sided convex polygon—where the triangle $v_1 v_i v_{n+1}$ is one of the triangles in the triangulation. [When $i = 2$, there is no region to the left of triangle $v_1 v_2 v_{n+1}$—there is *one* way this can happen. The region to the right of triangle $v_1 v_2 v_{n+1}$ is the interior of the n-sided convex polygon $v_2 v_3 v_4 \cdots v_n v_{n+1}$, which can be triangulated in $T_{(n+1)-(2-1)} = T_n$ ways. So triangle $v_1 v_2 v_{n+1}$ appears among the T_{n+1} triangulations of the given $(n+1)$-sided convex polygon in $1 \cdot T_n = T_2 T_n = T_2 T_{(n+1)-(2-1)}$ ways. Hence, we see why we define $T_2 = 1$. A similar situation arises for when $i = n$—and we get $T_n T_{(n+1)-(n-1)} = T_n T_2$ of the T_{n+1} possible triangulations.] Consequently,

$$T_{n+1} = T_2 T_n + T_3 T_{n-1} + \cdots + T_{n-1} T_3 + T_n T_2, \quad T_2 = 1.$$

If we shift subscripts and assign $T_n = S_{n-2}$, this recurrence relation becomes

$$S_{n-1} = S_0 S_{n-2} + S_1 S_{n-3} + \cdots + S_{n-3} S_1 + S_{n-2} S_0, \quad S_0 = 1.$$

Consequently,

$$S_n = S_0 S_{n-1} + S_1 S_{n-2} + \cdots + S_{n-2} S_1 + S_{n-1} S_0, \quad S_0 = 1.$$

Comparing this result with Eq. (29.1), we see that $C_n = S_n = T_{n+2}$, in agreement with the result in Example 23.1, where we learned that the number of triangulations of a convex polygon with $n + 2$ sides is C_n, for $n \ge 1$.

EXERCISES FOR CHAPTER 30

1. Kathy draws a convex decagon on a blackboard and labels the vertices consecutively, in a clockwise manner, as $v_1, v_2, v_3, \ldots, v_9, v_{10}$. Jill then draws in the diagonal from v_1 to v_7. In how many ways can Kathy triangulate the interior of this convex decagon without erasing the diagonal that Jill drew?

2. Paula draws a convex 12-sided polygon and labels the vertices consecutively, in a clockwise manner, as $v_1, v_2, v_3, \ldots, v_{11}, v_{12}$. In how many ways can she triangulate the interior of this 12-sided polygon if she wants to include the diagonal from v_1 to v_5 and the diagonal from v_6 to v_{10} as part of the triangulation?

3. Tim draws a convex n-gon and labels the vertices consecutively, in a clockwise manner, as $v_1, v_2, v_3, \ldots, v_{n-1}, v_n$. Upon drawing a diagonal from v_1 to v_6, he finds that there are 823,004 ways to triangulate the interior of his n-gon, using this diagonal as part of the triangulation. What is the value of n?

CHAPTER 31

Rooted Ordered Binary Trees, Pattern Avoidance, and Data Structures

In this chapter we have the opportunity to examine some additional examples where the Catalan numbers arise. This time, however, we shall verify that the structures are counted by these numbers by using the nonlinear recurrence relation developed in Chapter 29.

Example 31.1: Consider the trees shown in Fig. 31.1, where the vertices labeled with r are the *roots*. These trees are called *binary* because each vertex has at most two edges descending from it. Furthermore, these trees are *ordered* in the sense that a left branch descending from a vertex is considered to be different from a right branch descending from that vertex. In Fig. 31.2(a) we see the one rooted ordered binary tree for when we have $n = 1$ vertex—namely, just the root. Part (b) of Fig. 31.2 provides the two trees of this type for $n = 2$ vertices. If we let t_n count the number of rooted ordered binary trees on n vertices, then at this point we have $t_1 = 1$ and $t_2 = 2$. The results in Fig. 31.3 on p. 240 show us that $t_3 = 5$.

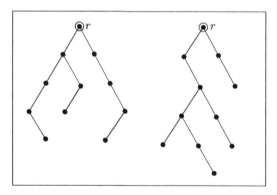

FIGURE 31.1

To count the number t_n of rooted ordered binary trees on $n \geq 0$ vertices, we shall assume that we know the values of t_i for $0 \leq i \leq n - 1$. In order to determine t_n, we

Fibonacci and Catalan Numbers: An Introduction, First Edition. Ralph P. Grimaldi.
© 2012 John Wiley & Sons, Inc. Published 2012 by John Wiley & Sons, Inc.

select one of the n vertices as the root, as in Fig. 31.4 on p. 240. Then we consider the substructures descending on the left and right sides of the root. These substructures are subtrees of the given tree, but more important, they themselves are two examples of rooted ordered binary trees having a total of $n - 1$ vertices. We realize it is possible for one of these subtrees to have 0 vertices, so we assign t_0 the value 1.

At this point we consider how the $n - 1$ vertices are divided up between the two subtrees. There are n cases to examine.

- (1) 0 vertices in the left subtree, $n - 1$ vertices in the right subtree: This gives us $t_0 t_{n-1}$ of the total number of possible rooted ordered binary trees on n vertices.
- (2) 1 vertex in the left subtree, $n - 2$ vertices in the right subtree: This case provides $t_1 t_{n-2}$ of the trees we are trying to count.

  ...

- $(i + 1)$ i vertices in the left subtree, $(n - 1) - i$ vertices in the right subtree: Here we obtain a count of $t_i t_{(n-1)-i}$ toward t_n.

  ...

- (n) $n - 1$ vertices in the left subtree, 0 vertices in the right subtree: This last case accounts for $t_{n-1} t_0$ of these trees.

Therefore, for all $n \geq 0$,

$$t_n = t_0 t_{n-1} + t_1 t_{n-2} + t_2 t_{n-3} + \cdots + t_{n-2} t_1 + t_{n-1} t_0 = \sum_{k=1}^{n} t_{k-1} t_{n-k}.$$

Since $t_0 = 1 \ (= C_0)$ and t_n satisfies the same (type of) nonlinear recurrence relation as in Eq. (29.1), it follows that for all $n \geq 0$,

$$t_n = C_n = \left(\frac{1}{n+1} \right) \binom{2n}{n}.$$

Before closing this example, let us look back at the complete binary trees in Fig. 24.6 of Example 24.3. Note that if we *prune* the trees for $n = 3$ by removing the four leaves for each tree in parts (1)–(5) of Fig. 24.6, the result is, respectively, the corresponding rooted ordered binary tree in parts (a)–(e) of Fig. 31.3.

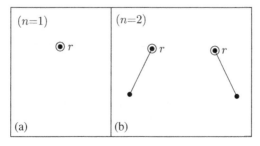

FIGURE 31.2

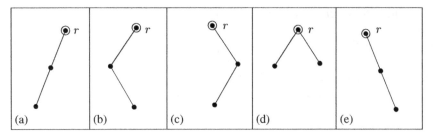

FIGURE 31.3

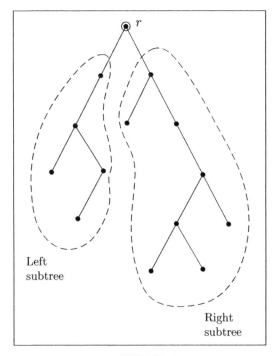

FIGURE 31.4

Example 31.2: Let us start with the permutations of $\{1, 2, 3\}$. There are six permutations of these integers—namely,

$$123 \quad 132 \quad 213 \quad 231 \quad 312 \quad 321.$$

Among these six permutations, the permutation 312 is the only one where the largest integer is first, the smallest integer is second, and the third integer is between the smallest and largest integers. Consequently, we say that this permutation exhibits the 312 *pattern*, while the other five permutations are said to be 312-*avoiding* permutations.

(This first example may seem somewhat trivial to the reader. That is why we better look at a larger example to see where this is leading.)

There are 24 permutations of $\{1, 2, 3, 4\}$. Let us examine four of these:

(1) The permutation 4132 is such that there are three integers—namely, 4, 1, and 3—where the first (namely, 4) is the largest, the second (namely, 1) is the smallest, and the third (namely, 3) is between these (smallest and largest) integers. Consequently, the permutation 4132 exhibits the 312 pattern. So we see that the integers involved in an instance of the 312 pattern need *not* be 3, 1, and 2.

Furthermore, the three integers 4, 1, and 2—though not consecutive within the permutation 4132—also exhibit the 312 pattern. So a permutation may have more than one instance where such a pattern arises.

(2) The permutation 3142 exhibits the 312 pattern. The integers here are those in the first, second, and fourth positions of the permutation, and they are precisely 3, 1, and 2.

(3) The permutation 2413 is the only permutation that starts with 2 and contains the 312 pattern—resulting from the integers 4, 1, and 3.

(4) Neither the permutation 1234 nor the permutation 2341 exhibits the 312 pattern. As a result, these permutations are said to be 312-avoiding permutations.

In total, there are 10 permutations that exhibit the 312 pattern:

(1) 1423	(2) 2413	(3) 3124	(4) 3142	(5) 3412
(6) 4123	(7) 4132	(8) 4213	(9) 4231	(10) 4312

Consequently, there are 14 permutations that are 312-avoiding permutations—namely,

(1) 1234	(2) 1243	(3) 1324	(4) 1342	(5) 1432
(6) 2134	(7) 2143	(8) 2314	(9) 2341	(10) 2431
(11) 3214	(12) 3241	(13) 3421	(14) 4321	

Even if the numbers 3, 1, and 2 are not in the set of integers we are permuting, we can still speak of the 312 pattern. For example, if we consider all 24 permutations of $\{5, 6, 8, 9\}$, there are still 10 permutations that exhibit the 312 pattern and 14 that are 312-avoiding permutations.

To determine the number of permutations of $\{1, 2, 3, 4, 5\}$ that are 312-avoiding permutations, we consider the location of 1 in the permutation:

(1) Suppose that 1 is in the first position of the permutation. Then the remaining integers—namely, 2, 3, 4, and 5—can be arranged in positions 2, 3, 4, and 5 of the permutation, so that the result avoids the 312 pattern, in 14 ($= C_4$) ways.

We let $1 (= C_0)$ account for the one way we can place nothing to the left of the 1 at the start of these permutations. So in this case we find $C_0 C_4$ permutations.

(2) Now consider the case where 1 is in position 2 of the permutation. If we place 3, 4, or 5 in position 1, then with 2 following 1, we have an occurrence of the 312 pattern. Consequently, these permutations start with 2, followed then by 1. Following that, we must place 3, 4, and 5 in the last three positions. This we can do—avoiding the 312 pattern—in $5 (= C_3)$ ways. Since the left side of 1 has been filled in $1 (= C_1)$ way and the right side in $5 (= C_3)$ ways, this case provides us with $C_1 C_3$ permutations.

(3) How can we avoid the 312 pattern when 1 is in position 3 of the permutation? In this case, we must place 2 and 3, in either order, in the first two positions. This can be done in $2 (= C_2)$ ways. Likewise, 4 and 5 must be placed, in either order, in the last two positions of the permutation. This too can be done in $2 (= C_2)$ ways. Therefore, this case provides $C_2 C_2$ permutations.

(4) When 1 is in position 4 of the permutation, in order to avoid the 312 pattern we must place 5 in the fifth position of the permutation. [This can be done in $1 (= C_1)$ way.] Then we can place 2, 3, and 4 in the first three positions, avoiding the 312 pattern, in $5 (= C_3)$ ways. So here we find $C_3 C_1$ permutations.

(5) Finally, if 1 is in position 5 of the permutation, then we can place 2, 3, 4, and 5 in the first four positions, and avoid the 312 pattern in $14 (= C_4)$ ways. There is $1 (= C_0)$ way in which we can place nothing to the right of 1 in these permutations, so here the number of permutations we obtain is $C_4 C_0$.

From the calculations in (1)–(5), it follows that the number of permutations of $\{1, 2, 3, 4, 5\}$ that avoid the 312 pattern is

$$C_0 C_4 + C_1 C_3 + C_2 C_2 + C_3 C_1 + C_4 C_0 = \sum_{i=0}^{4} C_i C_{4-i}.$$

From Eq. (29.1), we know that this is $C_5 = 42$ $[= 1 \cdot 14 + 1 \cdot 5 + 2 \cdot 2 + 5 \cdot 1 + 14 \cdot 1]$.

At this point, we know that the number of permutations of $\{1, 2, \ldots, n\}$ that avoid the 312 pattern is C_n, for $n = 3, 4, 5$. To establish this result for all $n \geq 3$, we apply the technique used above to determine the number of permutations of $\{1, 2, 3, 4, 5\}$ that avoid the 312 pattern.

For $n \geq 3$, let A_n count the number of permutations of $\{1, 2, \ldots, n\}$ that avoid the 312 pattern. Define $A_0 = 1$, $A_1 = 1$, and $A_2 = 2$. For $1 < i < n$, consider the permutations of $\{1, 2, \ldots, n\}$ where 1 is in position i. We place the integers 2, 3, $\ldots, i$ in the first $(i - 1)$ positions of the permutation. However, this cannot be done arbitrarily in one of the $(i - 1)!$ possible ways. To avoid the 312 pattern, there are only A_{i-1} ways in which this can be done. Then we must place the integers $i + 1, i + 2$, $\ldots, n$ in the final $n - i$ positions—avoiding the 312 pattern. There are A_{n-i} ways in which this can be accomplished. Consequently, there are $A_{i-1} A_{n-i}$ permutations of $\{1, 2, 3, \ldots, n\}$ that avoid the 312 pattern, when 1 is in position i of the permutation.

When $i = 1$ or $i = n$, then the numbers of permutations that avoid the 312 pattern in these two cases are $A_0 A_{n-1}$ and $A_{n-1} A_0$, respectively. So, for $n \geq 3$, the total number of permutations of $\{1, 2, 3, \ldots, n\}$ that avoid the 312 pattern is

$$A_0 A_{n-1} + A_1 A_{n-2} + \cdots + A_{n-2} A_1 + A_{n-1} A_0 = \sum_{i=1}^{n} A_{i-1} A_{n-i}, \quad A_0 = 1.$$

This is precisely the nonlinear recurrence relation of Eq. (29.1). Consequently, for all $n \geq 0$, $A_n = C_n$, and, for $n \geq 3$, the number of permutations of $\{1, 2, 3, \ldots, n\}$ that avoid the 312 pattern is the nth Catalan number C_n.

Although we have worked exclusively trying to avoid the 312 pattern, any permutation of this pattern leads to the same result. For instance, consider the 213 pattern, the *reverse* of the 312 pattern. Since the reverse of any permutation of $\{1, 2, 3, \ldots, n\}$ is likewise a permutation of $\{1, 2, 3, \ldots, n\}$, it follows that a permutation $p_1 p_2 p_3 \cdots p_{n-2} p_{n-1} p_n$ of $\{1, 2, 3, \ldots, n\}$ avoids the 312 pattern if and only if its reverse $p_n p_{n-1} p_{n-2} \cdots p_3 p_2 p_1$ avoids the 213 pattern. Consequently, for $n \geq 3$, the number of permutations of $\{1, 2, 3, \cdots, n\}$ that avoid the 213 pattern is also the nth Catalan number C_n.

It is also the case that, for $n \geq 3$, the number of permutations of $\{1, 2, 3, \ldots, n\}$ that avoid each of the patterns 132, 231, 123, and 321 is likewise counted by C_n. The cases for the patterns 132 and 231 are dealt with in the exercises for this chapter. The patterns 123 and 321 are examined in References [1, 2]. Further information on pattern avoidance for the permutations of $\{1, 2, 3, \ldots, n\}$ can be found in References [32,37,45].

Example 31.3 (The Stack Data Structure): This example is based on one given on page 86 of Reference [11].

The *stack* is an important abstract data structure that arises in computer science. This structure allows the storage of items of data according to the following rules:

(1) Each *insertion* takes place at *one* end of the structure. This is called the *top* of the stack. The insertion process is called the *push* procedure.

(2) Each *deletion* from the nonempty stack also takes place from the top of the stack. The deletion process is referred to as the *pop* procedure.

Since the *last* item pushed *onto* this data structure is the *first* item that can then be popped *out* of it, the stack is often referred to as a *last-in-first-out* (LIFO) structure.

Intuitive models for this data structure include a stack of poker chips on a table and a stack of trays in a cafeteria, where in both cases the stack can grow or shrink in size. Also, in both cases, one can only (1) insert a new entry at the top of the (possibly empty) stack or (2) delete the entry from the top of the nonempty stack.

Here we shall use this data structure, with its push and pop procedures, to permute the ordered list $1, 2, 3, \ldots, n$, where n is a positive integer. The diagram in Fig. 31.5 on p. 244 indicates how each integer of the input $1, 2, 3, \ldots, n$ is to be pushed onto

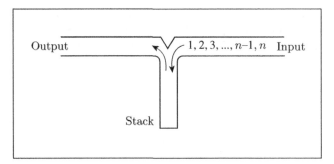

FIGURE 31.5

the top of the stack in the order given. However, an entry may be popped from the top of the nonempty stack at any time. But once an entry is popped from the stack , it may not be returned to either the top of the stack or the remaining input left to be pushed onto the stack. This process continues until the stack contains no entry. As a result, the ordered sequence of elements popped from the stack determines a permutation of $1, 2, 3, \ldots, n$.

We consider the following:

Case 1: Here $n = 1$ and the input list consists of only the integer 1. The 1 is inserted at the top of the empty stack and then popped out. This provides us with the permutation 1.

Case 2: This time $n = 2$ and we can obtain the two possible permutations of 1, 2, using the stack.

    (1) To arrive at the permutation 12, we place 1 at the top of the empty stack and then pop it. Then we place 2 at the top of the empty stack and pop it.

    (2) The permutation 21 is obtained when 1 is placed at the top of the empty stack and 2 is then pushed onto the top of this (nonempty) stack. Upon first popping the 2 from the top of the stack, and then the 1, the result is the permutation 21.

Case 3: Turning now to the case where $n = 3$, although there are $3! = 6$ possible permutations of $1, 2, 3$, we find that we can only obtain five of them, when using the stack. For example, the permutation 231 results when we take the following steps:

- Place 1 at the top of the empty stack.
- Push 2 onto the top of the stack—on top of 1.
- Pop 2 from the top of the stack.
- Push 3 onto the top of the stack—on top of 1.
- Pop 3 from the top of the stack.
- Pop 1 from the stack, leaving it empty.

However, suppose we try to generate the permutation 312 using the stack. In order to have 3 in the first position of the resulting permutation, we must build the stack by first pushing 1 onto the empty stack, then pushing 2 onto the top of the stack (on top of 1), and finally pushing 3 onto the stack (on top of 2). After 3 is popped from the top of the stack, we have 3 as the first entry in the resulting permutation. But now 2 is at the top of the stack, so we cannot pop 1 until after 2 has been popped. Consequently, the permutation 312 cannot be generated in this way, using the stack.

Case 4: When $n = 4$, there are 14 permutations of the ordered list 1, 2, 3, 4 that can be generated using the stack. These permutations are listed in Table 31.1, according to where 1 is located in the permutation.

TABLE 31.1

1 234	2 1 34	23 1 4	234 1
1 243	2 1 43	32 1 4	243 1
1 324			324 1
1 342			342 1
1 432			432 1

(1) There are five permutations with 1 in the first position, because after 1 is pushed onto the empty stack and then popped from it, there are five ways to permute 2, 3, 4 using the stack.

(2) If 1 is in the second position, then 2 must be in the first position. This follows because here we pushed 1 onto the empty stack, then we pushed 2 on top of 1, and then popped 2 and then 1. Since there are two ways 3, 4 can be permuted on the stack, this accounts for the two permutations in column 2 of Table 31.1.

(3) For column 3 of Table 31.1, we find 1 in the third position. We see that the only numbers that can precede 1 are 2 and 3. These two numbers can be permuted on the stack—with 1 on the bottom—in two ways. Then, after dealing with 2 and 3, 1 is popped. Finally, 4 is pushed onto the now empty stack and then popped from it.

(4) In the fourth column, we find five permutations: After we push 1 onto the empty stack, there are five ways we can permute 2, 3, 4 using the stack, with 1 on the bottom. After dealing with 2, 3, 4, we then pop 1 from the stack and complete the permutation.

On the basis of what we have seen in cases 1 through 4, for $n \geq 1$, let S_n count the number of ways to permute the integers $1, 2, 3, \ldots, n$ (or any list of n consecutive integers) using the stack. Furthermore, we set $S_0 = 1$ since there is only one way to permute nothing, using the stack. Then we find that

$$S_n = S_0 S_{n-1} + S_1 S_{n-2} + S_2 S_{n-3} + \cdots + S_{n-2} S_1 + S_{n-1} S_0,$$

where, for $1 \leq i \leq n$, the summand $S_{i-1}S_{n-i}$ accounts for the permutations of $1, 2, 3, \ldots, n-1, n$—using the stack—where 1 is in position i of the resulting permutation. Consequently, since $S_0 = 1$, it follows from Eq. (29.1) that

$$S_n = C_n, \quad n \geq 0.$$

Since the only permutation of 1, 2, 3 that fails to be generated in this way is 312, we could also have shown that $S_n = C_n$ by observing that a permutation of $1, 2, 3, \ldots, n$ fails to be generated using the stack when it contains the 312 pattern. For in Example 31.2, we showed that the number of such permutations that avoided the 312 pattern is C_n.

Furthermore, let us make one last observation about the permutations in Table 31.1. Consider, for instance, the permutation 3421. How did we generate this permutation of 1, 2, 3, 4, using the stack? First 1 is pushed onto the empty stack. Then 2 is pushed onto the stack, on top of 1. Now 3 is pushed onto the stack, on top of 2, and then 3 is popped from the top of the stack, leaving 2 and 1. At this point, 4 is pushed onto the stack, on top of 2. Then it is popped from the top of the stack, again leaving just 2 and 1 in the stack. Finally, 2 is popped from the top of the stack, and then 1 is popped from the top of the stack, now leaving the stack empty. So the permutation 3421 has been generated from 1, 2, 3, 4—using the stack—by means of the following sequence of four pushes and four pops:

push, push, push, pop, push, pop, pop, pop.

Now replace each "push" with an "1" and each "pop" with a "0." The result is the binary string

11101000.

Similarly, the permutation 1342 is determined by the procedures

push, pop, push, push, pop, push, pop, pop

and this corresponds with the binary string

10110100.

In fact, each of the permutations in Table 31.1 gives rise to a binary string of four 1's and four 0's. However, there are $8!/(4!4!) = 70$ ways to list four 1's and four 0's. Are these 14 binary strings special in any way? Yes, indeed! As each of these 14 binary strings of four 1's and four 0's is read from left to right, the number of 1's (pushes) is never exceeded by the number of 0's (pops), just like the binary strings in Table 20.1 in Example 20.2—one of our earlier situations that was counted by the Catalan numbers.

Along the same line, if $p = p_1 p_2 p_3 \cdots p_n$ is a permutation of $\{1, 2, 3, \ldots, n\}$, we call p *stack-sortable* if it is possible to generate the permutation $123 \cdots n$ from p, using the stack. The one permutation for $n = 1$ and the two permutations for $n = 2$ are all stack-sortable. When $n = 3$, the permutation 312 is stack-sortable. In this case, we first push 3 onto the empty stack. Then we push 1 onto the stack on top of 3 and then pop the 1. This is followed by pushing 2 onto the stack on top of 3. Then 2 is popped first, followed by 3 being popped. So the result is the permutation 123, proving that the permutation 312 is stack-sortable.

Are all permutations of 1, 2, 3 stack-sortable? Considering our earlier work with the stack, we probably feel that this is *not* the case. If we consider the permutation 231, we start by pushing 2 onto the empty stack. Then 3 is pushed onto the stack, on top of 2. This is followed by pushing 1 onto the stack on top of 3. Upon popping 1 from the top of the stack, we are now faced with 3 on top of 2 on the stack. There is no way that 2 can be popped from the stack at this point, so the permutation 231 is not stack-sortable.

In general, for $n \geq 3$, a permutation $p = p_1 p_2 p_3 \cdots p_n$ is stack-sortable if and only if it avoids the 231 pattern. Consequently, for $n \geq 1$, the number of permutations of $1, 2, 3, \ldots, n$ that are stack-sortable is C_n.

Example 31.4 (The Queue Data Structure): The *queue* is another example of an abstract data structure that arises in computer science. A queue is an ordered list wherein items are inserted at one end (called the *rear*) of the list and deleted at the other end (called the *front*). Whenever an item is added to the (rear of the) queue, all items that were added before it must be deleted before the new item can be deleted. The *first* item inserted *in* the queue is the first item that can be taken out from it. As a result, the queue is called a "first-in-first-out," (FIFO), data structure.

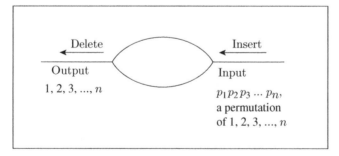

FIGURE 31.6

In Fig. 31.6 we find two parallel queues. We want to determine which of the permutations of $\{1, 2, 3, \ldots, n\}$ can be sorted into ascending order—namely, the permutation $123 \cdots n$—using these two parallel queues. When $n = 1$, there is only one such permutation—namely, 1. We simply insert 1 at the rear of either empty queue and then delete it from the front of that queue to arrive at 1. If $n = 2$, we can easily sort 1, 2 to get the permutation 12 by placing 1 at the rear of the first empty queue and 2 at the rear of the second empty queue. Then we delete 1 from the front of the first queue and 2 from the front of the second queue. For the permutation 21, the process is just as simple. For example, place 2 at the rear of the first empty queue and 1 at the rear of the second empty queue. Now delete 1 from the front of the second queue and follow that by deleting 2 from the front of the first queue. The result is the permutation 12.

What happens when $n = 3$? Are we able to sort the six possible permutations of 1, 2, 3 into ascending order—that is, into the permutation 123? For instance, suppose we consider the permutation 231. To sort this permutation into ascending order, we can start by inserting 2 at the rear of the empty top queue and then insert 3—following 2—at the rear of the top queue. Follow this by inserting 1 at the rear of the empty bottom queue. At this point, we delete 1 from the front of the bottom queue, leaving it empty. Then we delete 2 from the front of the top queue, followed by deleting 3 from the front of the top queue, now leaving the top queue empty. The result is the permutation 123.

Turning to the permutation 321, how can we get 1 as the first output? The only way possible is to insert 3 at the rear of one of the empty queues and follow that by inserting 2 at the rear of this queue, following 3. Then, at this point, 1 is inserted at the rear of the other queue (previously empty). Now 1 is deleted from the front of the queue where it was inserted. Unfortunately, 2 cannot be deleted from its present position since it is blocked by the 3 that precedes it (in the queue containing 3, 2). This turns out to be the only permutation of $\{1, 2, 3\}$ that cannot be sorted using the two parallel queues. In general, for $n \geq 3$, a permutation p of $\{1, 2, 3, \ldots, n\}$ can be sorted into ascending order (as the permutation $123 \cdots n$), using the two parallel queues, so long as p avoids the 321 pattern. Consequently, as a result of the remarks at the end of Example 31.2, for $n \geq 1$, the number of permutations of $\{1, 2, 3, \ldots, n\}$ that can be sorted into ascending order using two parallel queues is the nth Catalan number C_n. (This was first observed in Reference [43].)

EXERCISES FOR CHAPTER 31

1. For $n \geq 3$, let $p = p_1 p_2 p_3 \cdots p_{n-1} p_n$ be a permutation of $\{1, 2, 3, \cdots, n\}$. The *complement* of p is the permutation of $\{1, 2, 3, \ldots, n\}$ given by $\overline{p} = \overline{p_1} \; \overline{p_2} \; \overline{p_3} \; \ldots \overline{p_{n-1}} \; \overline{p_n}$, where $\overline{p_i} = n + 1 - p_i$, for $1 \leq i \leq n$. For example, if $n = 5$ and $p = 13452$, then $\overline{p} = 53214$. The permutation $q = 15243$ (of $\{1, 2, 3, 4, 5\}$) contains the 312 pattern, as exhibited by 524 and 523. The complement $\overline{q}$ of q is 51423, which contains the 132 pattern, as exhibited by 142 and 143. (Note that the complement of 312 is 132, the complement of 524 is 142, and the complement of 523 is 143.)

    (a) If p is a permutation of $\{1, 2, 3, \ldots, n\}$, what is $\overline{\overline{p}}$?

    (b) Let $p = 152643, q = 536412, r = 132456$, and $s = 351462$ be four of the 720 permutations of $\{1, 2, 3, 4, 5, 6\}$. Which of these permutations avoid the 312 pattern?

    (c) Find the complement of each permutation in part (b). Which of these complements avoid the 132 pattern?

(d) How many permutations of $\{1, 2, 3, 4, 5\}$ avoid the 132 pattern?

(e) For $n \geq 3$, how many permutations of $\{1, 2, 3, \ldots, n\}$ avoid the 132 pattern?

(f) For $n \geq 3$, how many permutations of $\{1, 2, 3, \ldots, n\}$ avoid the 231 pattern?

2. For $n \geq 3$, prove that the number of permutations of $\{1, 2, 3, \ldots, n\}$ that avoid both the 123 pattern and the 132 pattern is 2^{n-1}.

3. Which of the following permutations of $\{1, 2, 3, 4, 5, 6, 7, 8\}$ can be obtained from the permutation 12345678 using the stack of Example 31.3?

   (a) 42315678   (b) 54362187

   (c) 45321867   (d) 34217685

4. Suppose that the integers $1, 2, 3, 4, 5, 6, 7, 8$ are permuted using the stack of Example 31.3. (a) How many permutations are possible? (b) How many permutations have 1 in position 4 and 5 in position 8? (c) How many permutations have 1 in position 6? (d) How many permutations start with 321?

5. Which of the following permutations of $1, 2, 3, 4, 5, 6, 7$ is not stack-sortable?

   (a) 3145672   (b) 4123765

   (c) 5176423   (d) 4657321

6. For $n \geq 3$, let $p = p_1 p_2 p_3 \cdots p_{n-1} p_n$ be a permutation of $\{1, 2, 3, \ldots, n\}$. What pattern must p avoid so that if p is provided as the input to the stack, then the output is the permutation $n(n - 1) \cdots 321$? [Hence p is *reverse* stack-sortable.]

7. For $n \geq 3$, let $p = p_1 p_2 p_3 \cdots p_{n-1} p_n$ be a permutation of $\{1, 2, 3, \ldots, n\}$. What pattern must p avoid so that if p is provided as the input to a pair of parallel queues, then the output is the permutation $n(n - 1) \cdots 321$?

CHAPTER 32

Staircases, Arrangements of Coins, The Handshaking Problem, and Noncrossing Partitions

Even more examples, where the Catalan numbers come to the forefront, will be presented in this chapter. As we did in the previous chapter, here we shall use the nonlinear recurrence relation in Eq. (29.1) to verify that these new examples are counted by the Catalan numbers.

Example 32.1 (Staircases):   Consider the five-step staircase shown in Fig. 32.1 (a). We are interested in constructing this staircase using five rectangles. In parts (b), (c), and (d) of Fig. 32.1 we find three of the ways this can be accomplished.

Looking at the staircase in Fig. 32.1 (b), we find that if we remove the vertical rectangle with corner point $(1, 5)$, then the remaining four rectangles provide a way to construct a four-step staircase from four rectangles. Likewise, in Fig. 32.1 (c), upon removing the horizontal rectangle with corner point $(5, 1)$, the four rectangles pictured above it also provide a way to construct a four-step staircase from four rectangles. Finally, in Fig. 32.1 (d), we find a square with corner points $(0, 0)$, $(0, 3)$, $(3, 3)$, $(3, 0)$. To the right of this square is one of the (two) ways to construct a two-step staircase using two rectangles, and above this square is the other way to construct such a staircase. In general, we see how the rectangle with $(0, 0)$ as one of its corner points separates the remaining rectangles into smaller staircases—one of which may have (no rectangles and) no stairs.

To determine the number of ways one can construct an n-step staircase using n rectangles, we shall let St_n count the number of ways in which this can be done. From Fig. 32.2, we see that $St_1 = 1$, and $St_2 = 2$. Figure 32.3 (on p. 252) shows us that $St_3 = 5$. To account for the staircase with no stairs, we shall assign St_0 the value 1.

To determine St_n for $n \geq 4$, we start to construct the n-step staircase (using n rectangles) by considering the rectangle with $(0, 0)$ as one of its corner points. The other corner points of this rectangle are then found to be $(k, 0)$, $(k, n - k + 1)$, and $(0, n - k + 1)$, where $1 \leq k \leq n$.

Fibonacci and Catalan Numbers: An Introduction, First Edition. Ralph P. Grimaldi.
© 2012 John Wiley & Sons, Inc. Published 2012 by John Wiley & Sons, Inc.

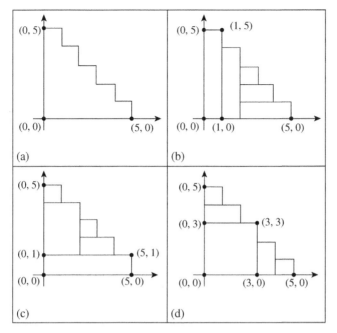

FIGURE 32.1

When $k = 1$, the rectangle with corner point $(0, 0)$ has its other corner points at $(1, 0)$, $(1, n)$, and $(0, n)$. We can place no other rectangles above this special rectangle in St_0 ways and $n - 1$ rectangles [forming $(n - 1)$-step staircases] to its right in St_{n-1} ways. In total, this case—where $k = 1$—provides $St_0 St_{n-1}$ of the staircases we are trying to count. In a similar way, when $k = n$, our special rectangle has corner points at $(0, 0)$, $(n, 0)$, $(n, 1)$, and $(0, 1)$. In this case, we can place $n - 1$ rectangles [forming $(n - 1)$-step staircases] above this special rectangle in St_{n-1} ways. Furthermore, we can place no other rectangles on the right of this special rectangle in St_0 ways. The result accounts for $St_{n-1} St_0$ more of the staircases to be counted.

Now consider the case for the positive integer k where $1 < k < n$. As mentioned previously, the corner points of the special rectangle are now located at $(0, 0)$, $(k, 0)$, $(k, n - k + 1)$, and $(0, n - k + 1)$. We can place $k - 1$ rectangles above the special rectangle, forming St_{k-1} $(k - 1)$-step staircases, and place $(n - k + 1) - 1 = n - k$

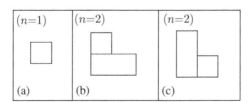

FIGURE 32.2

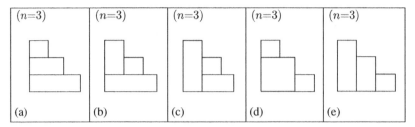

FIGURE 32.3

rectangles to the right of the special rectangle, forming St_{n-k} $(n - k)$-step staircases. This now accounts for $St_{k-1}St_{n-k}$ additional staircases. Therefore, we find that for $n \geq 1$,

$$St_n = \sum_{k=1}^{n} St_{k-1}St_{n-k}, \quad St_0 = 1.$$

Comparing this result with Eq. (29.1), we now know that, for $n \geq 1$, the number of ways we can construct an n-step staircase using n rectangles is counted by the nth Catalan number C_n.

A second way in which we can show that the number of these n-step staircases (made up from n rectangles) is counted by C_n is to associate a certain type of rooted tree with each staircase.

For $n = 1$, the tree consists of just the root—as shown in Fig. 32.4 (a). Parts (b) and (c) of Fig. 32.4 provide the rooted ordered binary trees that we correspond with the two 2-step staircases (made up from two rectangles).

In Fig. 32.5, we find the five 3-step staircases (made up from three rectangles)— each associated with the rooted ordered binary tree on its right. For example, the tree in part (c) is obtained by placing a circled dot in the special rectangle. (This is the special rectangle we used above to obtain the recurrence relation for St_n). We then draw the root r of the tree we want to construct. As we go to the *right* (away from the special rectangle), we draw an edge from the root r to the *right* child c_R of r. In going to the right (of the special rectangle), we encounter the 2-step staircase in Fig. 32.4 (b). The tree for this staircase becomes the subtree rooted at c_R. Going upward from the special rectangle, there are no additional rectangles—consequently, no

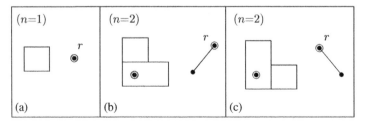

FIGURE 32.4

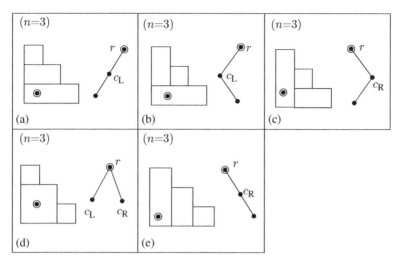

FIGURE 32.5

smaller staircase. As a result, we arrive at the rooted ordered binary tree at the right in Fig. 32.5 (c).

For the 3-step staircase in Fig. 32.5 (d), after placing a circled dot in the special rectangle, we then draw the root r of the tree we want to construct. As we go to the right, we draw an edge from the root r to the right child c_R. In going to the right of the special rectangle, we find the 1-step staircase (of one rectangle) for which the associated tree is a single vertex—as in Fig. 32.4 (a). This vertex becomes the subtree rooted at c_R. Going upward from the circled dot (in the special rectangle), we draw an edge from the root r to the left child c_L. In going up from the special rectangle, we encounter another 1-step staircase (of one rectangle). Its associated tree—namely, a single vertex—then becomes the subtree rooted at c_L. This results in the rooted ordered binary tree on the right of the 3-step staircase in Fig. 32.5 (d).

This process provides a one-to-one correspondence between the five 3-step staircases (made up from three rectangles) in parts(a)–(e) of Figs. 32.3 and 32.5 with the five rooted ordered binary trees in parts (a)–(e), respectively, of Figs. 31.3 and 32.5.

In general, for $n \geq 4$, if we have an n-step staircase (made up from n rectangles), we place a circled dot in the special rectangle for the staircase. We then draw the root r of the associated tree for the staircase. If a smaller staircase exists to the right of the special rectangle, then in going to the right we draw an edge from the root r to its right child c_R. The rooted tree for the smaller staircase to the right of the special rectangle then provides the subtree (perhaps, just a single vertex) at the right child c_R. This is the right subtree of r. If a smaller staircase exists above the special rectangle, then in going upward we draw an edge from the root r to its left child c_L. The smaller staircase above the special rectangle provides the subtree (perhaps, just a single vertex, once again) at the left child c_L. This is the left subtree of r. In this way, we construct a one-to-one correspondence between the n-step staircases (made up from n rectangles) and the rooted ordered binary trees on n vertices. It was shown in Example 31.1 that these trees are counted by C_n. Consequently, it now follows for the second time that $St_n = C_n$, $n \geq 0$.

Example 32.2 (Arranging Pennies):   This example is due to James G. Propp.

Abigail and Trevor want to arrange some pennies on a flat surface. They plan to start with a contiguous line of n pennies as a base on which they can place additional pennies. Each additional penny will then sit on the two pennies below it. In addition, neither Abigail nor Trevor is at all concerned with whether a given penny is placed heads up or heads down in any of their arrangements. The cases for $n = 1$ and $n = 2$ are shown in Fig. 32.6. In Fig. 32.7 we find their arrangements for when $n = 3$. Consequently, if we let a_n count the number of ways Abigail and Trevor can arrange pennies on a contiguous base of n pennies, we have $a_1 = 1$, $a_2 = 2$, and $a_3 = 5$. As we have done previously, we shall assign a_0 the value 1.

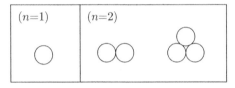

FIGURE 32.6

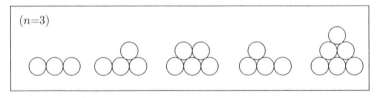

FIGURE 32.7

For $n \geq 4$, start with a base of n pennies. These pennies provide $n - 1$ possible positions for placing a penny on the second level. We shall consider two cases:

Case 1: As the second level is scanned from left to right, suppose that position k is the first position that is empty, where $1 \leq k \leq n - 1$. This happens when there are $k - 1$ pennies in the positions to the left of position k—and above the first k pennies in the contiguous bottom row. These $k - 1$ contiguous pennies (on the second level) provide a_{k-1} possible arrangements. The $(n - 1) - [(k - 1) + 1]$ positions (on the second level) to the right of position k are determined by a row of $(n - 1) - k + 1 \, (= n - k)$ contiguous pennies at the bottom level and these $n - k$ contiguous pennies provide a_{n-k} arrangements. As k varies from 1 to $n - 1$, we get a total of

$$a_0 a_{n-1} + a_1 a_{n-2} + \cdots + a_{n-2} a_1 = \sum_{k=1}^{n-1} a_{k-1} a_{n-k}$$

arrangements.

Case 2: The only situation not covered in Case 1 occurs when there is no empty position on the second level. So we have a contiguous row of n pennies on the bottom level and $n - 1$ contiguous pennies on the second level—and above this row of $n - 1$ pennies on the second level, we can have any one of a_{n-1} ($= a_{n-1}a_0$) possible arrangements.

From the results for the two exclusive cases above, we now know that the total number of ways Abigail and Trevor can arrange pennies on a contiguous base of n pennies is

$$a_n = a_0a_{n-1} + a_1a_{n-2} + \cdots + a_{n-2}a_1 + a_{n-1}a_0 = \sum_{k=1}^{n} a_{k-1}a_{n-k}.$$

With $a_0 = 1$, it now follows from Eq (29.1) that $a_n = C_n$, for $n \geq 0$.

Example 32.3: (a) (The Handshaking Problem) For $n \geq 1$, imagine that there are $2n$ people seated around a circular table. At a given time, all of these people decide to shake hands with someone else at the table. We would like to determine in how many ways this can be done so that we have n simultaneous handshakes with no set of two arms from one handshake crossing over another set of two arms from another handshake. If we let H_n count the number of ways these n handshakes can take place with the condition prescribed, we readily find that $H_1 = 1$ and $H_2 = 2$—as shown in parts (a) and (b), respectively, of Fig. 32.8. We also learn that $H_3 = 5$—as we see in Fig. 32.9.

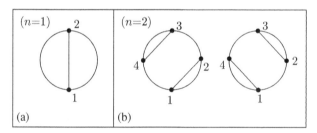

FIGURE 32.8

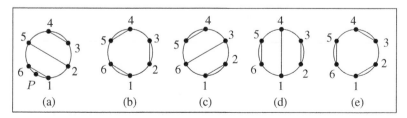

FIGURE 32.9

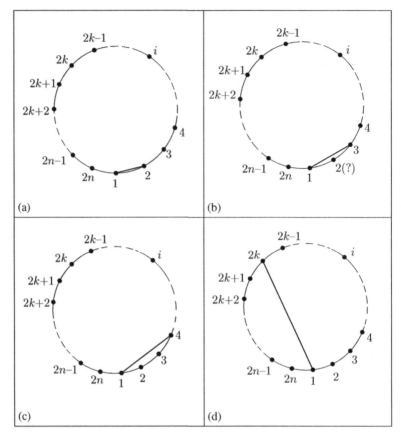

FIGURE 32.10

To determine H_n, for $n \geq 4$, consider the circular table shown in Fig. 32.10 (a). For $1 \leq i \leq 2n$, we shall let i represent the person seated at the location labeled i. We start by assuming that 1 and 2 shake hands. Then there is 1 ($= H_0$) way for the people seated between 1 and 2—namely, nobody—to shake hands. The remaining $2(n-1)$ people can then shake hands, with no pairs of arms crossing, in H_{n-1} ways. So in this case, where 1 and 2 shake hands, there is a total of $H_0 H_{n-1}$ possible ways for everyone to shake hands under the prescribed condition. Proceeding, we next observe in Fig. 32.10 (b) that 1 and 3 cannot shake hands under the prescribed condition—for then 2 would not be able to shake hands with anyone at the other $2n - 3$ locations without creating a set of two arms that crosses the set of two arms created by 1 and 3. In fact, 1 cannot shake hands with any person located at an odd position, without violating the condition prescribed for the problem. So next we consider what happens if 1 and 4 shake hands, as in Fig. 32.10 (c). Then 2 and 3 can shake hands in 1 ($= H_1$) way, while the remaining $2(n - 2)$ people can shake hands and follow the prescribed condition in H_{n-2} ways. Consequently, this case provides $H_1 H_{n-2}$ more of the situations we want to count. In general, for $1 \le k \le n$, when 1 shakes hands with $2k$, as in Fig. 32.10 (d), then 2, 3, ..., $2k - 1$ can shake hands, without violating the prescribed condition, in H_{k-1} ways, and $2k + 1$, $2k + 2$, ..., $2n - 1$, $2n$ can shake hands, according to the prescribed condition, in H_{n-k} ways. This accounts for $H_{k-1}H_{n-k}$ of the situations we are interested in counting. It now follows that the total number of ways we can have n simultaneous handshakes for these $2n$ people, with no two sets of two arms crossing, is

$$H_0 H_{n-1} + H_1 H_{n-2} + \cdots + H_{n-2} H_1 + H_{n-1} H_0 = \sum_{k=1}^{n} H_{k-1} H_{n-k}.$$

With $H_0 = 1$, it then follows—once again from Eq. (29.1)—that for $n \ge 0$, $H_n = C_n$.

(b) From the results in Figs. 32.8 and 32.9, it is clear that this "handshaking problem" can be stated geometrically—in terms of placing $2n$ labeled points on the circumference of a circle and then drawing n chords so that (i) every labeled point is an endpoint of one chord, and (ii) no two chords intersect within the circle. To show that these drawings are counted by the Catalan numbers, we do the following. For the circle in Fig. 32.9 (a) on p. 255, we place a point P on the arc from 6 to 1. Now we traverse the circumference of the circle in a counterclockwise manner starting and finishing at P. When we first encounter a chord we record a 1. When a chord is encountered for the second time, a -1 is recorded. So for the circle in Fig. 32.9 (a) we obtain the sequence 1, 1, 1, -1, -1, -1. Doing likewise for the circle in Fig. 32.9 (b), the sequence 1, 1, -1, 1, -1, -1 is recorded. Continuing in this way, we obtain a one-to-one correspondence between the configurations in parts (a)–(e) of Fig. 32.9 and the respective arrangements of three 1's and three -1's in parts (a)–(e) of Fig. 20.4 in Example 20.6. The number of these arrangements is C_3. In general, for $n \ge 1$, the number of ways to place $2n$ labeled points on the circumference of a circle and then draw n chords satisfying conditions (i) and (ii) above is C_n.

(c) Now let us take another look at the circles in part (b) of this example. Once again, for $n \ge 1$, let us start by placing $2n$ points about the circumference of a circle. We still want to draw n chords so that every point is used and no two chords intersect within the interior of the circle. This time, however, we shall find ourselves returning to an earlier example involving rooted trees.

In the case of $n = 3$, we redraw the configurations in Fig. 32.9 so that the arc 12 is at the top of the circle. Now consider the circle with chords 16, 25, and 34, as shown in Fig. 32.11 (a) on p. 258. Place a vertex in each of the four regions determined by the three chords. If the boundaries of two regions share a common chord, draw an edge connecting the vertices in these neighboring regions. The vertex in the region where the arc 12 is part of the boundary becomes the root r of the resulting tree. This ordered rooted tree (on $n + 1 = 4$ vertices) appears below the circle in Fig. 32.11 (a). Doing likewise for the other configurations where three non-intersecting chords are drawn, we obtain the resulting trees in parts (b)–(e) of Fig. 32.11. Now look back at the trees in Fig. 24.7. At this point we have a one-to-one correspondence between the configurations in parts (a)–(e) of Fig. 32.11 and the ordered rooted trees in parts

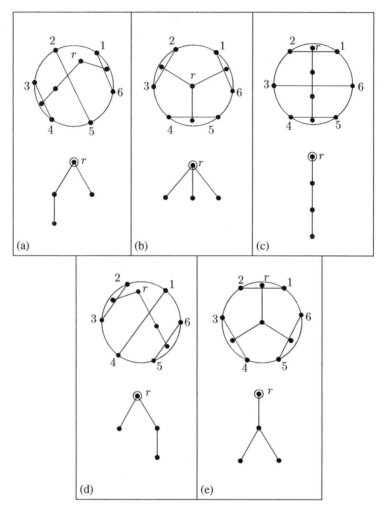

FIGURE 32.11

(1)–(5) of Fig. 24.7. Here the correspondence is (a) - (4), (b) - (5), (c) - (1), (d) - (3), (e) - (2).

The question that still needs to be answered, however, is how we go from an ordered rooted tree on $n + 1$ vertices to a circular configuration with n chords satisfying the stated conditions. To this end, consider the ordered rooted tree on seven vertices, as shown in Fig. 32.12 (a). Here $n = 6$, so we need a circular configuration with 12 points on the circumference of the circle and six chords, no two of which share a common point either within or on the circumference of the circle. We accomplish this by first drawing a circle about the tree—as in Fig. 32.12 (b). Then for each of the six edges of the tree, we place two points on the circumference of the circle—one on each side of the edge. This is demonstrated in Fig. 32.12 (b). Joining each of these pairs

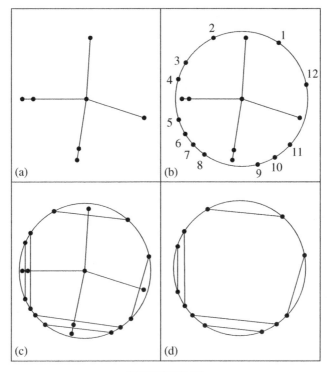

FIGURE 32.12

of points with a chord as in Fig. 32.12 (c), upon removing the tree we arrive at the configuration in Fig. 32.12 (d). Here we have the corresponding circular configuration with 12 points and six chords, satisfying the given conditions.

Now it is time to get away from the handshakes, the chords, and the trees, and introduce a new mathematical idea. We start with some basic definitions.

Definition 32.1:

(a) A *partition* of a finite set A consists of a collection B_1, B_2, $\ldots$, B_m of nonempty subsets of A such that (1) $A = B_1 \cup B_2 \cup \cdots \cup B_m$, and (2) $B_i \cap B_j = \varnothing$ for all $1 \le i < j \le m$. The subsets B_1, B_2, $\ldots$, B_m are called the *blocks* of the partition.

(b) A partition of $\{1, 2, 3, \ldots, n\}$ is called a *noncrossing partition* if whenever $a, b, c, d \in \{1, 2, 3, \ldots, n\}$ and $a < b < c < d$ with $a, c \in B_i$ and $b, d \in B_j$, then $B_i = B_j$.

Example 32.4: If $A = \{1, 2, 3, 4, 5, 6\}$, then

(1) the partition $B_1 = \{1, 2, 3\}$, $B_2 = \{4, 5\}$, $B_3 = \{6\}$ is a noncrossing partition of A.

(2) the partition $B_1 = \{1, 2\}$, $B_2 = \{3, 5\}$, $B_3 = \{4, 6\}$, however, is *not* a non-crossing partition of A because $3 < 4 < 5 < 6$ with $3, 5 \in B_2$ and $4, 6 \in B_3$, and $B_2 \neq B_3$.

Let $A = \{1, 2, 3, \ldots, n\}$. For $n = 1$, A has only one partition—namely, $\{1\}$. There are two partitions of A when $n = 2$. They are $\{1\}, \{2\}$ and $\{1, 2\}$. When $n = 3$, we find the five partitions of $A = \{1, 2, 3\}$ listed in Table 32.1.

TABLE 32.1

(a) $\{1\}, \{2\}, \{3\}$	(b) $\{1, 2\}, \{3\}$	(c) $\{1\}, \{2, 3\}$	(d) $\{1, 3\}, \{2\}$	(e) $\{1, 2, 3\}$

In all three cases—that is, $n = 1, 2$, or 3—all the partitions are noncrossing. For $n = 4$, however, we find that $\{1, 3\}, \{2, 4\}$ is not a noncrossing partition of $\{1, 2, 3, 4\}$. In fact, it is the only partition of $\{1, 2, 3, 4\}$ that is not noncrossing. There are 15 partitions of $\{1, 2, 3, 4\}$ in total, of which 14 are noncrossing. We list the 14 noncrossing partitions of $\{1, 2, 3, 4\}$ in Table 32.2.

TABLE 32.2

$\{1\}, \{2\}, \{3\}, \{4\}$	$\{1, 2\}, \{3\}, \{4\}$	$\{1, 3\}, \{2\}, \{4\}$	$\{1, 4\}, \{2\}, \{3\}$
$\{1\}, \{2, 3\}, \{4\}$	$\{1, 2\}, \{3, 4\}$	$\{1, 2, 3\}, \{4\}$	$\{1, 2, 4\}, \{3\}$
$\{1\}, \{2\}, \{3, 4\}$			$\{1, 4\}, \{2, 3\}$
$\{1\}, \{2, 4\}, \{3\}$			$\{1, 3, 4\}, \{2\}$
$\{1\}, \{2, 3, 4\}$			$\{1, 2, 3, 4\}$

Considering the results for $n = 1, 2, 3, 4$, the reader may feel, and rightfully so, that these partitions are counted by the Catalan numbers. Our next example addresses that issue.

Example 32.5 (Noncrossing Partitions): For $n \geq 5$, we want to determine the number of noncrossing partitions that exist for $\{1, 2, 3, \ldots, n\}$. This will be denoted by NC_n. (At this point, we know that $NC_1 = 1$, $NC_2 = 2$, $NC_3 = 5$, and $NC_4 = 14$. As we have done in previous similar situations, we assign NC_0 the value 1.) To count these partitions for a given $n \geq 5$, let B^* be the block of the noncrossing partition that contains 1 and suppose that k is the largest element in B^*.

If $k = 1$, then $B^* = \{1\}$. There is $1 = NC_0$ noncrossing partition of 1. There are NC_{n-1} noncrossing partitions of $\{2, 3, \ldots, n\}$. Any such partition together with $\{1\}$ ($= B^*$) provides a noncrossing partition of $\{1, 2, 3, \ldots, n\}$. So we have $NC_0 NC_{n-1}$ noncrossing partitions in this case.

Now suppose that $1 \in B^*$ and $k \in B^*$ for some $1 < k < n$. Consider the set $\{1, 2, \ldots, k - 1\}$. There are NC_{k-1} noncrossing partitions of $\{1, 2, \ldots, k - 1\}$ and NC_{n-k} noncrossing partitions of $\{k + 1, k + 2, \ldots, n\}$. If P is a noncrossing partition of $\{1, 2, \ldots, k - 1\}$, upon placing k in the block containing 1—thus forming B^*—the result is a noncrossing partition of $\{1, 2, \ldots, k - 1, k\}$. If not, then we have a, b, c, k, where $a < b < c < k$, with a, c in one block B and b, k in the block B^*, and $B \neq B^*$.

However, since $1 < a < b < c$, with $1, b \in B^*$ and $a, c \in B$, and $B \neq B^*$, we contradict P being a noncrossing partition of $\{1, 2, \ldots, k - 1\}$. Also, we realize that we cannot have an integer greater than k in the same block as an integer less than k. For if there were integers s, t with $s < k < t$, then $s \neq 1$ because 1 is in the same block as k. But then this leads to $1 < s < k < t$ with s, t in one block and $1, k$ in a different block—violating the definition for a noncrossing partition. With NC_{k-1} noncrossing partitions for $\{1, 2, \ldots, k - 1\}$ and NC_{n-k} noncrossing partitions for $\{k + 1, \ldots, n\}$, we see that there are $NC_{k-1} NC_{n-k}$ noncrossing partitions of $\{1, 2, 3, \ldots, n\}$, where 1 and k are in the same block, and $1 < k < n$.

Finally, suppose that $1, n$ are in the same block B^* of a noncrossing partition P of $\{1, 2, 3, \ldots, n\}$. Upon removing n from B^*, we obtain a noncrossing partition of $\{1, 2, 3, \ldots, n - 1\}$. Conversely, if we start with a noncrossing partition P' of $\{1, 2, 3, \ldots, n - 1\}$, after placing n in the block containing 1, we arrive at a noncrossing partition P of $\{1, 2, 3, \ldots, n\}$. If not, then we have integers a, b, c, n, where $a < b < c < n$, with a, c in one block and b, n in another block. So it follows that $1 < a < b < c$ with $1, b$ in one block (namely, B^*) and a, c in another block and this contradicts the fact that P' is a noncrossing partition of $\{1, 2, 3, \ldots, n - 1\}$. This accounts for the remaining NC_{n-1} ($= NC_{n-1} NC_0$) noncrossing partitions of $\{1, 2, 3, \ldots, n\}$.

From the preceding results, we now know that $NC_0 = 1$ and, for $n \geq 1$, NC_n, the number of noncrossing partitions of $\{1, 2, 3, \ldots, n\}$, satisfies

$$NC_n = NC_0 NC_{n-1} + NC_1 NC_{n-2} + \cdots + NC_{n-2} NC_1 + NC_{n-1} NC_0$$
$$= \sum_{i=1}^{n} NC_{i-1} NC_{n-i}.$$

Consequently, it follows from Eq. (29.1) that the number of noncrossing partitions of $\{1, 2, 3 \ldots, n\}$ is the nth Catalan number C_n.

Example 32.6:   We find that the notion of noncrossing partitions arises in a related geometric situation introduced by Robert Steinberg. This situation uses the following idea.

Let S be a subset of the plane. We say that S is *convex* if whenever points P and Q are in S, then the line segment made up of P and Q and all points between them is also in S. If A is a subset of the plane, the smallest convex subset of the plane that contains A is called the *convex hull* of A.

For example, if P and Q are points in the plane, then the convex hull of this set of two points is the line segment joining them (and this includes P and Q). If P, Q, and R are three noncollinear points in the plane, then the convex hull of this set of three points consists of all points on the perimeter and within the triangle PQR.

To demonstrate the connection between the notions of noncrossing partitions and the convex hull of a set, consider a circle with 10 labeled points on its circumference, as in Fig. 32.13 on p. 262. In Fig. 32.13 (a) we have the noncrossing partition $\{1, 5, 6, 10\}$, $\{2, 3, 4\}$, $\{7\}$, $\{8, 9\}$ of $\{1, 2, 3, 4, 5, 6, 7, 8, 9, 10\}$. The convex hull for the block $\{1, 5, 6, 10\}$ is a quadrilateral (and its interior), for the block $\{2, 3, 4\}$ it is a

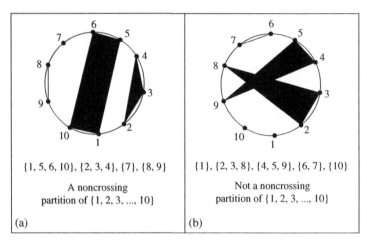

{1, 5, 6, 10}, {2, 3, 4}, {7}, {8, 9}

A noncrossing
partition of {1, 2, 3, ..., 10}

(a)

{1}, {2, 3, 8}, {4, 5, 9}, {6, 7}, {10}

Not a noncrossing
partition of {1, 2, 3, ..., 10}

(b)

FIGURE 32.13

triangle (and its interior), for the block {7} it is just the one point, and for the block {8, 9} it is the line segment joining these two points. When we have a noncrossing partition, for any two blocks of the partition the corresponding convex hulls have no point in common. The partition in Fig. 32.13 (b)—namely, {1}, {2, 3, 8}, {4, 5, 9}, {6, 7}, {10}—is not a noncrossing partition. For here $2 < 5 < 8 < 9$ with 2, 8 in one block and 5, 9 in another. Note how the convex hulls for the blocks {2, 3, 8} and {4, 5, 9} intersect within the circle.

We can generalize the situation demonstrated in Fig. 32.13 as follows. Let $B_1, B_2, \ldots, B_k$ be the blocks of a partition of $\{1, 2, 3, \ldots, n\}$. Place the labels $1, 2, 3, \ldots, n$ in a counterclockwise (or clockwise) manner on n consecutive points on the circumference of a circle. Then the blocks $B_1, B_2, \ldots, B_k$ determine a noncrossing partition of $\{1, 2, 3, \ldots, n\}$ if and only if the convex hulls of any two blocks of the partition do not intersect within the interior of the circle.

Example 32.7: In Fig. 32.14, we find the five *Murasaki* diagrams for $n = 3$. Here the vertical lines are connected by a horizontal line if and only if their respective labels are in the same block of the corresponding noncrossing partition. Consequently, we have a one-to-one correspondence between the noncrossing partitions in parts (a)–(e) of Table 32.1 and the respective Murasaki diagrams in parts (a)–(e) of Fig. 32.14.

These diagrams were used in *The Tale of Genji* to represent the 52 partitions of a five-element set. The noncrossing Murasaki diagrams are in a one-to-one correspondence with the noncrossing partitions of $\{1, 2, 3, 4, 5\}$. *The Tale of Genji* was written by Murasaki Shikibu (c. 976–1031), also known as Lady Murasaki, in the late Heian period during the reign of the empress Akiko. It provides an exquisite portrayal of country life in medieval Japan. In addition, it is widely regarded as the world's first psychological novel.

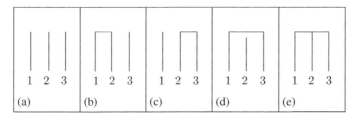

FIGURE 32.14

That such diagrams are enumerated by the Catalan numbers was first observed by Henry W. Gould.

Example 32.8: For our last example on noncrossing partitions, let us start with a positive integer n. We would like to count the number of noncrossing partitions of $\{1, 2, 3, \ldots, 2n + 1\}$ into $n + 1$ blocks where no block contains two consecutive integers. In Reference [29], the authors set up a one-to-one correspondence between these structures and the rooted trees for $n = 3$ in Fig. 24.6 of Example 24.3. The author in Reference [33] provides a one-to-one correspondence with the triangulations of a convex polygon, as developed in Examples 23.1 and 30.1. The approach given here follows the method developed in 1997 by Adrian Vetta.

Earlier, in Example 24.3, we showed that for $n \geq 0$, the number of complete binary trees on $2n + 1$ vertices is counted by the nth Catalan number C_n. In Fig. 32.15 on p. 264, we find the same five complete binary trees for the case of $n = 3$. Now, however, we want to show how we can partition the $2 \cdot 3 + 1 = 7$ vertices—this time labeled with the integers $1, 2, 3, 4, 5, 6, 7$—into $3 + 1 = 4$ blocks, so that there are no consecutive integers in any block.

Starting with a complete binary tree, we list the vertices (and label them consecutively) as we visit them in preorder. This is shown for each of the five trees in Fig. 32.15. If x and y are two integers between 1 and 7, and y is the right child of x, then we place x and y in the same block of our partition. Below each complete binary tree, we find the corresponding noncrossing partition and we see that there are no cases where a block contains consecutive integers.

From the results in Fig. 32.15, we make the following observations:

(1) The vertex labeled 1 is the root of the complete binary tree.
(2) The vertex labeled 2 is the left child of the root.
(3) The block containing 1 indicates the edges descending from the right child of the root or one of its right descendants—that is, if x and y are in the same block as 1 and there is no w such that $x < w < y$, then there is an edge from x to its right child y.
(4) A block containing a single label k corresponds with a leaf that is the left child of the vertex with label $k - 1$.
(5) Since the left child of a vertex is visited before the right child, two consecutive integers (labels) cannot occur within the same block.

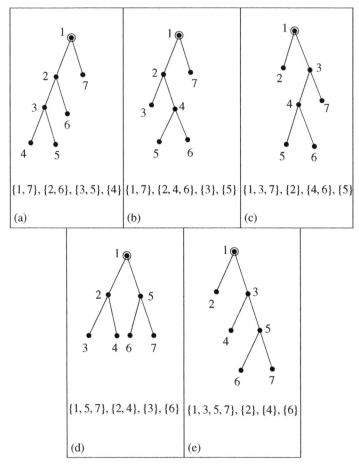

FIGURE 32.15

To go in the reverse direction, we use the prior five observations to go from a non-crossing partition of $\{1, 2, 3, \ldots, 2n + 1\}$ into $n + 1$ blocks, where no block contains two consecutive integers, to a complete binary tree on $2n + 1$ vertices. For example, when $n = 4$, the successive parts of Fig. 32.16 show how we can go from the noncrossing partition $\{1, 5, 9\}, \{2, 4\}, \{3\}, \{6, 8\}, \{7\}$ of $\{1, 2, 3, 4, 5, 6, 7, 8, 9\}$ to the complete binary tree in Fig. 32.16 (e).

As a result, we now have another situation counted by the Catalan numbers. Hence we now know that for $n \geq 0$, the number of noncrossing partitions of $\{1, 2, 3, \ldots, 2n + 1\}$ into $n + 1$ blocks, with no block containing two consecutive integers, is C_n.

At this point, we have introduced the reader to many situations where the Catalan numbers arise. Hopefully, this introduction has not been too overwhelming. Yet, there are many other situations where these numbers keep reappearing. The interested reader

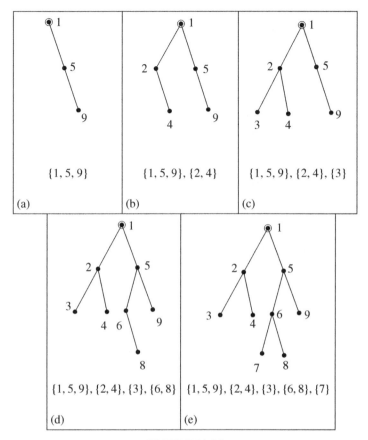

FIGURE 32.16

who wants to further investigate such situations will find other fascinating examples in References [5, 15, 16, 22, 24, 39] and, especially, in References [41, 42].

EXERCISES FOR CHAPTER 32

1. At a convention for superheroes (who can extend their right arms to rather long lengths), a party of 24 such superheroes are seated around a circular table, where the seats are numbered consecutively from 1 to 24—counterclockwise around the table.

   (a) In how many ways can they all shake hands simultaneously so that no pairs of arms cross?

   (b) Two of the superheroes seated at the table are "Michael the Marvel" and "Rebecca the Radiant." Michael is at seat 1 while Rebecca is occupying seat

      8. In how many ways can all the superheroes shake hands simultaneously so that Michael and Rebecca shake hands and no pairs of arms cross?

  (c) Two other superheroes seated at this table are "Cara the Courageous" and "Alberto the Avenger." Cara is at seat 15 while Alberto is occupying seat 22. In how many ways can all the superheroes shake hands simultaneously so that Michael and Rebecca shake hands, Alberto and Cara shake hands, and no pairs of arms cross?

2. At the same convention, as in the previous exercise, another circular table of 24 superheroes finds "Patti the Powerful" at seat 1 and "Brendan the Brave" at seat 7.

  (a) What is the maximum number of handshakes that can take place simultaneously so that Patti and Brendan shake hands and there are no pairs of arms crossing?

  (b) In how many ways can this maximum number of simultaneous handshakes take place, so that Patti and Brendan shake hands and there are no pairs of arms crossing ?

3. For the convention mentioned in the previous two exercises, suppose there is one more circular table, occupied by an even number of superheroes, with "Betty the Beautiful" at seat 1 and "Paul the Protector" at seat 10. If there are 235,144 ways in which all of the superheroes at this table can shake hands simultaneously, with no pairs of arms crossing and with Betty and Paul shaking hands, how many superheroes are seated at this table?

4. (a) Determine x, y so that $\{1, 2, x\}, \{y, 5, 6\}, \{7, 8\}$ is not a noncrossing partition of $\{1, 2, 3, 4, 5, 6, 7, 8\}$.

  (b) Determine x, y, z so that $\{1, 2, x\}, \{y\}, \{z, 6\}, \{7, 8\}$ is not a noncrossing partition of $\{1, 2, 3, 4, 5, 6, 7, 8\}$.

5. If A, B, C, D are four points in the plane, what can you say about the convex hull of $\{A, B, C, D\}$?

6. In Chapter 22, we examined situations involving vertices along a horizontal (base) line and arcs connecting pairs of these vertices, subject to certain conditions. Here is another such example. This one is due to Nate Kube and is related to his research on network testing.

    For $n \geq 1$, we have n vertices along a horizontal (base) line. These vertices are labeled consecutively from 1 to n. Arcs connecting any pair of these vertices must lie above the horizontal (base) line and cannot intersect (except, possibly at a vertex). If two arcs do share a common vertex, then that vertex is the *left* vertex (endpoint) of both arcs. For $n = 3$, there are five possible configurations, as shown in Fig. 32.17. How many such configurations are there for each $n \geq 1$?

7. How can one construct the 14 staircases for the case where $n = 4$ in Example 32.1?

8. How many of the staircases for $n = 7$ include the rectangle with corner points (a) $(0, 0)$, $(6, 0)$, $(6, 2)$, and $(0, 2)$? (b) $(0, 0)$, $(4, 0)$, $(4, 4)$, and $(0, 4)$?

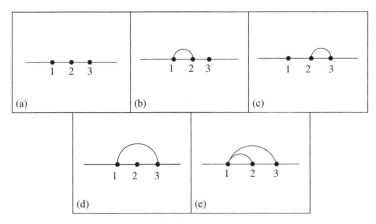

FIGURE 32.17

9. (a) For the arrangements in Example 32.2, suppose that Abigail and Trevor contiguously arrange seven pennies to serve as the bottom row of the arrangements. Then, according to the conditions prescribed in Example 32.2, they will be able to arrange pennies above this bottom row in C_7 $(= 429)$ ways. (i) How many of their arrangements will contain eight pennies? (ii) How many will contain nine pennies? (iii) How many will contain ten pennies?

   (b) For the arrangements in Example 32.2, now suppose that Abigail and Trevor contiguously arrange n pennies, for $n \geq 3$, to serve as the bottom row of the arrangements. Then, according to the conditions prescribed in Example 32.2, they will be able to arrange pennies above this bottom row in C_n ways. (i) How many of their arrangements will contain $n + 1$ pennies? (ii) How many will contain $n + 2$ pennies? (iii) How many will contain $n + 3$ pennies?

10. For the arrangements in Example 32.2, suppose there are $n \geq 4$ pennies in the bottom row of the arrangements. Under these conditions, how many arrangements will contain $n + 4$ pennies?

CHAPTER 33

The Narayana Numbers

In Example 5.6(a), we learned how each Fibonacci number can be expressed as a sum of binomial coefficients. A somewhat similar situation arises for the Catalan numbers by way of the Narayana numbers. Named after the Indian mathematician Tadepalli Venkata Narayana, the Narayana numbers are defined as follows:

$$N(0, 0) = 1,$$
$$N(n, 0) = 0, \quad n \geq 1,$$
$$N(n, k) = \frac{1}{n} \binom{n}{k} \binom{n}{k-1}, \quad n \geq k \geq 1.$$

Table 33.1 provides the values of $N(n, k)$ for $1 \leq k \leq n \leq 7$.

TABLE 33.1

$n \backslash k$	1	2	3	4	5	6	7	Row Sum
1	1							1
2	1	1						2
3	1	3	1					5
4	1	6	6	1				14
5	1	10	20	10	1			42
6	1	15	50	50	15	1		132
7	1	21	105	175	105	21	1	429

The last column of Table 33.1 suggests that for $n \geq 1$,

$$N(n, 1) + N(n, 2) + \cdots + N(n, n) = \sum_{k=1}^{n} N(n, k) = C_n.$$

We shall prove that this is indeed the case! Before we do so, however, we would like to know if the individual summands—namely, $N(n, k)$—have any combinatorial significance. The answer is a resounding "Yes!" The following will provide interpretations

Fibonacci and Catalan Numbers: An Introduction, First Edition. Ralph P. Grimaldi.
© 2012 John Wiley & Sons, Inc. Published 2012 by John Wiley & Sons, Inc.

of the Narayana numbers for some of the examples we have encountered for the Catalan numbers.

Example 33.1:

(1) In Fig. 33.1, we have the five lattice paths from (0, 0) to (3, 3) that never rise above the line $y = x$. We first encountered these paths in Fig. 19.2 of Example 19.1. Here we see that there is $1 = N(3, 1)$ such path with one turn—where there is an R followed by a U. Also, there are $3 = N(3, 2)$ paths with two such turns, and $1 = N(3, 3)$ path with three such turns. Figure 33.2 on pp. 270, 271, demonstrates the case for $n = 4$, for here we find $1 = N(4, 1)$ path with one turn, $6 = N(4, 2)$ paths with two turns, $6 = N(4, 3)$ paths with three turns, and $1 = N(4, 4)$ path with four turns.

(2) Figure 33.3 on p. 272 shows us how the Narayana numbers count the Dyck paths (or mountain ranges) in Fig. 21.1 of Example 21.1. In Fig. 33.3, we find

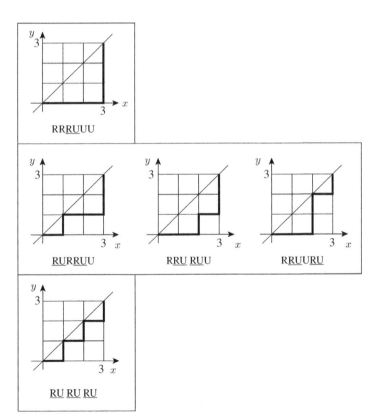

FIGURE 33.1

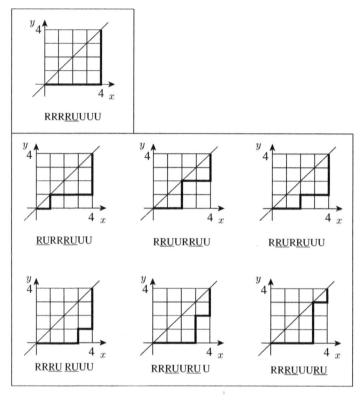

FIGURE 33.2

that there is one Dyck path with $1 = N(3, 1)$ peak—where there is a D ($\nearrow$) followed by a D* ($\searrow$), as in Example 21.6. Furthermore, there are $3 = N(3, 2)$ Dyck paths with two peaks, and $1 = N(3, 3)$ Dyck path with three peaks. The results in Fig. 33.4 on pp. 272, 273 settle the case for $n = 4$. Here we find $1 = N(4, 1)$ Dyck path with one peak, $6 = N(4, 2)$ Dyck paths with two peaks, $6 = N(4, 3)$ Dyck paths with three peaks, and $1 = N(4, 4)$ Dyck path with four peaks.

(3) In Example 20.1, we encountered the balanced strings made up of n left parentheses and n right parentheses. For $n = 3$, these were

   (a) ((()))   (b) (()())   (c) ()(())   (d) (())()   (e) ()()().

Here, we see that there is $N(3, 1) = 1$ balanced string with one occurrence of "()"—namely, the string in part (a). There are $N(3, 2) = 3$ balanced strings with two occurrences of "()"—those in parts (b), (c), and (d)—and $N(3, 3) = 1$ balanced string with three occurrences of "()"—namely, the balanced string in part (e).

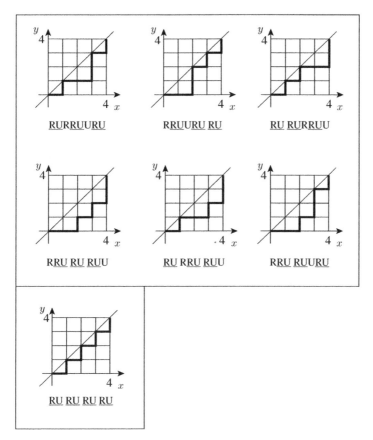

FIGURE 33.2 (*Continued*)

The case for $n = 4$ is demonstrated in the following:

$$((()))$$

$$((())() \quad (())() \quad ()(()) \quad (())() \quad (()()) \quad (()))$$

$$()(()) \quad ()(()) \quad ()(()) \quad (())() \quad (())() \quad (()))$$

$$()()()$$

Here we have partitioned the 14 balanced strings (with four left parentheses and four right parentheses) and find $N(4, 1) = 1$ balanced string with one occurrence of "()," $N(4, 2) = 6$ balanced strings with two occurrences of "()," $N(4, 3) = 6$ balanced strings with three occurrences of "()," and $N(4, 4) = 1$ balanced string with four occurrences of "()."

(4) In Example 26.3, we found that

$$w(w + x)(w + x + y) = w^3 + 2w^2x + wx^2 + w^2y + wxy,$$

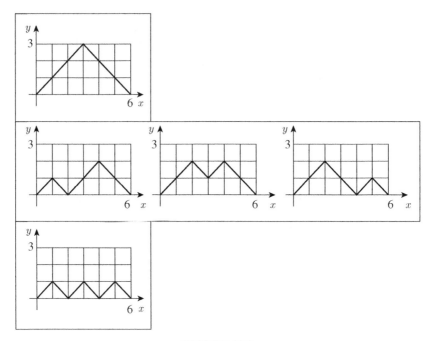

FIGURE 33.3

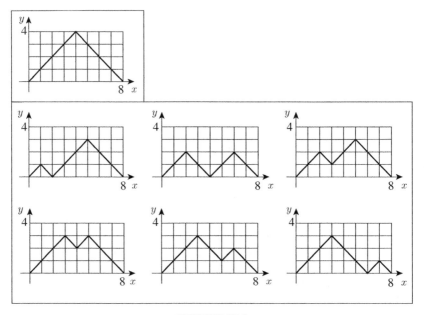

FIGURE 33.4

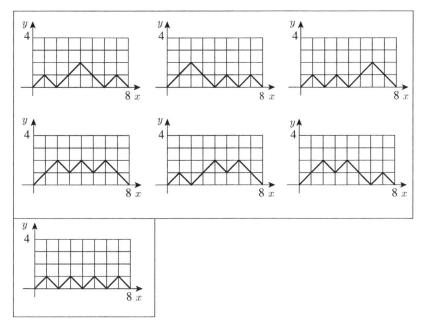

FIGURE 33.4 (*Continued*)

where (i) there is $N(3, 1) = 1$ term—namely, w^3—that involves one variable, (ii) each of the $N(3, 2) = 3$ terms—namely, $2w^2x$, wx^2, w^2y—involves two variables, and (iii) $N(3, 3) = 1$ term—namely, wxy—that involves all three variables.

When dealing with four variables, one finds that

$$w(w + x)(w + x + y)(w + x + y + z)$$
$$= w^4 + 3w^3x + 2w^3y + w^3z + 3w^2z^2 + 4w^2xy + 2w^2xz$$
$$+ w^2y^2 + w^2yz + wx^3 + 2wx^2y + wx^2z + wxy^2 + wxyz.$$

Table 33.2 provides an inventory of the 14 terms in the expansion, according to the number of variables involved.

TABLE 33.2

n = The Number of Variables Involved	The Terms Involving n Variables	The Number of Terms Involving n Variables
$n = 1$	w^4	$1 = N(4, 1)$
$n = 2$	$3w^3x$, $2w^3y$, w^3z, $3w^2z^2$, w^2y^2, wx^3	$6 = N(4, 2)$
$n = 3$	$4w^2yz$, $2w^2xz$, w^2yz, $2wx^2y$, wx^2z, wxy^2	$6 = N(4, 3)$
$n = 4$	$wxyz$	$1 = N(4, 4)$

Once again we see the presence of the Narayana numbers in yet another instance where the Catalan numbers arise.

(5) For our last example, let us look back at the noncrossing partitions of $\{1, 2, 3, \ldots, n\}$, which we investigated in Example 32.5. In Table 32.1, we find the five noncrossing partitions of $\{1, 2, 3\}$ and we see that there is $1 = N(3, 1)$ such partition with one block, $3 = N(3, 2)$ such partitions with two blocks, and $1 = N(3, 3)$ with three blocks. Turning to Table 32.2, we have the 14 noncrossing partitions of $\{1, 2, 3, 4\}$. Here we find that there is $1 = N(4, 1)$ such partition with one block, $6 = N(4, 2)$ such partitions with two blocks, $6 = N(4, 3)$ such partitions with three blocks, and $1 = N(4, 4)$ with four blocks.

In addition to what we have just learned about the combinatorial significance of the Narayana numbers, these numbers satisfy the following properties:

(a) $$N(n, k) = N(n, n + 1 - k), \quad n \geq k \geq 1$$
(The Symmetry Property)

(b) $$\binom{k + 1}{2} N(n + 1, k + 1) = \binom{n + 1}{2} N(n, k), \quad n \geq k \geq 0$$
(The Absorption Property)

(c) $$\binom{n}{k - 1} N(n, k + 1) = \binom{n}{k + 1} N(n, k), \quad n \geq k \geq 1$$

(d) $$\binom{n - k + 2}{2} N(n + 1, k) = \binom{n + 1}{2} N(n, k), \quad n \geq k \geq 1$$

(e) $$(n + 1)N(n, k) = (n - 1)[N(n - 1, k - 1) + N(n - 1, k)]$$
$$+ 2\binom{n - 1}{k - 1}^2, \quad n \geq k \geq 2$$

(f) $$N(n + 1, k + 1) = \binom{n}{k}^2 - \binom{n}{k - 1}\binom{n}{k + 1}, \quad n \geq k \geq 1$$

The proofs of properties (a)–(d) are requested in the exercises for this chapter. We shall prove properties (e) and (f). Property (f) will prove useful in establishing that $N(n, 1) + N(n, 2) + \cdots + N(n, n) = C_n$.

Property (e): The proof of this property provides us with an opportunity to use a result we saw earlier at the end of Example 5.6(a)—namely,

$$\binom{n + 1}{r} = \binom{n}{r} + \binom{n}{r - 1}, \quad \text{for } n \geq r \geq 1.$$

We find that

$$(n-1)[N(n-1,k-1)+N(n-1,k)]+2\binom{n-1}{k-1}^2$$

$$= (n-1)\left[\frac{1}{n-1}\binom{n-1}{k-1}\binom{n-1}{k-2}+\frac{1}{n-1}\binom{n-1}{k}\binom{n-1}{k-1}\right]+2\binom{n-1}{k-1}^2$$

$$= \binom{n-1}{k-1}\left[\binom{n-1}{k-2}+\binom{n-1}{k}\right]+2\binom{n-1}{k-1}^2$$

$$= \binom{n-1}{k-1}\left[\binom{n-1}{k-2}+2\binom{n-1}{k-1}+\binom{n-1}{k}\right]$$

$$= \binom{n-1}{k-1}\left\{\left[\binom{n-1}{k-1}+\binom{n-1}{k-2}\right]+\left[\binom{n-1}{k}+\binom{n-1}{k-1}\right]\right\}$$

$$= \binom{n-1}{k-1}\left[\binom{n}{k-1}+\binom{n}{k}\right]=\binom{n-1}{k-1}\binom{n+1}{k}=\binom{n+1}{k}\binom{n-1}{k-1}$$

$$= \frac{(n+1)!}{k!(n+1-k)!}\frac{(n-1)!}{(k-1)!(n-k)!}$$

$$= (n+1)\frac{n!}{k!(n+1-k)!}\frac{n}{n}\frac{(n-1)!}{(k-1)!(n-k)!}$$

$$= (n+1)\left(\frac{1}{n}\right)\frac{n!}{k!(n-k)!}\frac{n!}{(k-1)!(n+1-k)!}$$

$$= (n+1)\left(\frac{1}{n}\right)\binom{n}{k}\binom{n}{k-1}=(n+1)N(n,k), \quad n\geq k\geq 2.$$

Property (f): First we observe that

$$N(n+1,k+1)=\frac{1}{n+1}\binom{n+1}{k+1}\binom{n+1}{k}$$

$$= \frac{1}{n+1}\frac{(n+1)!}{(k+1)!(n-k)!}\frac{(n+1)!}{k!(n+1-k)!}$$

$$= \frac{n!}{(k+1)k!(n-k)!}\frac{(n+1)n!}{k!(n+1-k)(n-k)!}$$

$$= \frac{n!}{k!(n-k)!}\frac{n!}{k!(n-k)!}\frac{(n+1)}{(k+1)(n+1-k)}$$

$$= \binom{n}{k}^2\frac{(n+1)}{(k+1)(n+1-k)}.$$

Then

$$\binom{n}{k}^2 - \binom{n}{k-1}\binom{n}{k+1}$$

$$= \binom{n}{k}^2 - \frac{n!}{(k-1)!(n-k+1)!} \frac{k}{k} \frac{n!}{(k+1)!(n-k-1)!}$$

$$= \binom{n}{k}^2 - \frac{n!k}{k!(n-k+1)(n-k)!} \frac{n!(n-k)}{(k+1)k!(n-k)!}$$

$$= \binom{n}{k}^2 - \binom{n}{k}^2 \frac{k(n-k)}{(n-k+1)(k+1)}$$

$$= \binom{n}{k}^2 \left[1 - \frac{k(n-k)}{(n-k+1)(k+1)}\right]$$

$$= \binom{n}{k}^2 \left[\frac{(n-k+1)(k+1)}{(n-k+1)(k+1)} - \frac{k(n-k)}{(n-k+1)(k+1)}\right]$$

$$= \binom{n}{k}^2 \left[\frac{nk - k^2 + k + n - k + 1 - kn + k^2}{(n-k+1)(k+1)}\right]$$

$$= \binom{n}{k}^2 \frac{n+1}{(n-k+1)(k+1)} = \binom{n}{k}^2 \frac{n+1}{(k+1)(n-k+1)}.$$

Consequently,

$$N(n+1, k+1) = \binom{n}{k}^2 - \binom{n}{k-1}\binom{n}{k+1}, \quad n \geq k \geq 1.$$

Before we can prove that $N(n, 1) + N(n, 2) + \cdots + N(n, n) = C_n$, we need to examine the following two lemmas.

Lemma 33.1 (Lagrange's Identity): This result is named after the renowned French mathematician Joseph-Louis Lagrange (1736–1813).

For $n \geq 0$,

$$\sum_{k=0}^{n} \binom{n}{k}^2 = \binom{2n}{n}.$$

Proof: From the Binomial theorem, we know that the coefficient of x^n in $(1 + x)^{2n}$ is $\binom{2n}{n}$. However, it is also the case that

$$(1 + x)^{2n} = (1 + x)^n (1 + x)^n$$

$$= \left[\binom{n}{0} + \binom{n}{1} x + \binom{n}{2} x^2 + \cdots + \binom{n}{n} x^n \right]$$

$$\cdot \left[\binom{n}{0} + \cdots + \binom{n}{n-2} x^{n-2} + \binom{n}{n-1} x^{n-1} + \binom{n}{n} x^n \right],$$

where the coefficient of x^n is $\binom{n}{0}\binom{n}{n} + \binom{n}{1}\binom{n}{n-1} + \binom{n}{2}\binom{n}{n-2} + \cdots + \binom{n}{n}\binom{n}{0}$. Since $\binom{n}{k} = \binom{n}{n-k}$ for $0 \leq k \leq n$, it follows that the coefficient of x^n in $(1 + x)^{2n}$ is $\binom{n}{0}\binom{n}{0} + \binom{n}{1}\binom{n}{1} + \binom{n}{2}\binom{n}{2} + \cdots + \binom{n}{n}\binom{n}{n} = \sum_{k=0}^{n} \binom{n}{k}^2$. Consequently,

$$\sum_{k=0}^{n} \binom{n}{k}^2 = \binom{2n}{n}.$$

$\square$

A comparable result is found in the following.

Lemma 33.2: For $n \geq 0$,

$$\sum_{k=0}^{n} \binom{n}{k}\binom{n}{k-2} = \sum_{k=2}^{n} \binom{n}{k}\binom{n}{k-2} = \binom{2n}{n+2}.$$

Proof: The first equality follows because $\binom{n}{-2} = \binom{n}{-1} = 0$. To establish the second equality, we once again use the Binomial theorem. In the expansion of $(1 + x)^{2n}$, the coefficient of x^{n+2} is $\binom{2n}{n+2}$. However, since

$$(1 + x)^{2n} = (1 + x)^n (1 + x)^n$$

$$= \left[\binom{n}{0} + \binom{n}{1} x + \binom{n}{2} x^2 + \cdots + \binom{n}{n} x^n \right]$$

$$\cdot \left[\binom{n}{0} + \cdots + \binom{n}{n-2} x^{n-2} + \binom{n}{n-1} x^{n-1} + \binom{n}{n} x^n \right],$$

it is also the case that the coefficient of x^{n+2} is

$$\binom{n}{2}\binom{n}{n} + \binom{n}{3}\binom{n}{n-1} + \cdots + \binom{n}{n}\binom{n}{2}$$

$$= \binom{n}{2}\binom{n}{0} + \binom{n}{3}\binom{n}{1} + \cdots + \binom{n}{n}\binom{n}{n-2}$$

$$= \sum_{k=2}^{n} \binom{n}{k}\binom{n}{k-2}.$$

Consequently,

$$\sum_{k=2}^{n} \binom{n}{k}\binom{n}{k-2} = \binom{2n}{n+2}.$$

$\square$

The results in Lemmas 33.1 and 33.2, along with property (f), now help us to provide a proof for the last result in this chapter.

Theorem 33.1: For $n \geq 0$,

$$\sum_{k=0}^{n} N(n, k) = C_n.$$

Proof: The result follows for $n = 0$ because $N(0, 0) = 1 = C_0$. So from this point on, we only consider $n \geq 1$ and realize that $\sum_{k=0}^{n} N(n, k) = \sum_{k=1}^{n} N(n, k)$ since $N(n, 0) = 0$ for $n \geq 1$. From property (f), it follows that

$$N(n, k) = \binom{n-1}{k-1}^2 - \binom{n-1}{k-2}\binom{n-1}{k},$$

so

$$\sum_{k=1}^{n} N(n, k) = \sum_{k=1}^{n} \binom{n-1}{k-1}^2 - \sum_{k=1}^{n} \binom{n-1}{k-2}\binom{n-1}{k}$$

$$= \sum_{k=0}^{n-1} \binom{n-1}{k}^2 - \sum_{k=2}^{n-1} \binom{n-1}{k-2}\binom{n-1}{k},$$

because $\binom{n-1}{-1} = \binom{n-1}{n} = 0$, and this implies that $\sum_{k=1}^{n} \binom{n-1}{k-1}^2 = \sum_{k=0}^{n} \binom{n-1}{k-1}^2 = \sum_{k=0}^{n-1} \binom{n-1}{k}^2$. From Lemma 33.1 we have $\sum_{k=0}^{n-1} \binom{n-1}{k}^2 = \binom{2(n-1)}{n-1}$, while from Lemma 33.2 it follows that $\sum_{k=2}^{n-1} \binom{n-1}{k-2}\binom{n-1}{k} = \sum_{k=2}^{n-1} \binom{n-1}{k}\binom{n-1}{k-2} = \binom{2(n-1)}{(n-1)+2} =$

$\binom{2(n-1)}{n+1}$. Consequently,

$$\sum_{k=1}^{n} N(n, k) = \binom{2n - 2}{n - 1} - \binom{2n - 2}{n + 1}$$

$$= \frac{(2n - 2)!}{(n - 1)!(n - 1)!} - \frac{(2n - 2)!}{(n + 1)!(n - 3)!}$$

$$= \frac{(n + 1)(n)(2n - 2)!}{(n + 1)(n)(n - 1)!(n - 1)!}$$
$$- \frac{(n - 1)(n - 2)(2n - 2)!}{(n + 1)!(n - 1)(n - 2)(n - 3)!}$$

$$= \frac{(n + 1)(n)(2n - 2)!}{(n + 1)!(n - 1)!} - \frac{(n - 1)(n - 2)(2n - 2)!}{(n + 1)!(n - 1)!}$$

$$= \frac{(2n - 2)!}{(n + 1)!(n - 1)!} \left[n^2 + n - (n^2 - 3n + 2) \right]$$

$$= \frac{(2n - 2)!}{(n + 1)!(n - 1)!} (4n - 2) = \frac{2(2n - 1)(2n - 2)!}{(n - 1)!(n + 1)!}$$

$$= \frac{2(2n - 1)!}{(n - 1)!(n + 1)!} = \frac{(2n)(2n - 1)!}{n(n - 1)!(n + 1)!} = \frac{(2n)!}{n!(n + 1)!}$$

$$= \frac{1}{n + 1} \frac{(2n)!}{n!n!} = \frac{1}{n + 1} \binom{2n}{n} = C_n.$$

$\square$

(More on the Narayana numbers can be found in References [1, 41].)

EXERCISES FOR CHAPTER 33

1. Starting with two 1's and one 0, there are three ways in which these three symbols can be arranged as binary strings (in a linear manner)—namely, 101, 110, and 011. However, if we start with 101 and move the 1 at the (right) end of this binary string to the front of the string, then we obtain the binary string 110. In this case, we say that 110 was obtained from 101 by a single cyclic shift. In like manner, we could obtain the binary string 011 from the binary string 110 by a single cyclic shift or from the binary string 101 by two (single) cyclic shifts. In either case, we would say that 011 is a cyclic shift of either 110 or 101. If we consider two binary strings of three symbols to be the "same" if one is a cyclic shift of the other, then our original three binary strings become one!

Should we start with three 1's and two 0's, we should now find that there are $5!/(3!2!) = 10$ possible binary strings made up of these five symbols. They are

$$11100, \ 01110, \ 00111, \ 10011, \ 11001$$
$$10100, \ 01010, \ 00101, \ 10010, \ 01001.$$

If we do not distinguish two of these binary strings when one is a cyclic shift of the other, then these ten strings reduce to just two. Simply choose one string in each of the two rows.

(a) How many binary strings can one make up from $n + 1$ 1's and n 0's?

(b) For the binary strings in part (a), consider two strings the "same" if one can be obtained from the other by a cyclic shift. How many distinguishable binary strings are there under this condition?

(c) Interpret the Narayana numbers $N(n, k)$ in terms of the number of binary strings made up of $n + 1$ 1's and n 0's, where no binary string is a cyclic shift of another.

2. (a) Consider the rooted ordered trees in parts (1) - (5) of Fig. 24.7 in Example 24.4. How many of these trees have (i) one leaf; (ii) two leaves; and (iii) three leaves?

(b) Interpret the Narayana number $N(n, k)$ in the context of this example.

3. How many terms in the expansion of

$$\prod_{i=1}^{8} \left(\sum_{j=1}^{i} x_j \right)$$

contain (exactly) six of the variables $x_1, x_2, x_3, x_4, x_5, x_6, x_7, x_8$?

4. (a) Consider the graphs for $n = 3$ in Fig. 27.2 of Example 27.1. How many of these graphs have (i) one maximal clique; (ii) two maximal cliques; and, (iii) three maximal cliques?

(b) Now consider the graphs for $n = 4$ in Fig. 27.2 of Example 27.1. How many of these graphs have (i) one maximal clique; (ii) two maximal cliques; (iii) three maximal cliques; and, (iv) four maximal cliques?

(c) For $n \geq k \geq 1$, how many of the graphs of Example 27.1 have vertex set $\{1, 2, 3, \ldots, n\}$ and (exactly) k maximal cliques?

5. Among the 16, 796 Dyck paths from $(0, 0)$ to $(20, 0)$, how many have seven peaks?

6. For $n \geq k \geq 1$, prove that

$$N(n, k) = \frac{1}{k} \binom{n}{k-1} \binom{n-1}{k-1}.$$

7. For $n \geq k \geq 1$, prove that

$$N(n, k) = N(n, n + 1 - k).$$

(This is known as the Symmetry Property for the Narayana numbers.)

8. For $n \geq k \geq 0$, prove that

$$\binom{k + 1}{2} N(n + 1, k + 1) = \binom{n + 1}{2} N(n, k).$$

(This is known as the Absorption Law for the Narayana numbers.)

9. For $n \geq k \geq 1$, prove that

$$\binom{n}{k - 1} N(n, k + 1) = \binom{n}{k + 1} N(n, k).$$

10. For $n \geq k \geq 1$, prove that

$$\binom{n - k + 2}{2} N(n + 1, k) = \binom{n + 1}{2} N(n, k).$$

11. (a) For $k \geq 1$, determine $N(k, k)$ and $N(k + 1, k)$.

(b) For $k \geq 2$ and $p \geq 1$, with p fixed, prove that

$$N(k + p, k) < N(p + k + 1, k).$$

Related Number Sequences: The Motzkin Numbers, The Fine Numbers, and The Schröder Numbers

When dealing with the Fibonacci numbers we learned, in Example 12.5, about a closely related sequence of numbers called the Lucas numbers. These numbers are often considered "a first cousin" of the Fibonacci numbers.

Turning to the Catalan numbers, now we find several "first cousins" for this sequence of numbers.

Example 34.1 (The Motzkin Numbers): This sequence of numbers first appeared in Reference [28] and has consequently come to be called the sequence of Motzkin numbers, after the (German-born) American mathematician Theodore Samuel Motzkin (1908–1970).

For $n \geq 0$, the nth Motzkin number M_n can be given by any of the following three formulas:

(1) $M_n = \sum_{k \geq 0} \binom{n}{2k} C_k$

(2) $M_n = M_{n-1} + \sum_{k=0}^{n-2} M_k M_{n-2-k}, \quad M_0 = 1$

(3) $M_n = \frac{3(n-1)M_{n-2} + (2n+1)M_{n-1}}{n+2}, \quad M_0 = M_1 = 1$

The first of these provides an immediate connection with the sequence of Catalan numbers.

One finds that the first 11 Motzkin numbers are 1, 1, 2, 4, 9, 21, 51, 127, 323, 835, and 2188.

Two situations where the Motzkin numbers arise are as follows:

(a) For $n \geq 0$, M_n counts the number of ways one can draw chords between n points on the circumference of a circle, so that no two chords intersect

---

Fibonacci and Catalan Numbers: An Introduction, First Edition. Ralph P. Grimaldi.
© 2012 John Wiley & Sons, Inc. Published 2012 by John Wiley & Sons, Inc.

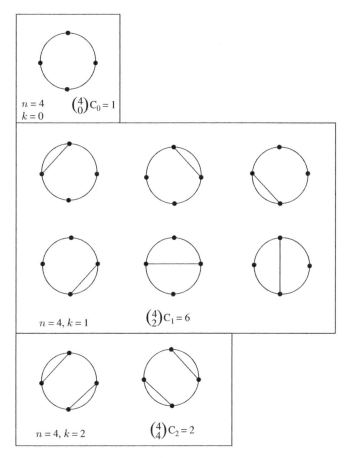

FIGURE 34.1

on, or within the interior of, the circle. Note here that, unlike the situation in Example 32.3 (b) for the Catalan numbers, we have not specified that n must be even, nor have we mentioned anything about the number of chords that are to be drawn. The configurations in Fig. 34.1 demonstrate the case for $n = 4$. Here we find the total number of configurations to be $\binom{4}{0}C_0 + \binom{4}{2}C_1 + \binom{4}{4}C_2 = \sum_{k=0}^{2} \binom{4}{2k} C_k = 1 \cdot 1 + 6 \cdot 1 + 1 \cdot 2 = 9 = M_4$.

(b) Now let us return to the Dyck paths of Example 21.1. For $n \geq 0$, we now want to determine the number of paths from $(0, 0)$ to $(n, 0)$, where we never go below the x-axis, but this time we are allowed to use the step R: $(x, y) \longrightarrow (x + 1, y)$, in addition to the diagonal steps D: $(x, y) \nearrow (x + 1, y + 1)$ and D*: $(x, y) \searrow (x + 1, y - 1)$. So in this situation, each such path is made up of an equal number of D steps and D* steps, with the remaining steps being R steps. The number of R steps is even or odd, depending on whether n is even or odd.

Suppose, for example, we want to count the number of these paths from $(0, 0)$ to $(17, 0)$ that contain five R steps. There are $\binom{17}{5}$ ways in which we can locate the five R's among the 17 steps. Upon removing these five steps, we coalesce the right endpoint of a D step with the left endpoint of a D* step, where this pair of D and D* steps was originally separated by one or more R steps. The result is one of the C_6 Dyck paths from $(0, 0)$ to $(12, 0)$. Consequently, there are $\binom{17}{5}C_6 = \binom{17}{12}C_6$ such paths from $(0, 0)$ to $(17, 0)$ that contain five R steps. In total, there are $\sum_{k=0}^{8} \binom{17}{2k}C_k = M_{17}$ such paths.

In Fig. 34.2 we see the case for $n = 4$. Here we can have 0, 2, or 4 R steps. There is $1 = 1 \cdot 1$ $\left[= \binom{4}{4}C_0 = \binom{4}{0}C_0\right]$ path with four R steps. There are $6 = 6 \cdot 1$ $\left[= \binom{4}{2}C_1\right]$ paths with two R steps and $2 = 1 \cdot 2$ $\left[= \binom{4}{4}C_2\right]$ paths with no R steps. In total there are $\binom{4}{0}C_0 + \binom{4}{2}C_1 + \binom{4}{4}C_2 = \sum_{k=0}^{2} \binom{4}{2k}C_k$ $= 1 \cdot 1 + 6 \cdot 1 + 1 \cdot 2 = 9$ such paths from $(0, 0)$ to $(4, 0)$.

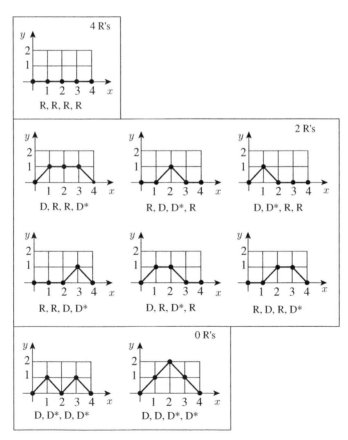

FIGURE 34.2

Further examples of the Motzkin numbers in different combinatorial settings are provided in References [9, 39].

Example 34.2 (The Fine Numbers):   Another sequence of numbers that is closely related to the Catalan numbers is the sequence of Fine numbers. This sequence is named after Terrence Leon Fine since they first appeared in Reference [12].

These numbers can be obtained from the recurrence relation

$$C_n = 2 \mathcal{F}_n + \mathcal{F}_{n-1}, \quad n \geq 1, \; \mathcal{F}_0 = 1.$$

The first seven Fine numbers are $1, 0, 1, 2, 6, 18, 57$.

Four situations where we find these numbers are given in the following:

(a) For a given Dyck path, a consecutive pair of steps, consisting of a D step: $(x, 0) \nearrow (x + 1, 1)$ followed by a D* step: $(x + 1, 1) \searrow (x + 2, 0)$, is called a *hill*. [So a hill is a consecutive pair of steps with height 1 (above the x-axis).]

  When $n = 4$, upon examining the 14 Dyck paths from $(0, 0)$ to $(8, 0)$ in Fig. 33.4, we find that there are $6 (= \mathcal{F}_4)$ of these paths that have no hills.

(b) Returning to the Young Tableaux of Example 22.1, we find that for $n = 4$ there are $C_4 = 14$ such arrangements of $1, 2, 3, \dots, 8$. These are presented in Table 34.1.

TABLE 34.1

(1) $\begin{bmatrix} 1 & 2 & 3 & 4 \\ 5 & 6 & 7 & 8 \end{bmatrix}$	(2) $\begin{bmatrix} 1 & 2 & 3 & 5 \\ 4 & 6 & 7 & 8 \end{bmatrix}$	(3) $\begin{bmatrix} 1 & 2 & 3 & 6 \\ 4 & 5 & 7 & 8 \end{bmatrix}$	
(4) $\begin{bmatrix} 1 & 2 & 4 & 5 \\ 3 & 6 & 7 & 8 \end{bmatrix}$	(5) $\begin{bmatrix} 1 & 2 & 4 & 6 \\ 3 & 5 & 7 & 8 \end{bmatrix}$	(6) $\begin{bmatrix} 1 & 2 & 5 & 6 \\ 3 & 4 & 7 & 8 \end{bmatrix}$	
(7) $\begin{bmatrix} 1 & 2 & 3 & 7 \\ 4 & 5 & 6 & 8 \end{bmatrix}$	(8) $\begin{bmatrix} 1 & 2 & 4 & 7 \\ 3 & 5 & 6 & 8 \end{bmatrix}$	(9) $\begin{bmatrix} 1 & 2 & 5 & 7 \\ 3 & 4 & 6 & 8 \end{bmatrix}$	
(10) $\begin{bmatrix} 1 & 3 & 4 & 7 \\ 2 & 5 & 6 & 8 \end{bmatrix}$	(11) $\begin{bmatrix} 1 & 3 & 5 & 7 \\ 2 & 4 & 6 & 8 \end{bmatrix}$	(12) $\begin{bmatrix} 1 & 3 & 4 & 6 \\ 2 & 5 & 7 & 8 \end{bmatrix}$	
(13) $\begin{bmatrix} 1 & 3 & 4 & 5 \\ 2 & 6 & 7 & 8 \end{bmatrix}$	(14) $\begin{bmatrix} 1 & 3 & 5 & 6 \\ 2 & 4 & 7 & 8 \end{bmatrix}$		

Here we see that there are $6 (= \mathcal{F}_4)$ of these tableaux—namely, (1)–(6)—that do *not* contain a column of the form $\begin{bmatrix} k \\ k + 1 \end{bmatrix}$.

(c) At the end of Example 31.2 in, we learned that there are $C_4 = 14$ permutations of $\{1, 2, 3, 4\}$ that avoid the pattern 321. They are

$$
\begin{array}{ccc}
1234 & 2134 & 3124 \\
1243 & 2143 & 3142 \\
1324 & 2314 & 3412 \\
1342 & 2341 & \\
1423 & 2413 & 4123.
\end{array}
$$

Of these, how many have no fixed points—that is, 1 is not in the first position, 2 is not in the second position, 3 is not in the third position, and 4 is not in the fourth position? Those with no fixed points are found to be

(1) 2143   (2) 2341   (3) 2413   (4) 3142   (5) 3412   (6) 4123,

and we see that they number 6 $(= \mathcal{F}_4)$.

(d) In Table 32.2 of Chapter 32 we found the 14 noncrossing partitions of $\{1, 2, 3, 4\}$—namely,

$$
\begin{array}{llll}
\{1\}, \{2\}, \{3\}, \{4\} & \{1, 2\}, \{3\}, \{4\} & \{1, 3\}, \{2\}, \{4\} & \{1, 4\}, \{2\}, \{3\} \\
\{1\}, \{2, 3\}, \{4\} & \{1, 2\}, \{3, 4\} & \{1, 2, 3\}, \{4\} & \{1, 2, 4\}, \{3\} \\
\{1\}, \{2\}, \{3, 4\} & & & \{1, 4\}, \{2, 3\} \\
\{1\}, \{2, 4\}, \{3\} & & & \{1, 3, 4\}, \{2\} \\
\{1\}, \{2, 3, 4\} & & & \{1, 2, 3, 4\}.
\end{array}
$$

Of these 14 noncrossing partitions, how many are such that the size of the block containing 1 is even? We find that the noncrossing partitions of $\{1, 2, 3, 4\}$ with this property are

(1) $\{1, 2\}, \{3\}, \{4\}$ (2) $\{1, 2\}, \{3, 4\}$ (3) $\{1, 3\}, \{2\}, \{4\}$
(4) $\{1, 4\}, \{2\}, \{3\}$ (5) $\{1, 4\}, \{2, 3\}$ (6) $\{1, 2, 3, 4\}$,

which number 6 $(= \mathcal{F}_4)$.

More on this number sequence can be found in References [8, 39].

Example 34.3 (The Schröder Numbers):   This sequence of numbers is named after the German logician Friedrich Wilhelm Karl Ernst Schröder (1841–1902). The first seven numbers in the sequence are 1, 2, 6, 22, 90, 394, and 1806. Further members of the sequence can be obtained from the recurrence relation

$$
R_n = R_{n-1} + \sum_{k=0}^{n-1} R_k R_{n-1-k}, \quad R_0 = 1, \ n \geq 1.
$$

Two situations, where these numbers arise and show a resemblance to situations we studied for the Catalan numbers, are given in the following.

(a) For $n \geq 0$, the Schröder numbers count the number of paths from $(0, 0)$ to (n, n) made up from the steps R : $(x, y) \longrightarrow (x + 1, y)$, U: $(x, y) \uparrow (x, y + 1)$, and D: $(x, y) \nearrow (x + 1, y + 1)$, where the path never rises above the line $y = x$. The case for $n = 2$ is demonstrated in Fig. 34.3.

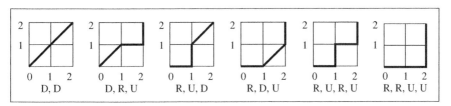

FIGURE 34.3

(b) This time we consider the paths from $(0, 0)$ to $(2n, 0)$, for $n \geq 0$, where the allowable steps are D: $(x, y) \nearrow (x + 1, y + 1)$, D*: $(x, y) \searrow (x + 1, y - 1)$, and R*: $(x, y) \longrightarrow (x + 2, y)$, and where the path cannot fall below the x-axis. Note that here R* is a double step—namely, two consecutive R steps. Figure 34.4 provides the six paths for the case where $n = 2$.

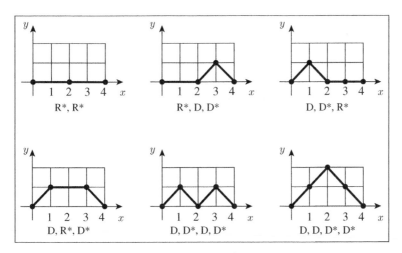

FIGURE 34.4

The reader may be wondering why, unlike the cases for the Motzkin numbers and the Fine numbers, we did not use S_n or s_n to denote the Schröder numbers. The numbers we have introduced as the Schröder numbers are more precisely called (in the literature) the *large* Schröder numbers. This distinguishes our sequence from the *small* Schröder numbers, the first seven of which are

$$s_1 = 1, \quad s_2 = 1, \quad s_3 = 3, \quad s_4 = 11, \quad s_5 = 45, \quad s_6 = 197, \quad s_7 = 903.$$

Further members of this sequence can be obtained from the recurrence relation

$$(n + 2)s_{n+2} - 3(2n + 1)s_{n+1} + (n - 1)s_n = 0, \quad n \geq 1, \ s_1 = 1, \ s_2 = 1.$$

The large and small Schröder numbers are related as follows:

$$R_n = 2s_{n+1}, \quad n \geq 1.$$

The following provides a situation where the small Schröder numbers arise. The result is due to David Callan.

Example 34.4: In Example 22.2, we considered pairs of compositions of a positive integer n. For a given n, each pair was of the form

$$a_1 + a_2 + \cdots + a_k \quad \text{and} \quad b_1 + b_2 + \cdots + b_k,$$

where $a_i > 0$ and $b_i > 0$, for $1 \leq i \leq k$, and

$$a_1 \geq b_1, \quad a_1 + a_2 \geq b_1 + b_2, \ldots, \quad \text{and}$$
$$a_1 + a_2 + \cdots + a_k = b_1 + b_2 + \cdots + b_k = n$$

Now we shall do the same thing, except that this time we shall allow 0 to be a summand (or part) in the first composition. That is, for $1 \leq i \leq k$, we shall allow a_i to be 0.

For $n = 2$, we find $3 = s_3 = s_{2+1}$ such pairs of compositions—namely,

$$(2, 2), \quad (2 + 0, 1 + 1), \quad \text{and} \quad (1 + 1, 1 + 1).$$

For $n = 3$, we have $11 = s_4 = s_{3+1}$ pairs of compositions under these conditions. They are listed in Table 34.2.

TABLE 34.2

(1) (3, 3)	(2) $(3 + 0, 2 + 1)$	(7) $(3 + 0 + 0, 1 + 1 + 1)$
	(3) $(3 + 0, 1 + 2)$	(8) $(2 + 1 + 0, 1 + 1 + 1)$
	(4) $(2 + 1, 2 + 1)$	(9) $(2 + 0 + 1, 1 + 1 + 1)$
	(5) $(2 + 1, 1 + 2)$	(10) $(1 + 2 + 0, 1 + 1 + 1)$
	(6) $(1 + 2, 1 + 2)$	(11) $(1 + 1 + 1, 1 + 1 + 1)$

In general, for $n \geq 1$, there are s_{n+1} pairs of compositions of n that satisfy the given conditions.

More about these numbers and their actual origin can be found in References [4, 36 , 40]. Further examples where these numbers arise are given in References [39, 41].

EXERCISES FOR CHAPTER 34

1. Determine the value of the Motzkin number M_{11}.
2. Determine the Fine numbers $\mathcal{F}_7$ and $\mathcal{F}_8$.
3. (a) Determine the value of each of the following:

$$(\mathrm{i})\,(-1)^0\,|C_0|\,;\;(\mathrm{ii})\,\,(-1)^1\begin{vmatrix} C_0 & C_1 \\ 1 & C_0 \end{vmatrix};\;(\mathrm{iii})\,(-1)^2\begin{vmatrix} C_0 & C_1 & C_2 \\ 1 & C_0 & C_1 \\ 0 & 1 & C_0 \end{vmatrix};$$

$$(\mathrm{iv})\,(-1)^3\begin{vmatrix} C_0 & C_1 & C_2 & C_3 \\ 1 & C_0 & C_1 & C_2 \\ 0 & 1 & C_0 & C_1 \\ 0 & 0 & 1 & C_0 \end{vmatrix}$$

   (b) Conjecture a formula based on the results in part (a).
4. Determine the (large) Schröder numbers R_7 and R_8.
5. Determine the (small) Schröder numbers s_8 and s_9.

CHAPTER 35

Generalized Catalan Numbers

We know that for $n \geq 0$, the nth Catalan number is given by

$$C_n = \frac{1}{n+1}\binom{2n}{n}$$

which can be rewritten as

$$C_n = \frac{1}{(2-1)n+1}\binom{2n}{n}.$$

But why would we want to do this?

If we let k denote a fixed positive integer, then we can define the *generalized Catalan number* $C_k(n)$ as

$$C_k(n) = \frac{1}{(k-1)n+1}\binom{kn}{n}.$$

When $k = 2$ we find that $C_2(n) = C_n$, our familiar nth Catalan number. For $k = 1$ we have $C_1(n) = \binom{n}{n} = 1$, for $n \geq 0$.

Corresponding to the cases for $k = 3$ and $k = 4$, we find the sequences

$$(k = 3):\ 1,\ 1,\ 3,\ 12,\ 55,\ 273,\ 1428,\ 7752,\ \ldots,\ \frac{1}{2n+1}\binom{3n}{n},\ \ldots$$

$$(k = 4):\ 1,\ 1,\ 4,\ 22,\ 140,\ 969,\ 7084,\ 53820,\ \ldots,\ \frac{1}{3n+1}\binom{4n}{n},\ \ldots.$$

Based on what we have seen so far, it seems that these generalized Catalan numbers are all integers. This is indeed the case, as we shall now establish. Before doing so,

Fibonacci and Catalan Numbers: An Introduction, First Edition. Ralph P. Grimaldi.
© 2012 John Wiley & Sons, Inc. Published 2012 by John Wiley & Sons, Inc.

however, we need to recall the following facts about the system of integers:

(1) If a and b are positive integers, then the greatest common divisor of a and b, denoted $\gcd(a, b)$, is the smallest positive integer that can be written as a linear combination of a and b—so there exist integers s and t with $\gcd(a, b) = as + bt$ and no smaller positive integer can be expressed in this manner.

(2) If a and b are positive integers with $\gcd(a, b) = 1$, then we say that a and b are *relatively prime*.

(3) If a, b, and c are positive integers and a divides the product bc, then if a and b are relatively prime, it follows that a divides c.

Theorem 35.1: For $n \geq 0$ and $k \geq 1$, the generalized Catalan number $C_k(n) = \frac{1}{(k-1)n+1}\binom{kn}{n}$ is an integer.

Proof: We see that

$$\binom{kn}{n} - \binom{kn}{n-1} = \frac{(kn)!}{n!(kn-n)!} - \frac{(kn)!}{(n-1)!(kn-n+1)!}$$

$$= \frac{(kn)!}{n!(kn-n)!} - \frac{(kn)!(n)}{n!(kn-n)!(kn-n+1)}$$

$$= \frac{(kn)!}{n!(kn-n)!}\left(1 - \frac{n}{kn-n+1}\right)$$

$$= \binom{kn}{n}\left(\frac{kn-n+1-n}{kn-n+1}\right)$$

$$= \frac{1}{(k-1)n+1}\binom{kn}{n}[(k-2)n+1].$$

Since $\binom{kn}{n} - \binom{kn}{n-1}$ is an integer, it follows that $(1/((k-1)n+1))\binom{kn}{n}[(k-2)n+1]$ is also an integer. Consequently, $(k-1)n+1$ divides $\binom{kn}{n}[(k-2)n+1]$. Since $(k-1)[(k-2)n+1] + (-1)(k-2)[(k-1)n+1] = (k-1) - (k-2) = 1$, we see that $1 = \gcd((k-1)n+1, (k-2)n+1)$, so $(k-1)n+1$ and $(k-2)n+1$ are relatively prime. Therefore, $(k-1)n+1$ does not divide $(k-2)n+1$, so $(k-1)n+1$ does divide $\binom{kn}{n}$. Consequently, for $n \geq 0$ and $k \geq 1$,

$$C_k(n) = \frac{1}{(k-1)n+1}\binom{kn}{n}$$

is an integer. $\square$

We shall now examine a few examples where these generalized Catalan numbers arise. For more examples and results on these sequences, we refer the reader to References [21, 39].

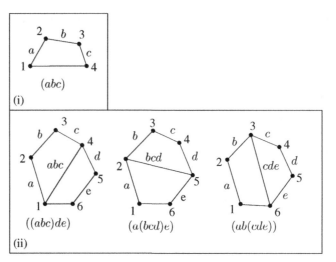

FIGURE 35.1

Example 35.1:   Once again we consider convex polygons, as we did in Example 23.1. For the case where $k = 3$, this time we shall let P_n be a convex polygon with $2n + 2$ sides. We want to count the number of ways $n - 1$ diagonals can be drawn within the interior of P_n, so that no two of these diagonals intersect (within the interior of the polygon) and these $n - 1$ diagonals partition the interior of P_n into n (nonoverlapping) *quadrilaterals*. In Fig. 35.1(i), we find the one way in which we can start with a convex quadrilateral—this is the case for $n = 1$. There are no diagonals in this case and the one $[= C_3(1)]$ possible partition leaves the interior of P_1 intact.

For $n = 2$, the convex polygon P_2 has $2 \cdot 2 + 2 = 6$ sides, so it is a convex *hexagon*. In Fig. 35.1(ii) we find the three $[= C_3(2)]$ ways in which we can draw one diagonal and partition the interior of P_2 into two (nonoverlapping) quadrilaterals. In each of these three partitions, we have labeled the vertices of the hexagon with the labels 1, 2, 3, 4, 5, 6 in a clockwise manner. Likewise, in the same manner, we have labeled five of the sides of the hexagon with the labels a, b, c, d, and e, leaving the side connecting vertices 1 and 6 unlabeled. Whenever we draw a diagonal, it completes a quadrilateral with three labeled sides or previously labeled diagonals. These three labels, taken in a clockwise manner, then provide the label for the new diagonal. As this process continues, it takes us to the final label that is placed on the side connecting vertices 1 and 6. In this way, we arrive at the expressions in Fig 35.1(ii)—namely,

$$((abc)de) \quad (a(bcd)e) \quad (ab(cde)).$$

These are the three $[= C_3(2)]$ orders in which we can combine a, b, c, d, and e for a *ternary* operation. [We have seen a similar situation for a binary operation in Table 20.2 of Example 20.3. Here we are counting the number of ways we can parenthesize the product $x_1 x_2 \cdots x_{2n+1}$ of the $2n + 1$ symbols x_1, x_2, ..., x_{2n+1} using n pairs of parentheses under the following condition: Within each pair of parentheses, there are (i) three symbols; (ii) two symbols and a parenthesized expression; (iii) one symbol and two parenthesized expressions; or (iv) three parenthesized expressions.]

These observations are instances of the following general results.

For $n \geq 1$, $C_3(n)$ counts

(i) the number of ways one can draw $n - 1$ diagonals within the interior of a convex polygon with $2n + 2$ sides, so that no two of the diagonals intersect within the interior of the polygon and the interior is partitioned into n (nonoverlapping) quadrilaterals.

(ii) the number of ways one can insert n pairs of parentheses into a product of $2n + 1$ variables, in order to show the order in which these variables are combined under a ternary operation.

Example 35.2: Analogous to the rooted ordered binary trees in Figs. 31.2 and 31.3 of Example 31.1, in Fig. 35.2 we find the rooted ordered *ternary* trees on n vertices for $n = 1$, 2, 3. In each case the root is labeled r. Each vertex of such a tree has at most three children—distinguished as the left, middle, or right child. For $n = 1$, we see that there is one [$= C_3(1)$] such rooted tree—consisting of just the root r (with no children). In the case of two vertices, there are three [$= C_3(2)$] such rooted ordered ternary trees, each with one child (which is also a leaf). Finally, when $n = 3$, we find 12 [$= C_3(3)$] such rooted ordered ternary trees—nine with one leaf and three with two leaves.

In general, for $n \geq 1$, there are $C_3(n)$ rooted ordered ternary trees on n vertices.

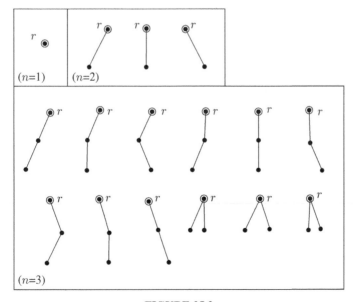

FIGURE 35.2

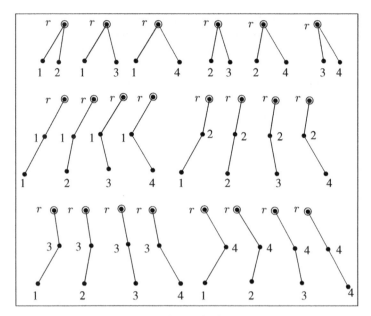

FIGURE 35.3

Example 35.3: In Fig. 35.3 we have the 22 [= $C_4(3)$] rooted ordered *quaternary* trees on $n = 3$ vertices. In this case, the root is labeled r and each vertex has at most four children, which we identify with the labels 1, 2, 3, and 4—for the first, second, third, and fourth child, respectively. We see that six of these trees have two leaves (both at level 1), while the other 16 trees each have only one leaf (at level 2).

In general, for $n \geq 1$, there are $C_4(n)$ rooted ordered quaternary trees on n vertices.

EXERCISES FOR CHAPTER 35

1. For $n > 0$ and $k > 0$, prove that

$$C_k(n) = \frac{1}{n}\binom{kn}{n-1}.$$

2. For $n > 0$ and $k > 0$, prove that

$$C_k(n) = \frac{1}{kn+1}\binom{kn+1}{n}.$$

3. (a) Suppose that you are in a box, as in Example 27.3, and tennis balls labeled 1, 2, and 3 are tossed into the box. If you are allowed to throw out one ball

onto the lawn beside the box, what are the possible labels for the ball you decide to throw out?

(b) After throwing out one of the tennis balls labeled 1, 2, and 3, the tennis balls labeled 4, 5, and 6 are tossed into the box. You are then allowed to throw out one of the five balls presently in the box onto the lawn beside the box. What are the possible labels for the two tennis balls you have now thrown out onto the lawn?

(c) How many sets of three labels are possible if the tennis balls labeled 7, 8, and 9 are then tossed into the box and you are allowed to throw out onto the lawn one of the seven balls presently in the box with you?

(d) Conjecture the number of possibilities if this process is repeated n times, for $n \geq 1$.

4. (a) Returning to the situation in the previous exercise, suppose once again that you are in a box, as in Example 27.3, and tennis balls labeled 1, 2, and 3 are tossed into the box. This time, however, you are allowed to throw out two balls onto the lawn beside the box. What are the possible labels for the balls you decide to throw out?

(b) After throwing out two of the tennis balls labeled 1, 2, and 3, the tennis balls labeled 4, 5, and 6 are tossed into the box. You are then allowed to throw out two of the four balls presently in the box onto the lawn beside the box. What are the possible labels for the four tennis balls you have now thrown out onto the lawn?

(c) How many sets of six labels are possible if the tennis balls labeled 7, 8, and 9 are then tossed into the box and you are allowed to throw out onto the lawn two of the five balls presently in the box with you?

(d) Conjecture the number of possibilities if this process is repeated n times, for $n \geq 1$.

5. (a) One last time, let us return to the box in the preceding two exercises and Example 27.3. As before, you are in the box, but this time tennis balls labeled 1, 2, 3, and 4 are tossed into the box. You are allowed to throw out one ball onto the lawn beside the box. What are the possible labels for the ball you decide to throw out?

(b) After throwing out one of the tennis balls labeled 1, 2, 3, and 4, the tennis balls labeled 5, 6, 7, and 8 are tossed into the box. You are then allowed to throw out one of the seven balls presently in the box onto the lawn beside the box. What are the possible labels for the two tennis balls you have now thrown out onto the lawn?

(c) How many sets of three labels are possible if the tennis balls labeled 9, 10, 11, and 12 are then tossed into the box and you are allowed to throw out onto the lawn one of the ten balls presently in the box with you?

(d) Conjecture the number of possibilities if this process is repeated n times, for $n \geq 1$.

CHAPTER 36

One Final Example?

As we did in Chapter 17 at the end of Part One, we now want to consider an example where the beginning pattern of results suggests the Catalan numbers—when it is not! This will emphasize one last time that in order to claim that a new example is truly counted by the Catalan numbers, we must either (i) set up a one-to-one correspondence with an example that is known to be counted by the Catalan numbers, or (ii) obtain a recurrence relation like the one in Eq. (29.1), remembering to check the initial condition.

To drive this point home, consider the following:

Example 36.1: For $n \geq 0$, start with n *distinct objects* and distribute them among at most n *identical containers*. Do this, however, while adhering to the following conditions:

(i) Do not place more than three objects in any one container.

(ii) Do not be concerned about how the objects in any given container are arranged.

We shall let d_n count the number of these distributions, for $n \geq 0$. From the results in Fig. 36.1, we find that

$$d_0 = 1, \quad d_1 = 1, \quad d_2 = 2, \quad d_3 = 5, \quad d_4 = 14$$

So it appears that we have come upon another situation that is counted by the Catalan numbers. Not so fast! Unfortunately, the suggested pattern does *not* continue and one finds, for instance, that

$$d_5 = 46 \neq 42 \,(= C_5) \quad \text{and} \quad d_6 = 166 \neq 132 \,(= C_6).$$

The distributions given here were studied in Reference [27]. Other such counterexamples can be found in Reference [20].

Fibonacci and Catalan Numbers: An Introduction, First Edition. Ralph P. Grimaldi.
© 2012 John Wiley & Sons, Inc. Published 2012 by John Wiley & Sons, Inc.

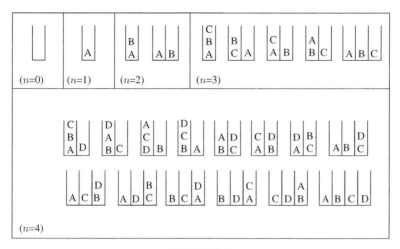

FIGURE 36.1

REFERENCES

1. Bóna, Miklós. *Combinatorics of Permutations*. Boca Raton, Florida: CRC Press - Chapman Hall, 2004.
2. Bóna, Miklós. *A Walk Through Combinatorics*, Second Edition. Hackensack, New Jersey: World Scientific Publishing Co., 2006.
3. Brown, William G. Historical Note on a Recurrent Combinatorial Problem, *American Mathematical Monthly*, Volume 72, 1965, Pp. 973 - 977.
4. Brualdi, Richard A. *Introductory Combinatorics*, fifth edition. Upper Saddle River, New Jersey: Pearson Prentice-Hall, 2010.
5. Conway, John Horton, and Guy, Richard Kenneth. *The Book of Numbers*. New York: Copernicus (Springer-Verlag), 1996.
6. de Mier, Anna, and Noy, Marc. A Solution to the Tennis Ball Problem, *Theoretical Computer Science*, Volume 346, Issues 2, 3, November, 2005, Pp. 254 - 264.
7. Deutsch, Emeric. Solution to Problem 11170, *American Mathematical Monthly*, Volume 114, May, 2007, Pp. 456 - 457.
8. Deutsch, Emeric, and Shapiro, Louis W. A Survey of Fine Numbers, *Discrete Mathematics*, Volume 241, No. 1 - 3, October 2001, Pp. 241 - 265.
9. Donaghey, Robert, and, Shapiro, Louis W. Motzkin Numbers, *Journal of Combinatorial Theory, Series A*, Volume 23, 1977, Pp. 291 - 301.
10. Edelman, Paul H., and Simion, Rodica. Chains in the Lattice of Noncrossing Partitions, *Discrete Mathematics*, Volume 126, 1994, Pp. 107 - 119.
11. Even, Shimon. *Graph Algorithms*. Rockville, Maryland: Computer Science Press, 1979.
12. Fine, Terrence Leon. Extrapolation When Very Little Is Known about the Source, *Information and Control*, Volume 16, 1970, Pp. 331 - 359.
13. Forder, Henry George. Some Problems in Combinatorics, *The Mathematical Gazette*, Volume 45, October, 1961, Pp. 199 - 201.
14. Gallian, Joseph A. *Contemporary Abstract Algebra*, Sixth Edition. Boston, Massachusetts: Houghton Mifflin, 2006.

15. Gould, Henry W. *Bell and Catalan Numbers: Research Bibliography of Two Special Number Sequences*, Revised Edition. Morgantown, West Virginia: Combinatorial Research Institute, 1978.

16. Graham, Ronald Lewis, Knuth, Donald Ervin, and Patashnik, Oren. *Concrete Mathematics*, Second Edition. Reading, Massachusetts: Addison-Wesley, 1994

17. Grimaldi, Ralph P. The Catalan Numbers via a Partition, *Congressus Numerantium*, Volume 102, 1994, Pp. 237 - 242.

18. Grimaldi, Ralph P., and Moser, Joseph. The Catalan Numbers and a Tennis Ball Problem, *Congressus Numerantium*, Volume 125, 1997, Pp. 65 - 72.

19. Grimaldi, Ralph P. *Discrete and Combinatorial Mathematics*, Fifth Edition. Boston, Massachusetts: Pearson Addison-Wesley, 2004.

20. Guy, Richard K. The Second Strong Law of Large Numbers. *The Mathematics Magazine*, Volume 63, No. 1, February, 1990, Pp. 3 - 20.

21. Hilton, Peter, and Pedersen, Jean. Catalan Numbers, Their Generalizations, and Their Uses, *The Mathematical Intelligencer*, Volume 13, No. 2, 1991, Pp. 64 - 75.

22. Koshy, Thomas. *Catalan Numbers and Applications*. New York: Oxford University Press, 2009.

23. Koshy, Thomas, and Salmassi, Mohammad. Parity and Primality of Catalan Numbers, *College Mathematics Journal*, Volume 37, January, 2006, Pp. 52 - 53.

24. Kuchinski, M. J. *Catalan Structures and Correspondences*. Master's Thesis, West Virginia University. Morgantown, West Virginia, 1977.

25. Mallows, Colin L., and Shapiro, Louis W. Balls on the Lawn. *The Journal of Integer Sequences*, Volume 2, 1999, Article 99.1.5.

26. Merlini, Donatella, Rogers, D. G., Sprugnoli, Renzo, and Verri, M. Cecilia. The Tennis Ball Problem, *Journal of Combinatorial Theory, Series A*, Volume 99, 2002, Pp. 307 - 344.

27. Miksa, F. L., Moser, Leo, and Wyman, Max. Restricted Partitions of Finite Sets, *Canadian Mathematical Bulletin*, Volume 1, 1958, Pp. 87 - 96.

28. Motzkin, Theodore Samuel. Relations Between Hypersurface Cross Ratios, and a Combinatorial Formula for Partitions of a Polygon, for Permanent Preponderance, and for Non-Associative Products. *Bulletin of the American Mathematical Society*, Volume 54, 1948, Pp. 352 - 360.

29. Mullin, Ronald C., and Stanton, Ralph G. A Map-Theoretic Approach to Davenport-Schinzel Sequences. *Pacific Journal of Mathematics*, Volume 40, No. 1, 1972, Pp. 167 - 172.

30. Netto, Eugen Otto Erwin. *Lehrbuch der Combinatorik*, Teubner, Leipzig, 1900.

31. Olds, Carl Douglas. Problem E 760. *American Mathematical Monthly*, Volume 54, February, 1947, P. 108.

32. Richards, D. Ballot Sequences and Restricted Permutations. *Ars Combinatoria*, Volume 25, 1988, Pp. 83 - 86.

33. Roselle, D. P. An Algorithmic Approach to Davenport-Schinzel Sequences. *Utilitas Mathematica*, Volume 6, 1974, Pp. 91 - 93.

34. Rosen, Kenneth H. *Elementary Number Theory and Its Applications*, Fifth Edition. Boston, Massachusetts: Pearson Addison-Wesley, 2005.

35. Shapiro, Louis W., and Hamilton, Wallace. The Catalan Numbers Visit the World Series, *The Mathematics Magazine*, Volume 66, February, 1993, Pp. 20 - 22.

36. Shapiro, Louis W., and Sulanke, Robert A. Bijections for Schröder Numbers. *Mathematics Magazine*, Volume 73, 2000, Pp. 369 - 376.

37. Simion, Rodica, and Schmidt, Frank W. Restricted Permutations. *European Journal of Combinatorics*, Volume 6, 1985, Pp. 383 - 406.

38. Simion, Rodica, and Ullman, Daniel. On the Structure of the Lattice of Noncrossing Partitions, *Discrete Mathematics*, Volume 98, 1991, Pp. 193 - 206.

39. Sloane, Neil James Alexander. The On-Line Encyclopedia of Integer Sequences.
http://www.research.att.com/~njas/sequences/

40. Stanley, Richard Peter. Hipparchus, Plutarch, Schröder, and Hough. *American Mathematical Monthly*, Volume 104, 1997, Pp. 344 - 350.

41. Stanley, Richard Peter. *Enumerative Combinatorics*, Volume 2. Cambridge, United Kingdom: Cambridge University Press, 1999.

42. Stanley, Richard Peter. *Catalan Addendum*.
http://www-math.mit.edu/~rstan/ec/catadd.pdf.

43. Tarjan, Robert Endre. Sorting Using Networks of Queues and Stacks. *Journal of the Association of Computing Machinery*, Volume 19, 1972, Pp. 341 - 346.

44. Tymoczko, Thomas, and Henle, Jim. *Sweet Reason: A Field Guide to Modern Logic*. Freeman Publishing, 1995, P. 304.

45. West, Julian. Generating Trees and the Catalan and Schröder Numbers. *Discrete Mathematics*, Volume 146, 1995, Pp. 247 - 262.

46. Wigner, Eugene Paul. Characteristic Vectors of Bordered Matrices with Infinite Dimension. *Annals of Mathematics*, Volume 62, 1955, Pp. 548 - 564.

Solutions for the Odd-Numbered Exercises

PART ONE: THE FIBONACCI NUMBERS

Exercises for Chapter 4

1. **Proof:** If not, there is a first integer $r > 0$ such that $\gcd(F_r, F_{r+2}) > 1$. But $\gcd(F_{r-1}, F_{r+1}) = 1$. Consequently, there exists a positive integer d, where $d > 1$ and d divides both F_r and F_{r+2}. Since $F_{r+2} = F_{r+1} + F_r$, it follows that d divides F_{r+1}. But this contradicts $\gcd(F_r, F_{r+1}) = 1$, the result in Property 4.1.

3. **Proof:** $F_{2(n+1)} = F_{2n+2} = F_{2n+1} + F_{2n} = (F_{2n} + F_{2n-1}) + F_{2n} = 2\,F_{2n} + F_{2n-1}$.

5. **Proof:** Simply add F_n to each side of the result in Exercise 4.

7. **Proof:** $F_{3n} = F_{3n-1} + F_{3n-2} = (F_{3n-2} + F_{3n-3}) + (F_{3n-3} + F_{3n-4}) = F_{3n-2} + 2F_{3n-3} + F_{3n-4} = (F_{3n-3} + F_{3n-4}) + 2F_{3n-3} + F_{3n-4} = 3F_{3n-3} + 2F_{3n-4} = 3F_{3n-3} + 2(F_{3n-3} + F_{3n-5}) = 5F_{3n-3} - 2F_{3n-5} = 4F_{3n-3} + (F_{3n-4} + F_{3n-5}) - 2F_{3n-5} = 4F_{3n-3} + (F_{3n-5} + F_{3n-6} + F_{3n-5}) - 2F_{3n-5} = 4F_{3n-3} + F_{3n-6}$.

9. **Proof (By Mathematical Induction):** We see that $F_0 = \sum_{r=0}^{0} F_r = 0 = 1 - 1 = F_2 - 1$, so the given statement holds in this first case. This provides the basis step of the proof. For the inductive step, we assume the truth of the result for $n = k$ (≥ 0)—that is, that $\sum_{r=0}^{k} F_r = F_{k+2} - 1$. Now we consider what happens for $n = k + 1$. In this case, we see that

$$\sum_{r=0}^{k+1} F_r = \left(\sum_{r=0}^{k} F_r \right) + F_{k+1} = F_{k+2} - 1 + F_{k+1} = (F_{k+2} + F_{k+1}) - 1$$
$$= F_{k+3} - 1 = F_{(k+1)+2} - 1,$$

so the truth of the statement at $n = k$ implies the truth at $n = k + 1$. Consequently, $\sum_{r=0}^{n} F_r = F_{n+2} - 1$ for all $n \geq 0$—by the Principle of Mathematical Induction.

Fibonacci and Catalan Numbers: An Introduction, First Edition. Ralph P. Grimaldi.
© 2012 John Wiley & Sons, Inc. Published 2012 by John Wiley & Sons, Inc.

11. **Proof:** Although we could use the Principle of Mathematical Induction to establish this property, instead we consider the following:

$$F_1 = F_2 - F_0$$
$$F_3 = F_4 - F_2$$
$$F_5 = F_6 - F_4$$
$$\vdots \quad \vdots \quad \vdots$$
$$F_{2n-3} = F_{2n-2} - F_{2n-4}$$
$$F_{2n-1} = F_{2n} - F_{2n-2}.$$

When we add up these n equations, the result on the left-hand side gives us $\sum_{r=1}^{n} F_{2r-1}$, while on the right-hand side we obtain $F_{2n} + (F_{2n-2} - F_{2n-2}) + (F_{2n-4} - F_{2n-4}) + \cdots + (F_4 - F_4) + (F_2 - F_2) - F_0 = F_{2n} - F_0 = F_{2n} - 0 = F_{2n}$.

13. **Proof:**

$$n = 3: \quad F_2 F_4 - F_3^2 = 1 \cdot 3 - 2^2 = 3 - 4 = -1 = (-1)^3$$
$$n = 4: \quad F_3 F_5 - F_4^2 = 2 \cdot 5 - 3^2 = 10 - 9 = 1 = (-1)^4$$
$$n = 5: \quad F_4 F_6 - F_5^2 = 3 \cdot 8 - 5^2 = 24 - 25 = -1 = (-1)^5$$
$$n = 6: \quad F_5 F_7 - F_6^2 = 5 \cdot 13 - 8^2 = 65 - 64 = 1 = (-1)^6.$$

15. **Proof:** $F_n F_{n+1} F_{n+2} = F_n F_{n+1}(F_{n+1} + F_n) = F_n F_{n+1}^2 + F_n^2 F_{n+1} = (F_{n+1} - F_{n-1})F_{n+1}^2 + F_n^2 F_{n+1} = F_{n+1}^3 - F_{n-1}F_{n+1}^2 + F_n^2 F_{n+1} = F_{n+1}^3 + F_{n+1}(F_n^2 - F_{n-1}F_{n+1}) = F_{n+1}^3 + F_{n+1}(-1)^{n+1}$, by the Cassini property.

17. **Proof (By Mathematical Induction):** For $n = 1$, we see that $\sum_{r=1}^{1} rF_r = 1F_1 = 1$, while $nF_{n+2} - F_{n+3} + 2 = 1F_3 - F_4 + 2 = 1(2) - 3 + 2 = 1$, so the result is true in this first case. This establishes the basis case for our inductive proof. Next, we assume the result is true for some $n = k$ (≥ 1)—that is, we assume that $\sum_{r=1}^{k} rF_r = kF_{k+2} - F_{k+3} + 2$. Now, for $n = k + 1$ (≥ 2), we have

$$\sum_{r=1}^{k+1} rF_r = \sum_{r=1}^{k} rF_r + (k + 1) F_{k+1}$$
$$= (kF_{k+2} - F_{k+3} + 2) + (k + 1) F_{k+1}$$
$$= [(k + 1)F_{k+2} - F_{k+2} - F_{k+3} + 2] + (k + 1) F_{k+1}$$
$$= [(k + 1)F_{k+2} + (k + 1) F_{k+1}] - (F_{k+2} + F_{k+3}) + 2$$
$$= (k + 1)(F_{k+2} + F_{k+1}) - (F_{k+2} + F_{k+3}) + 2$$
$$= (k + 1)F_{k+3} - F_{k+4} + 2 = (k + 1)F_{(k+1)+2} - F_{(k+1)+3} + 2.$$

So the result for $n = k + 1$ follows from the result for $n = k$. Therefore, by the Principle of Mathematical Induction, this weighted sum formula holds for all $n \geq 1$.

Alternate Solution Using Property 4.5:

$$\sum_{r=1}^{n} rF_r = 1F_1 + 2F_2 + 3F_3 + \cdots + (n-1)F_{n-1} + nF_n$$

$$= \sum_{r=1}^{n} F_r + \sum_{r=2}^{n} F_r + \sum_{r=3}^{n} F_r + \cdots + \sum_{r=n-1}^{n} F_r + \sum_{r=n}^{n} F_r.$$

For this sum, note, for example, that F_2 occurs only in the first two summations, while F_3 occurs only in the first three summations. Continuing we find that

$$\sum_{r=1}^{n} F_r + \sum_{r=2}^{n} F_r + \sum_{r=3}^{n} F_r + \cdots + \sum_{r=n-1}^{n} F_r + \sum_{r=n}^{n} F_r$$

$$= \sum_{r=1}^{n} F_r + \left(\sum_{r=1}^{n} F_r - \sum_{r=1}^{1} F_r \right) + \left(\sum_{r=1}^{n} F_r - \sum_{r=1}^{2} F_r \right) + \cdots$$

$$+ \left(\sum_{r=1}^{n} F_r - \sum_{r=1}^{n-2} F_r \right) + \left(\sum_{r=1}^{n} F_r - \sum_{r=1}^{n-1} F_r \right)$$

$$= n \sum_{r=1}^{n} F_r - \left(\sum_{r=1}^{1} F_r + \sum_{r=1}^{2} F_r + \cdots + \sum_{r=1}^{n-1} F_r \right) = n \sum_{r=1}^{n} F_r - \sum_{m=1}^{n-1} \left(\sum_{r=1}^{m} F_r \right)$$

$$= n(F_{n+2} - 1) - \sum_{m=1}^{n-1} (F_{m+2} - 1), \quad \text{by Property 4.5}$$

$$= nF_{n+2} - n - \sum_{m=1}^{n-1} F_{m+2} + \sum_{m=1}^{n-1} 1$$

$$= nF_{n+2} - n - \sum_{k=3}^{n+1} F_k + (n-1)$$

$$= nF_{n+2} - \left(\sum_{k=1}^{n+1} F_k - F_1 - F_2 \right) - 1$$

$$= nF_{n+2} - ((F_{n+3} - 1) - 2) - 1, \quad \text{also by Property 4.5}$$

$$= nF_{n+2} - F_{n+3} + 2.$$

19. **Proof (By Mathematical Induction):** For $n = 1$, we find that $\sum_{i=1}^{n}(-1)^{i+1}F_{i+1} = \sum_{i=1}^{1}(-1)^{i+1}F_{i+1} = (-1)^{1+1}F_{1+1} = F_2 = 1$, while $(-1)^{n-1}F_n = (-1)^{1-1}F_1 = 1$. So the result is true in this first case and this establishes the basis for our inductive argument. Now assume the result true for some $n = k \ (\geq 1)$—that is, assume that $\sum_{i=1}^{k}(-1)^{i+1}F_{i+1} = (-1)^{k-1}F_k$. Then for $n = k+1 \ (\geq 2)$, we have

$$\sum_{i=1}^{k+1}(-1)^{i+1}F_{i+1} = \sum_{i=1}^{k}(-1)^{i+1}F_{i+1} + (-1)^{(k+1)+1}F_{(k+1)+1}$$

$$= (-1)^{k-1}F_k + (-1)^{k+2}F_{k+2}$$

$$= (-1)^{k-1}F_k + (-1)^{k}F_{k+2}$$

$$= (-1)^{k}[F_{k+2} - F_k] = (-1)^{k}F_{k+1}.$$

Therefore, the result is true for all $n \geq 1$, by the Principle of Mathematical Induction.

21. **Proof:**

$$
\begin{aligned}
F_n^2 + F_{n+4}^2 &= (F_{n+2} - F_{n+1})^2 + (F_{n+2} + F_{n+3})^2 \\
&= F_{n+2}^2 - 2F_{n+2}F_{n+1} + F_{n+1}^2 + F_{n+2}^2 + 2F_{n+2}F_{n+3} + F_{n+3}^2 \\
&= F_{n+1}^2 + F_{n+3}^2 + 2F_{n+2}^2 + 2F_{n+2}(F_{n+3} - F_{n+1}) \\
&= F_{n+1}^2 + F_{n+3}^2 + 2F_{n+2}^2 + 2F_{n+2}F_{n+2} \\
&= F_{n+1}^2 + F_{n+3}^2 + 4F_{n+2}^2.
\end{aligned}
$$

23. **Proof:** For the even case, the result takes the form $\sum_{i=1}^{2n} F_i F_{i+1} = F_{2n+1}^2 - (1/2)[1 + (-1)^{2n}] = F_{2n+1}^2 - 1$. This is the result due to K. S. Rao, in the preceding exercise.

For the odd case, we have to show that for $n \geq 1$, $\sum_{i=1}^{2n-1} F_i F_{i+1} = F_{2n}^2$. The proof will be by Mathematical Induction. For $n = 1$, $\sum_{i=1}^{2(1)-1} F_i F_{i+1} = \sum_{i=1}^{1} F_i F_{i+1} = F_1 F_2 = 1 \cdot 1 = 1$, while $F_{2(1)}^2 = F_2^2 = 1^2 = 1$. Consequently, the result is true for this first case and this establishes the basis step for our proof by induction. Now assume the result true for some $n = k \, (\geq 1)$—that is, assume that $\sum_{i=1}^{2k-1} F_i F_{i+1} = F_{2k}^2$. Then for $n = k + 1 \, (\geq 2)$, we find that

$$
\begin{aligned}
\sum_{i=1}^{2(k+1)-1} F_i F_{i+1} &= \sum_{i=1}^{2k+1} F_i F_{i+1} \\
&= \sum_{i=1}^{2k-1} F_i F_{i+1} + F_{2k} F_{2k+1} + F_{2k+1} F_{2k+2} \\
&= F_{2k}^2 + F_{2k} F_{2k+1} + F_{2k+1} F_{2k+2} \\
&= F_{2k}(F_{2k} + F_{2k+1}) + F_{2k+1} F_{2k+2} \\
&= F_{2k} F_{2k+2} + F_{2k+1} F_{2k+2} \\
&= (F_{2k} + F_{2k+1}) F_{2k+2} \\
&= F_{2k+2}^2 = F_{2(k+1)}^2.
\end{aligned}
$$

Therefore, this result is true for all $n \geq 1$, by the Principle of Mathematical Induction.

25. **Proof (By the Alternative, or Strong, Form of the Principle of Mathematical Induction):** For $n = 0$, $F_{n+5} - 3F_n = F_5 - 3F_0 = 5 - 3 \cdot 0 = 5$. So the result is true for $n = 0$. For $n = 1$, $F_{n+5} - 3F_n = F_6 - 3F_1 = 8 - 3 \cdot 1 = 5$. So the result is also true for $n = 1$. These two results form the basis for this inductive argument. Now assume the result true for $n = 0, 1, 2, \ldots, k \, (\geq 1)$. That is, assume that $F_{n+5} - 3F_n$ is divisible by 5 for $n = 0, 1, 2, \ldots, k \, (\geq 1)$. For $n = k + 1 \, (\geq 2)$, we find that $F_{(k+1)+5} - 3F_{k+1} = [F_{k+5} + F_{(k-1)+5}] - 3[F_k + F_{k-1}] = [F_{k+5} - 3F_k] + [F_{(k-1)+5} - 3F_{(k-1)}]$. By the induction hypothesis, each of the summands $[F_{k+5} - 3F_k]$ and $[F_{(k-1)+5} - 3F_{(k-1)}]$ is divisible by 5, so their sum is divisible by 5. Consequently, the result is true for

all $n \geq 0$, by the Alternative, or Strong, form of the Principle of Mathematical Induction.

27. (a): $(F_3 + F_4 + F_5 + F_6) + F_4 = (2 + 3 + 5 + 8) + 3 = 18 + 3 = 21 = F_8$.

   (b): $(F_4 + F_5 + F_6 + F_7 + F_8) + F_5 = (3 + 5 + 8 + 13 + 21) + 5 = 50 + 5 = 55 = F_{10}$.

   (c):
   $$(F_n + F_{n+1} + F_{n+2} + \cdots + F_{n+m}) + F_{n+1}$$
   $$= \left[\sum_{r=1}^{n+m} F_r - \sum_{r=1}^{n-1} F_r \right] + F_{n+1}$$
   $$= \left[(F_{n+m+2} - 1) - (F_{(n-1)+2} - 1) \right] + F_{n+1}, \quad \text{using Property 4.5}$$
   $$= \left[(F_{n+m+2} - 1) - (F_{n+1} - 1) \right] + F_{n+1} = F_{n+m+2}.$$

   [*Note*: We could have gone directly from $(F_n + F_{n+1} + F_{n+2} + \cdots + F_{n+m}) + F_{n+1}$ to $(F_{n+m+2} - F_{n+1}) + F_{n+1}$, using the result in Exercise 26.]

Exercises for Chapter 5

1. Here it is better to solve part (c) first. Let a_n count the number of sequences of heads and tails until the $(n-1)$st and nth flips are the first two consecutive flips resulting in heads. Then there are (i) a_{n-1} possibilities if the first flip (by Tanya) was a tail and (ii) a_{n-2} possibilities if Tanya's first flip was a head, but Greta's first flip was a tail. Since these two cases have nothing in common and cover all of the possibilities, we have

$$a_n = a_{n-1} + a_{n-2}, \quad n \geq 3, \quad a_1 = 0, \quad a_2 = 1.$$

   Consequently, $a_n = F_{n-1}, n \geq 1$.
   For part (a), the answer is $F_6 (= 8)$. The answer for part (b) is $F_{11} (= 89)$.

3. (a) This is the same as the number of subsets of $\{1, 2, 3, \ldots, 13, 14\}$ that contain no consecutive integers. From Example 5.1, this number is $F_{14+2} = F_{16} = 987$.

   (b) This is the same as the number of subsets of $\{1, 2, 3, \ldots, n, n+1\}$ that contain no consecutive integers. This number is $F_{(n+1)+2} = F_{n+3}$.

   (c) Since $55 = F_{10} = F_{8+2}$, there are eight elements in U. Consequently, $U = \{31, 32, 33, \ldots, 37, 38\}$, where the largest element is 38.

   (d) Since $377 = F_{14} = F_{12+2}$, there are 12 elements in W. So $W = \{-4, -3, -2, \ldots, 6, 7\}$, where the smallest element is -4.

5. Suppose that n is even and $x_n = 1$. Then $x_1, x_2, \ldots, x_{n-2}, x_{n-1}$ can be any of the sequences counted by a_{n-1}. However, if $x_n = 0$, we have $x_{n-1} = 0$, and since $n - 1$ is odd, it then follows that $x_1, x_2, \ldots, x_{n-2}$ can be any of the sequences counted by a_{n-2}. Hence, $a_n = a_{n-1} + a_{n-2}$.

Meanwhile, if n is odd and $x_n = 1$, then $x_{n-1} = 0$ and $x_1, x_2, \ldots, x_{n-2}$ can be any of the sequences counted by a_{n-2}. However, if $x_n = 0$, then $x_1, x_2, \ldots, x_{n-2}, x_{n-1}$ can be any of the sequences counted by a_{n-1}. Once again we have $a_n = a_{n-1} + a_{n-2}$.

When $n = 1$, we have $a_1 = 2$ for the binary sequences 0 and 1. For $n = 2$, we have the binary sequences 0, 0; 0, 1; 1, 1; so $a_2 = 3$. Consequently, $a_n = F_{n+2}$, for $n \geq 1$.

7. (a) Keep the elements of A ordered as 1, 2, 3. Assign each subset S of A a binary string xyz of length 3, where

$$x = \begin{cases} 0, \text{ if } 1 \notin S \\ 1, \text{ if } 1 \in S \end{cases}, \quad y = \begin{cases} 0, \text{ if } 2 \notin S \\ 1, \text{ if } 2 \in S \end{cases}, \quad z = \begin{cases} 0, \text{ if } 3 \notin S \\ 1, \text{ if } 3 \in S \end{cases}.$$

When we compare two subsets S_i and S_{i+1} of A, we compare the individual components of their corresponding length 3 binary strings. Similar to the solution of the previous exercise, here the x_i's, y_i's, and z_i's are independent of each other, so the answer is F_{n+2}^3.

(b) F_{n+2}^m

Exercises for Chapter 6

1. (a) **Proof (By Mathematical Induction):** For $n = 1$, there is only one composition—namely, 1. So the number of compositions of 1 is $1 = 2^{1-1} = 2^{n-1}$, and the result is true in this first case. This establishes the basis case for our inductive proof. Next we assume the result is true for some $n = k \, (\geq 1)$—that is, we assume that the number of compositions of k is 2^{k-1}. Now, for $n = k + 1 \, (\geq 2)$, we obtain the compositions of $k + 1$ from the compositions of k (i) by appending "+1" to each of the 2^{k-1} compositions of k, and (ii) by adding 1 to the last summand of each of the 2^{k-1} compositions of k. Consequently, there are $2^{k-1} + 2^{k-1} = 2(2^{k-1}) = 2^k$ compositions of $k + 1$. Therefore, by the Principle of Mathematical Induction, it follows that the number of compositions of n is 2^{n-1}, for all $n \geq 1$.

(b) Suppose that n is even, with $n = 2k$, for $k \geq 1$. There is one palindrome with center summand n. For $1 \leq r \leq k - 1$, there are 2^{r-1} palindromes with center summand $n - 2r$. (One palindrome for each of the 2^{r-1} compositions of r.) In addition, there are 2^{k-1} palindromes of n with a plus sign as the center symbol. (One palindrome for each of the $2^{(n/2)-1} = 2^{k-1}$ compositions of $n/2 = k$.) So in total, the number of palindromes for n is $1 + (1 + 2^1 + 2^2 + 2^3 + \cdots + 2^{k-2} + 2^{k-1}) = 1 + (2^k - 1) = 2^k = 2^{n/2}$.

Turning to the case were n is odd, we find the following. For $n = 1$, there is just one palindrome—namely, 1. So suppose now that $n = 2k + 1$, for $k \geq 1$. As in the even case, there is one palindrome with center summand

n. For $1 \leq s \leq k$, there are 2^{s-1} palindromes of n with center summand $n - 2s$. (One palindrome for each of the 2^{s-1} compositions of s.) So the total number of palindromes for n is $1 + (1 + 2^1 + 2^2 + 2^3 + \cdots + 2^{k-2} + 2^{k-1}) = 1 + (2^k - 1) = 2^k = 2^{(n-1)/2}$.

The two cases can be combined into the one result $2^{\lfloor n/2 \rfloor}$, $n \geq 1$.

3. (a) Start with a composition of 24, where the only summands are 1's and 2's. Multiply each summand by 2. The result is a composition of 48, where the only summands are 2's and 4's. So from the result in Example 6.2, it follows that the number of compositions of 48, where the only summands are 2's and 4's, is $F_{24+1} = F_{25} (= 75,025)$.

   (b) Now start with a composition of 16, where the only summands are 1's and 2's. Multiply each summand by 3. The result is a composition of 48, where the only summands are 3's and 6's. So from the result in Example 6.2, it follows that the number of compositions of 48, where the only summands are 3's and 6's, is $F_{16+1} = F_{17} (= 1597)$.

5. (a) The answer here can be thought of as the number of compositions of 10 where the only summands are 1's and 2's. From Example 6.2, this is $F_{10+1} = F_{11} = 89$.

   (b) Due to independence, Anthony can ascend the eight staircases in $F_{11}^8 = 89^8 = 3,936,588,805,702,081$ ways.

7. The number of compositions of $n - 6 - 3 = n - 9$, where all the summands are odd, is $610 = F_{15}$. From Example 6.6, it follows that $F_{n-9} = F_{15}$, so $n = 24$.

Exercises for Chapter 7

1. **Proof:** Using Property 7.1 from Example 7.3,

$$F_n = F_{(n-m)+m} = F_m F_{(n-m)+1} + F_{m-1} F_{n-m} = F_m F_{n-m+1} + F_{m-1} F_{n-m}.$$

3. **Proof:** From the previous exercise, we have $F_{n+m} = F_{m+1} F_n + F_m F_{n-1}$. Continuing from this, we find that

$$\begin{aligned}
F_{m+1} F_n + F_m F_{n-1} &= F_{m+1}(F_{n+1} - F_{n-1}) + F_{n-1}(F_{m+1} - F_{m-1}) \\
&= F_{m+1} F_{n+1} - F_{m+1} F_{n-1} + F_{n-1} F_{m+1} - F_{n-1} F_{m-1} \\
&= F_{n+1} F_{m+1} - F_{n-1} F_{m-1}.
\end{aligned}$$

5. **Proof:** Using Property 7.1 from Example 7.3,

$$\begin{aligned}
F_{2n-1} &= F_{n+(n-1)} = F_{n-1} F_{n+1} + F_{n-2} F_n \\
&= F_{n+1}(F_n - F_{n-2}) + F_{n-2} F_n \\
&= F_n F_{n+1} + F_{n-2}(F_n - F_{n+1}) \\
&= F_n F_{n+1} - F_{n-2}(F_{n+1} - F_n) \\
&= F_n F_{n+1} - F_{n-1} F_{n-2}.
\end{aligned}$$

7. **Proof:** This follows from Property 7.2 (in Example 7.4) because $F_4 = 3$.

9. **Proof:** Since $F_6 = 8$, it follows from Property 7.2 (in Example 7.4) that F_n is a multiple of 8. Consequently, F_n is a multiple of 4.

11. **Proof:** Yes! If not, then $n = n_1 n_2$ where $n_1 \geq 3$ or $n_2 \geq 3$. Otherwise, $1 \leq n_1 \leq 2$ and $1 \leq n_2 \leq 2$, so $1 \leq n \leq 4$. Assume that $n_1 \geq 3$. Then $F_{n_1} \geq F_3 = 2$, so $F_n \geq F_5 \ (= 5)$ and F_n is divisible by 2 $(= F_3)$ by Property 7.2 (in Example 7.4). This contradicts F_n being prime, so it now follows that n is prime.

13. **Proof:** This is a special case of the preceding exercise. Let $p = q = r = n$.

15. **Proof (By Mathematical Induction):** For $n = 0$, $F_{60n+r} - F_r = F_r - F_r = 0$, which is divisible by 10. Consequently, the result is true for $n = 0$ and this establishes the basis step for our inductive proof. Assuming the result true for some arbitrary, but fixed, value $n = k \ (\geq 0)$, we have $F_{60k+r} - F_r$ divisible by 10. Now for $n = k + 1 \ (\geq 1)$, we find that

$$
\begin{aligned}
F_{60(k+1)+r} - F_r &= F_{(60k+r)+60} - F_r \\
&= \left(F_{(60k+r)-1} F_{60} + F_{60k+r} F_{60+1} \right) - F_r, \quad \text{by Property 7.1} \\
&= F_{(60k+r)-1} F_{60} + F_{60k+r}((F_{61} - 1) + 1) - F_r \\
&= F_{(60k+r)-1} F_{60} + F_{60k+r}(F_{61} - 1) + (F_{60k+r} - F_r).
\end{aligned}
$$

Since the units digit of F_{60} is 0, F_{60} and, consequently, $F_{(60k+r)-1} F_{60}$ are both divisible by 10. The units digit of F_{61} is 1, so $(F_{61} - 1)$ and, consequently, $F_{60k+r}(F_{61} - 1)$ are both divisible by 10. Finally, from the induction hypothesis we know that $(F_{60k+r} - F_r)$ is divisible by 10. Since all three summands are divisible by 10, it follows that $F_{60(k+1)+r} - F_r$ is divisible by 10. Therefore, the result follows for all $n \geq 0$ and $r \geq 0$, with r fixed, by the Principle of Mathematical Induction.

Exercise for Chapter 9

1. For example, when $n = 3$, consider the five paths shown in Fig. 9.1(g). All five emerge through Face 4, so the last reflection occurs at either Face 1 or Face 3.

   (i) Consider the three paths where the last reflection occurs at Face 1. Here we can take any of the paths for $n = 2$ and, instead of letting the ray emerge through Face 1, we add a reflection at that face and have the ray emerge through Face 4.

   (ii) For the other two paths emerging through Face 4, the last reflection occurs at Face 3. The previous reflection occurs at Face 4—it cannot occur at Face 2, as can be seen from Fig. 9.1(c). (A reflection at Face 4 can be preceded by a reflection at either Face 1 or Face 3.) So now we start with any path for $n = 1$, but before we let the ray emerge through Face 4, we add a reflection at Face 4 followed by a reflection at Face 3.

   To generalize the preceding observations, let n be odd with $n \geq 3$. Then the rays emerge through Face 4 and the last reflection occurs at Face 1 or Face 3.

(i) For the paths where the last reflection is at Face 1, the previous reflection occurs at Face 2 or Face 4. In this case, take any of the s_{n-1} paths for the (even) case $n-1$ and add a reflection at Face 1 before allowing the ray to emerge through Face 4.

(ii) Otherwise, the last reflection occurs at Face 3. (The previous reflection occurs at Face 4, since it cannot occur at Face 2—as we see in Fig. 9.1(c).) Now for any of s_{n-2} paths for the (odd) case $n-2$, we add a reflection at Face 4, then one at Face 3 before allowing the ray to emerge through Face 4.

In this way we arrive at the result

$$s_n = s_{n-1} + s_{n-2}, \quad n \geq 2, \ n \text{ odd}.$$

Exercises for Chapter 10

1. Substituting $a_n = Ar^n$, with $A \neq 0$ and $r \neq 0$, into the recurrence relation, we find that $Ar^n - 10Ar^{n-1} + 21Ar^{n-2} = 0$. Upon dividing by A and r^{n-2}, we arrive at the characteristic equation $r^2 - 10r + 21 = 0$. As $r^2 - 10r + 21 = (r-3)(r-7)$, the characteristic roots are 3 and 7. So the solution has the form $c_1 3^n + c_2 7^n$, $n \geq 0$. From $1 = a_0 = c_1 + c_2$ and $1 = a_1 = 3c_1 + 7c_2$, it follows that $c_1 = 3/2$ and $c_2 = -(1/2)$. Consequently, $a_n = (3/2) 3^n - (1/2) 7^n$, $n \geq 0$.

3. For $n \geq 3$, consider the nth (last) column of the $3 \times n$ chessboard. If that column is covered with three square 1×1 tiles, then the preceding $(n-1)$ columns of the chessboard can be tiled in a_{n-1} ways. There are two other ways in which the nth (last) column can be covered—namely, (i) place two square 1×1 tiles in the top row (and last two columns) and a square 2×2 tile below them, or (ii) place two square 1×1 tiles in the bottom row (and last two columns) and a square 2×2 tile above them. In either case, the preceding $(n-2)$ columns can be covered in a_{n-2} ways. This leads us to the following recurrence relation with initial conditions:

$$a_n = a_{n-1} + 2a_{n-2}, \quad a_1 = 1, \ a_2 = 3, \ n \geq 3.$$

Substituting $a_n = Ar^n$, with $A \neq 0$ and $r \neq 0$, into the recurrence relation, we find that $Ar^n - Ar^{n-1} - 2Ar^{n-2} = 0$. Upon dividing by A and r^{n-2}, we arrive at the characteristic equation $r^2 - r - 2 = 0$. Since $r^2 - r - 2 = (r-2)(r+1)$, the characteristic roots are 2 and -1. So the solution has the form $c_1 2^n + c_2 (-1)^n$, $n \geq 1$. From $1 = a_1 = 2c_1 - c_2$ and $3 = a_2 = 4c_1 + c_2$, it follows that $c_1 = 2/3$ and $c_2 = 1/3$, so $a_n = (2/3)2^n + (1/3)(-1)^n = (1/3)[2^{n+1} - (-1)^{n+1}]$, $n \geq 1$.

(One can also solve this exercise by using the recurrence relation with initial conditions given by

$$a_n = a_{n-1} + 2a_{n-2}, \quad a_0 = 1, \ a_1 = 1, \ n \geq 2.)$$

(The sequence of numbers $a_0 = 1$, $a_1 = 1$, $a_2 = 3$, $a_3 = 5$, $a_4 = 11$, ... is known as the *Jacobsthal* numbers.)

5. **Proof:**

(a) (i) $\alpha + \beta = ((1 + \sqrt{5})/2) + ((1 - \sqrt{5})/2) = (1 + \sqrt{5} + 1 - \sqrt{5})/2 = 2/2 = 1$.

(ii) $\alpha - \beta = ((1 + \sqrt{5})/2) - ((1 - \sqrt{5})/2) = (1 + \sqrt{5} - 1 + \sqrt{5})/2 = (2\sqrt{5})/2 = \sqrt{5}$.

(b) (i) $\alpha^2 + \beta^2 = ((1 + \sqrt{5})/2)^2 + ((1 - \sqrt{5})/2)^2 = ((6 + 2\sqrt{5})/4) + ((6 - 2\sqrt{5})/4) = (6 + 2\sqrt{5} + 6 - 2\sqrt{5})/4 = 12/4 = 3$.

(ii) From part (a) we see that $\alpha^2 - \beta^2 = (\alpha + \beta)(\alpha - \beta) = 1 \cdot \sqrt{5} = \sqrt{5}$.

(c) (i) $\alpha^2 = \alpha + 1 \Rightarrow \alpha^3 = \alpha^2 + \alpha$ and $\beta^2 = \beta + 1 \Rightarrow \beta^3 = \beta^2 + \beta$. Consequently, from parts (a) (i) and (b) (i), it follows that $\alpha^3 + \beta^3 = (\alpha^2 + \alpha) + (\beta^2 + \beta) = (\alpha^2 + \beta^2) + (\alpha + \beta) = 3 + 1 = 4$.

(ii) $\alpha^2 = \alpha + 1 \Rightarrow \alpha^3 = \alpha^2 + \alpha$ and $\beta^2 = \beta + 1 \Rightarrow \beta^3 = \beta^2 + \beta$. Consequently, from parts (a) (ii) and (b) (ii), it follows that $\alpha^3 - \beta^3 = (\alpha^2 + \alpha) - (\beta^2 + \beta) = (\alpha^2 - \beta^2) + (\alpha - \beta) = \sqrt{5} + \sqrt{5} = 2\sqrt{5}$.

7. **Proof:**

(i) $\alpha^2 = \alpha + 1 \Rightarrow \alpha^2/\alpha^{n+2} = \alpha/\alpha^{n+2} + 1/\alpha^{n+2} \Rightarrow 1/\alpha^n = 1/\alpha^{n+1} + 1/\alpha^{n+2}$.

(ii) $\beta^2 = \beta + 1 \Rightarrow \beta^2/\beta^{n+2} = \beta/\beta^{n+2} + 1/\beta^{n+2} \Rightarrow 1/\beta^n = 1/\beta^{n+1} + 1/\beta^{n+2}$.

9. **Proof:** This is a geometric series with ratio $1/\alpha^2$. Since $\left|1/\alpha^2\right| = (1/\alpha^2) < 1$, the series converges to

$$\left(\frac{1}{\alpha^2}\right)\left[\frac{1}{1 - \frac{1}{\alpha^2}}\right] = \left(\frac{1}{\alpha^2}\right)\left(\frac{\alpha^2}{\alpha^2 - 1}\right) = \frac{1}{\alpha^2 - 1} = \frac{1}{\alpha}$$

11. **Proof:** $\sqrt{3 - \beta} = \sqrt{3 - ((1 - \sqrt{5})/2)} = \sqrt{(6 - (1 - \sqrt{5}))/2} = \sqrt{(5 + \sqrt{5})/2} = \sqrt{(10 + 2\sqrt{5})/4} = (1/2)\sqrt{10 + 2\sqrt{5}}$.

13. Dividing the numerator and denominator of $2n/(n + 1 + \sqrt{5n^2 - 2n + 1})$ by n, we have

$$\lim_{n\to\infty} \frac{2n}{n + 1 + \sqrt{5n^2 - 2n + 1}} = \lim_{n\to\infty} \frac{2}{1 + \frac{1}{n} + \sqrt{5 - \frac{2}{n} + \frac{1}{n^2}}} = \frac{2}{1 + \sqrt{5}} = \frac{1}{\alpha}.$$

15. **Proof:** Since $\beta(\alpha - \beta) = \alpha\beta - \beta^2 = -1 - (\beta + 1) = -2 - \beta = -(2 + \beta)$, it follows that $\beta/(2 + \beta) = -1/(\alpha - \beta)$.

17. When $1 + (1/x) = x$, then $x + 1 = x^2$, or $x^2 - x - 1 = 0$. So $x = (1 \pm \sqrt{5})/2$ and the points of intersection are $((1 + \sqrt{5})/2, (1 + \sqrt{5})/2) = (\alpha, \alpha)$ and $((1 - \sqrt{5})/2, (1 - \sqrt{5})/2) = (\beta, \beta)$.

19. $\sum_{k=0}^{n} \binom{n}{k} \alpha^{3k-2n} = \alpha^{-2n} \sum_{k=0}^{n} \alpha^{3k} = \alpha^{-2n} \sum_{k=0}^{n} (\alpha^3)^k 1^{n-k} = \alpha^{-2n} (1 + \alpha^3)^n$.

Now $\alpha^2 = \alpha + 1 \Rightarrow \alpha^3 = \alpha^2 + \alpha = (\alpha + 1) + \alpha = 2\alpha + 1 \Rightarrow \alpha^3 + 1 = 2\alpha + 2 = 2(\alpha + 1) = 2(\alpha^2) = 2\alpha^2$. Consequently,

$$\sum_{k=0}^{n} \binom{n}{k} \alpha^{3k-2n} = \alpha^{-2n} (2\alpha^2)^n = 2^n \alpha^{-2n} \alpha^{2n} = 2^n.$$

21. (a) (i) $a/3 = 12/a \Rightarrow a^2 = 36 \Rightarrow a = \pm 6$. Since $a > 0$, the mean proportion for 3 and 12 is 6.

(ii) $a/4 = 25/a \Rightarrow a^2 = 100 \Rightarrow a = \pm 10$. Since $a > 0$, the mean proportion for 4 and 25 is 10.

(iii) $a/5 = 10/a \Rightarrow a^2 = 50 \Rightarrow a = \pm\sqrt{50}$. Since $a > 0$, the mean proportion for 5 and 10 is $\sqrt{50} = 5\sqrt{2}$.

(b) Since r is the mean proportion for $r + s$ and s, it follows that $r/(r + s) = s/r$ or $r^2 = s(r + s) = sr + s^2$. Dividing through by s^2, we arrive at the equation

$$\frac{r^2}{s^2} - \frac{sr}{s^2} - \frac{s^2}{s^2} = 0 \quad \text{or} \quad \left(\frac{r}{s}\right)^2 - \left(\frac{r}{s}\right) - 1 = 0.$$

Solving the quadratic equation $(r/s)^2 - (r/s) - 1 = 0$ for r/s, we find that

$$\frac{r}{s} = \frac{-(-1) \pm \sqrt{(-1)^2 - 4(1)(-1)}}{2(1)} = \frac{1 \pm \sqrt{5}}{2}.$$

Since $r > 0$ and $s > 0$, we find that

$$\frac{r}{s} = \frac{1 + \sqrt{5}}{2} = \alpha, \quad \text{the golden ratio.}$$

23. We realize that $r \neq 1$, for if $r = 1$, then we have an equilateral triangle, which cannot be a right triangle.

(a) If s is the length of the hypotenuse, then $r < 1$ and, in this case,

$$s^2 = (rs)^2 + (r^2 s)^2,$$

so

$$r^4 + r^2 - 1 = 0, \quad \text{a quadratic in } r^2,$$

and

$$r^2 = \frac{-1 \pm \sqrt{(1)^2 - 4(1)(-1)}}{2(1)} = \frac{-1 \pm \sqrt{5}}{2}.$$

Since $r^2 > 0$, it follows that $r^2 = (-1 + \sqrt{5})/2 = -((1 - \sqrt{5})/2) = -\beta = 1/\alpha$ and $r = \sqrt{1/\alpha} = 1/\sqrt{\alpha}$.

(b) If r^2s is the length of the hypotenuse, then $r > 1$ and in this case

$$(r^2s)^2 = s^2 + (rs)^2,$$

so

$$r^4 - r^2 - 1 = 0, \quad \text{a quadratic in } r^2,$$

and

$$r^2 = \frac{-(-1) \pm \sqrt{(-1)^2 - 4(1)(-1)}}{2(1)} = \frac{1 \pm \sqrt{5}}{2}.$$

Since $r^2 > 0$, it follows that $r^2 = (1 + \sqrt{5})/2 = \alpha$ and $r = \sqrt{\alpha}$.

25. Since the three right triangles all have the same area, it follows that

$$\frac{1}{2} bc = \frac{1}{2} a(c + d) = \frac{1}{2} (a + b)d$$

or

$$bc = a(c + d) = (a + b)d$$

Then $a(c + d) = (a + b)d \Rightarrow ac + ad = ad + bd \Rightarrow ac = bd \Rightarrow$
$b/a = c/d$. Consequently,

$$\frac{b}{a}c = d + c \Rightarrow \frac{c}{d}c = d + c \Rightarrow c^2 = d^2 + cd$$

$$\Rightarrow \frac{c^2}{d^2} = 1 + \frac{c}{d} \Rightarrow \left(\frac{c}{d}\right)^2 - \frac{c}{d} - 1 = 0$$

So $c/d > 0$ and is a solution of $x^2 - x - 1 = 0$. Therefore, $c/d = b/a = \alpha = (1 + \sqrt{5})/2$.

27. **Proof:** Using the Binet form for the Fibonacci numbers, we find that

$$F_n F_{n+1} - F_{n-1} F_{n-2}$$
$$= \frac{\alpha^n - \beta^n}{\alpha - \beta} \cdot \frac{\alpha^{n+1} - \beta^{n+1}}{\alpha - \beta} - \frac{\alpha^{n-1} - \beta^{n-1}}{\alpha - \beta} \cdot \frac{\alpha^{n-2} - \beta^{n-2}}{\alpha - \beta}$$
$$= \frac{\left[\alpha^{2n+1} - \alpha(-1)^n - \beta(-1)^n + \beta^{2n+1}\right] - \left[\alpha^{2n-3} - \alpha(-1)^{n-2} - \beta(-1)^{n-2} + \beta^{2n-3}\right]}{(\alpha - \beta)^2},$$

because $\alpha\beta = -1$,

$$= \frac{\left(\alpha^{2n+1} - \alpha^{2n-3}\right) + \left(\beta^{2n+1} - \beta^{2n-3}\right)}{(\alpha - \beta)^2}$$

$$= \frac{\alpha^{2n-1}(\alpha^2 - \alpha^{-2}) + \beta^{2n-1}(\beta^2 - \beta^{-2})}{(\alpha - \beta)^2}$$

$$= \frac{\alpha^{2n-1}(\alpha - \beta) + \beta^{2n-1}(\beta - \alpha)}{(\alpha - \beta)^2},$$

because $\alpha^2 - \alpha^{-2} = \alpha^2 - (-\beta)^2 = \alpha^2 - \beta^2 = (\alpha + \beta)(\alpha - \beta) = \alpha - \beta$ and $\beta^2 - \beta^{-2} = \beta^2 - (-\alpha)^2 = \beta^2 - \alpha^2 = (\beta + \alpha)(\beta - \alpha) = (\beta - \alpha)$, since $\alpha + \beta = 1$. But then

$$\frac{\alpha^{2n-1}(\alpha - \beta) + \beta^{2n-1}(\beta - \alpha)}{(\alpha - \beta)^2}$$

$$= \frac{\alpha^{2n-1}(\alpha - \beta) - \beta^{2n-1}(\alpha - \beta)}{(\alpha - \beta)^2}$$

$$= \frac{\alpha^{2n-1} - \beta^{2n-1}}{(\alpha - \beta)} = F_{2n-1}.$$

29. **Proof:** Using the Binet form for F_n and the fact that $\alpha\beta = -1$, we find that

$$F_{n+1}^2 + F_n^2 = \left(\frac{\alpha^{n+1} - \beta^{n+1}}{\alpha - \beta}\right)^2 + \left(\frac{\alpha^n - \beta^n}{\alpha - \beta}\right)^2$$

$$= \frac{\alpha^{2n+2} - 2(-1)^{n+1} + \beta^{2n+2} + \alpha^{2n} - 2(-1)^n + \beta^{2n}}{(\alpha - \beta)^2}$$

$$= \frac{\alpha^{2n+1}(\alpha + (1/\alpha)) + \beta^{2n+1}(\beta + (1/\beta))}{(\alpha - \beta)^2}$$

$$= \frac{\alpha^{2n+1}(\alpha - \beta) + \beta^{2n+1}(\beta - \alpha)}{(\alpha - \beta)^2}$$

$$= \frac{\alpha^{2n+1} - \beta^{2n+1}}{\alpha - \beta} = F_{2n+1}.$$

31. **Proof (By Mathematical Induction):** For $n = 1$, $\sum_{i=1}^{1} F_{4i-2} = F_{4(1)-2} = F_2 = 1$, while $F_{2(1)}^2 = F_2^2 = 1^2 = 1$. So, the result is true for this first case and this establishes the basis step for the inductive proof.

Assume the result true for some $n = k$ (≥ 1)—that is, assume that $\sum_{i=1}^{k} F_{4i-2} = F_{2k}^2$. For $n = k + 1$ (≥ 2), we find that

$$\sum_{i=1}^{k+1} F_{4i-2} = \sum_{i=1}^{k} F_{4i-2} + F_{4(k+1)-2}$$

$$= F_{2k}^2 + F_{4k+2} = F_{2k}^2 + F_{2(2k+1)}$$

$$= F_{2k}^2 + \left[F_{2k+2}^2 - F_{2k}^2 \right], \quad \text{by the result in part (a)}$$

of the previous exercise

$$= F_{2k+2}^2 = F_{2(k+1)}^2.$$

Consequently, the result follows for all $n \geq 1$ by the Principle of Mathematical Induction.

33. **Proof:** Using the Binet form for the Fibonacci numbers and the fact that $\alpha\beta = -1$, we find that

$$F_n^2 + F_{n+2k+1}^2 = \left(\frac{\alpha^n - \beta^n}{\alpha - \beta} \right)^2 + \left(\frac{\alpha^{n+2k+1} - \beta^{n+2k+1}}{\alpha - \beta} \right)^2$$

$$= \frac{\alpha^{2n} + 2(-1)^n + \beta^{2n} + \alpha^{2n+4k+2} + 2(-1)^{n+2k+1} + \beta^{2n+4k+2}}{(\alpha - \beta)^2}$$

$$= \frac{\alpha^{2n} + \beta^{2n} + \alpha^{2n+4k+2} + \beta^{2n+4k+2}}{(\alpha - \beta)^2},$$

since $(-1)^{n+2k+1} = (-1)^{n+1}$ and, consequently, $2(-1)^n + 2(-1)^{n+2k+1} = 2(-1)^n + 2(-1)^{n+1} = 2(-1)^n[1 - 1] = 0$.

In addition,

$$F_{2k+1} F_{2n+2k+1} = \frac{\alpha^{2k+1} - \beta^{2k+1}}{\alpha - \beta} \frac{\alpha^{2n+2k+1} - \beta^{2n+2k+1}}{\alpha - \beta}$$

$$= \frac{\alpha^{2n+4k+2} - \alpha^{2n}(-1)^{2k+1} - \beta^{2n}(-1)^{2k+1} + \beta^{2n+4k+2}}{(\alpha - \beta)^2}$$

$$= \frac{\alpha^{2n} + \beta^{2n} + \alpha^{2n+4k+2} + \beta^{2n+4k+2}}{(\alpha - \beta)^2}.$$

Alternate Proof: This result also follows from the previous exercise. In that exercise, replace m by $n + k$ and n by k.

35. **Proof:** Since

$$\lim_{n \to \infty} \frac{(F_{n+1}/2^{n+1})}{(F_n/2^n)} = \frac{1}{2} \lim_{n \to \infty} \frac{F_{n+1}}{F_n} = \frac{\alpha}{2} < 1,$$

this series converges by the Ratio Test. In particular,

$$\sum_{i=0}^{\infty} \frac{F_i}{2^i} = \frac{1}{\alpha - \beta} \left[\sum_{i=0}^{\infty} \frac{\alpha^i}{2^i} - \sum_{i=0}^{\infty} \frac{\beta^i}{2^i} \right]$$

$$= \frac{1}{\alpha - \beta} \left[\frac{1}{1 - (\alpha/2)} - \frac{1}{1 - (\beta/2)} \right]$$

$$= \frac{1}{\alpha - \beta} \left[\frac{2}{2 - \alpha} - \frac{2}{2 - \beta} \right]$$

$$= \frac{1}{\alpha - \beta} \left[\frac{2(2 - \beta) - 2(2 - \alpha)}{(2 - \alpha)(2 - \beta)} \right]$$

$$= \frac{1}{\alpha - \beta} \left[\frac{4 - 2\beta - 4 + 2\alpha}{4 - 2\alpha - 2\beta + \alpha\beta} \right]$$

$$= \frac{1}{\alpha - \beta} \left[\frac{2\alpha - 2\beta}{4 - 2\alpha - 2\beta - 1} \right]$$

$$= \frac{1}{\sqrt{5}} \left[\frac{2\sqrt{5}}{3 - 2} \right] \quad \text{(Since } \alpha + \beta = 1.)$$

$$= 2.$$

37. **Proof:** Using the Binet form for the Fibonacci numbers, the Binomial theorem, and the facts that $\alpha^2 = \alpha + 1$ and $\beta^2 = \beta + 1$, we find that $\sum_{i=0}^{n}(-1)^{n-i}\binom{n}{i}F_{2i} = (1/\sqrt{5})\sum_{i=0}^{n}(-1)^{n-i}\binom{n}{i}(\alpha^{2i} - \beta^{2i}) = (1/\sqrt{5})\sum_{i=0}^{n}(-1)^{n-i}\binom{n}{i}\alpha^{2i} - (1/\sqrt{5})\sum_{i=0}^{n}(-1)^{n-i}\binom{n}{i}\beta^{2i} = (1/\sqrt{5})(\alpha^2 - 1)^n - (1/\sqrt{5})(\beta^2 - 1)^n = (1/\sqrt{5})(\alpha)^n - (1/\sqrt{5})(\beta)^n = (1/\sqrt{5})(\alpha^n - \beta^n) = (\alpha^n - \beta^n)/(\alpha - \beta) = F_n$.

39. **Proof:** Using the Binet form for the Fibonacci numbers, the Binomial theorem, and the facts that $\alpha^2 = \alpha + 1$ and $\beta^2 = \beta + 1$, we find that $\sum_{i=0}^{n}\binom{n}{i}F_{i+j} = \alpha^j(1/\sqrt{5})\sum_{i=0}^{n}\binom{n}{i}\alpha^i - \beta^j(1/\sqrt{5})\sum_{i=0}^{n}\binom{n}{i}\beta^i = \alpha^j(1/\sqrt{5})(1 + \alpha)^n - \beta^j(1/\sqrt{5})(1 + \beta)^n = \alpha^j(1/\sqrt{5})(\alpha^2)^n - \beta^j(1/\sqrt{5})(\beta^2)^n = (1/\sqrt{5})(\alpha^{2n+j} - \beta^{2n+j}) = F_{2n+j}$.

41. **Proof:** Using the Binet form for the Fibonacci numbers, we have

$$\alpha'^m F_{n-m+1} + \alpha'^{m-1} F_{n-m}$$

$$= \alpha'^m \left[\frac{\alpha^{n-m+1} - \beta^{n-m+1}}{\alpha - \beta} \right] + \alpha'^{m-1} \left[\frac{\alpha^{n-m} - \beta^{n-m}}{\alpha - \beta} \right]$$

$$= \frac{\alpha^{n+1} - \alpha'^m \beta^{n-m+1} + \alpha^{n-1} - \alpha'^{m-1}\beta^{n-m}}{\alpha - \beta}$$

$$= \frac{\alpha^{n+1} - \alpha'^m \beta^{n-m+1} + \alpha^{n-1} - \alpha'^{m-1}\beta^{n-(m-1)-1}}{\alpha - \beta}$$

$$= \frac{\alpha^{n+1} + \alpha^{n-1} - \beta^{n+1}(\alpha/\beta)^m - \beta^{n-1}(\alpha/\beta)^{m-1}}{\alpha - \beta}$$

$$= \frac{\alpha^{n+1} + \alpha^{n-1} - \beta^{n+1}\left(-1/\beta^2\right)^m - \beta^{n-1}\left(-1/\beta^2\right)^{m-1}}{\alpha - \beta}$$

$$= \frac{\alpha^{n+1} + \alpha^{n-1} - \beta^{n+1}\left(1/\beta^{2m}\right)(-1)^m - \beta^{n-1}\left(1/\beta^{2(m-1)}\right)(-1)^{m-1}}{\alpha - \beta}$$

$$= \frac{\alpha^{n+1} + \alpha^{n-1} - \beta^{n+1-2m}(-1)^m - \beta^{n-1-2(m-1)}(-1)^{m-1}}{\alpha - \beta}$$

$$= \frac{\alpha^{n+1} + \alpha^{n-1} - \beta^{n+1-2m}(-1)^m - \beta^{n+1-2m}(-1)^{m-1}}{\alpha - \beta}$$

$$= \frac{\alpha^{n+1} + \alpha^{n-1} - \beta^{n+1-2m}[(-1)^m + (-1)^{m-1}]}{\alpha - \beta}$$

$$= \frac{\alpha^{n+1} + \alpha^{n-1} - \beta^{n+1-2m}[0]}{\alpha - \beta} = \frac{\alpha^{n+1} + \alpha^{n-1}}{\alpha - \beta}$$

$$= \frac{\alpha^n(\alpha + (1/\alpha))}{\alpha - \beta} = \frac{\alpha^n(\alpha - \beta)}{\alpha - \beta} = \alpha^n.$$

43. **Proof:**

(a) $AC = \alpha BC$ and $AC = AD + DC = \alpha DC + DC = (\alpha+1)DC = \alpha^2 DC$, because $\alpha^2 = \alpha+1$. Consequently, $\alpha BC = AC = \alpha^2 DC$, so $BC = \alpha DC$ and $BC/DC = \alpha$.

(b) We have $\angle ACB$ (of $\triangle ABC$) $= \angle BCD$ (of $\triangle CBD$) and $AB/BC = \alpha = BC/CD$ [from part (a)]. So the triangles ABC and CBD are similar— because two pairs of corresponding sides are in proportion and the included angles are the same.

(c) Since $\triangle ABC$ is isosceles and $\triangle CBD$ is similar to $\triangle ABC$, it follows that $\triangle CBD$ is isosceles with $BD = BC = \alpha DC$ and $\angle BDC = \angle BCD$. With $AD = \alpha DC$, we have $AD = BD$. So $\triangle ADB$ is also isosceles with $\angle DAB = \angle DBA$. Consequently, $\angle DBC = 180° - 2(\angle BCD) = 180° - (\angle BCD + \angle ABC) = \angle DAB = \angle DBA$, and BD is the bisector of $\angle ABC$.

45. From Exercises 43 and 44, we know that $\angle DAB = \angle ABD = \angle DBC$ and $\angle DCB = \angle ABC$, with the measure of $\angle ABC$ equal to twice the measure of $\angle DBC$. So $180° = 5\times$ the measure of $\angle DAB = 5\times$ the measure of $\angle A$. Consequently, the measure of $\angle A$ is $180°/5 = 36°$.

47. Draw the perpendicular bisector SE of RT, with E the point where SE intersects the circle. Since $\triangle RST$ is equilateral (hence, isosceles), SE bisects $\angle BSC$. So the measure of arc ET is $2 \cdot 30° = 60°$ and the measure of arc TS is twice the measure of $\angle SRT$ or $2 \cdot 60° = 120°$. Consequently, SE is a diameter in the circle.

Since B is the midpoint of SR and C is the midpoint of ST, $\triangle SBC$ is similar to $\triangle SRT$, and it follows that $BC = (1/2)\, RT$ and that BC is parallel to RT. So

SE is perpendicular to BC, and SE bisects BC for $\triangle SBC$ is equilateral (hence, isosceles).

Let P be the point where SE meets BC, so that $BP = PC$. Furthermore, since the chords AD and SE are perpendicular and SE is a diameter, SE bisects AD. Therefore, $AP = PD$ and $AB = CD$.

Let $x = AB$ and $y = BC$. Then $CD = x$ and $SC = CT = BC = y$. Now consider the intersecting chords ST and AD. Here we have $AC \cdot CD = SC \cdot CT$ or $(x + y) \cdot x = y \cdot y$, so $x^2 + xy = y^2$ or $1 + (y/x) = (y/x)^2$. Solving this quadratic equation [in (y/x)], we learn that $BC/AB = y/x = \alpha$.

49. **Proof:** Since

$$\lim_{i \to \infty} \left| \left(\frac{F_{i+1}}{3^{i+2}} \right) \Big/ \left(\frac{F_i}{3^{i+1}} \right) \right| = \lim_{i \to \infty} \left| \left(\frac{1}{3} \right) \frac{F_{i+1}}{F_i} \right| = \left(\frac{1}{3} \right) \lim_{n \to \infty} \frac{F_{i+1}}{F_i} = \frac{\alpha}{3} < 1,$$

this series converges by the Ratio Test. Using the Binet form for the Fibonacci numbers, we have

$$\sum_{i=0}^{\infty} \frac{F_i}{3^{i+1}} = \sum_{i=0}^{\infty} \frac{1}{3^{i+1}} \frac{\alpha^i - \beta^i}{\alpha - \beta} = \frac{1}{3\sqrt{5}} \left(\sum_{i=0}^{\infty} \left(\frac{\alpha}{3} \right)^i - \sum_{i=0}^{\infty} \left(\frac{\beta}{3} \right)^i \right)$$

$$= \frac{1}{3\sqrt{5}} \left(\frac{1}{1 - (\alpha/3)} - \frac{1}{1 - (\beta/3)} \right), \quad \text{since} \quad \left| \frac{\alpha}{3} \right| < 1 \text{ and } \left| \frac{\beta}{3} \right| < 1$$

$$= \frac{1}{3\sqrt{5}} \left(\frac{3}{3 - \alpha} - \frac{3}{3 - \beta} \right) = \frac{1}{\sqrt{5}} \left(\frac{(3 - \beta) - (3 - \alpha)}{(3 - \alpha)(3 - \beta)} \right)$$

$$= \frac{1}{\sqrt{5}} \left(\frac{\alpha - \beta}{(3 - \alpha)(3 - \beta)} \right) = \frac{1}{(3 - \alpha)(3 - \beta)} = \frac{1}{9 - 3(\alpha + \beta) + \alpha\beta}$$

$$= \frac{1}{9 - 3 - 1}, \quad \text{since } \alpha + \beta = 1 \text{ and } \alpha\beta = -1$$

$$= \frac{1}{5}.$$

51. (a)
$$F_{-n} = \frac{\alpha^{-n} - \beta^{-n}}{\alpha - \beta} = \frac{(\alpha^{-1})^n - (\beta^{-1})^n}{\alpha - \beta}$$

$$= \frac{(-\beta)^n - (-\alpha)^n}{\alpha - \beta}, \quad \text{since } \alpha\beta = -1$$

$$= \frac{(-1)^n \beta^n - (-1)^n \alpha^n}{\alpha - \beta} = (-1)^n \frac{\beta^n - \alpha^n}{\alpha - \beta}$$

$$= (-1)^{n+1} \frac{\alpha^n - \beta^n}{\alpha - \beta} = (-1)^{n+1} F_n$$

(b) $L_{-n} = \alpha^{-n} + \beta^{-n} = (\alpha^{-1})^n + (\beta^{-1})^n = (-\beta)^n + (-\alpha)^n = (-1)^n \beta^n + (-1)^n \alpha^n = (-1)^n(\alpha^n + \beta^n) = (-1)^n L_n$.

(c) $\alpha F_{-n} + F_{-(n+1)} = \alpha[(-1)^{n+1} F_n] + (-1)^{n+2} F_{n+1} = (-1)^{n+1}(\alpha F_n - F_{n+1}) = \alpha^{-n}$, the final equality following from Exercise 48.

Exercises for Chapter 11

1. (a) $\sin\frac{\pi}{5} = \sqrt{1 - \cos^2\frac{\pi}{5}} = \sqrt{1 - \frac{\alpha^2}{4}} = \frac{1}{2}\sqrt{4 - \alpha^2}$.

   (b) $\cos^2\frac{\pi}{10} = \frac{1 + \cos(\pi/5)}{2} = \frac{1 + (\alpha/2)}{2} = \frac{2 + \alpha}{4}$,   so $\cos\frac{\pi}{10} = \frac{1}{2}\sqrt{2 + \alpha}$.

   (c) $\sin\frac{\pi}{10} = \sqrt{1 - \cos^2(\pi/10)} = \sqrt{1 - \left(\frac{2+\alpha}{4}\right)} = \frac{1}{2}\sqrt{4 - (2 + \alpha)} = $
   $\frac{1}{2}\sqrt{2 - \alpha}$.

3. (a)
$$[2; 1, 4, 9] = 2 + \cfrac{1}{1 + \cfrac{1}{4 + \frac{1}{9}}} = 2 + \cfrac{1}{1 + \frac{1}{\frac{37}{9}}}$$
$$= 2 + \cfrac{1}{1 + \frac{9}{37}} = 2 + \cfrac{1}{\frac{46}{37}}$$
$$= 2 + \frac{37}{46} = \frac{92 + 37}{46} = \frac{129}{46}.$$

   (b)
$$[3; 3, 7, 2, 4] = 3 + \cfrac{1}{3 + \cfrac{1}{7 + \cfrac{1}{2 + \frac{1}{4}}}} = 3 + \cfrac{1}{3 + \cfrac{1}{7 + \frac{1}{\frac{9}{4}}}}$$
$$= 3 + \cfrac{1}{3 + \cfrac{1}{7 + \frac{4}{9}}} = 3 + \cfrac{1}{3 + \frac{1}{\frac{67}{9}}} = 3 + \cfrac{1}{3 + \frac{9}{67}}$$
$$= 3 + \cfrac{1}{\frac{210}{67}} = 3 + \frac{67}{210} = \frac{630 + 67}{210} = \frac{697}{210}.$$

5. From Example 11.1, we know that $\cos 36° = \cos(\pi/5) = \alpha/2$.

   (a) The radius of the base is the length of CB, which is $s\cos 36° = s(\alpha/2)$. Consequently, the circumference of the base is $2\pi(s\alpha/2) = s\pi\alpha$.

   (b) The area of the base is $\pi(s\alpha/2)^2 = (\pi s^2\alpha^2)/4$.

   (c) For a right circular cone with base radius r and height h, the volume is $(1/3)\pi r^2 h$. Here the volume is given by $(1/3)\pi(s\alpha/2)^2\sqrt{s^2 - (s\alpha/2)^2} = (\pi/3)(s^2\alpha^2/4)s\sqrt{1 - (\alpha^2/4)} = (\pi s^3\alpha^2/24)\sqrt{4 - \alpha^2} = (\pi s^3(\alpha + 1)/24) \cdot \sqrt{4 - (\alpha + 1)} = (\pi s^3(\alpha + 1)/24)\sqrt{3 - \alpha}$.

   (d) For a right circular cone with base radius r and height h, the lateral surface area is $\pi r\sqrt{r^2 + h^2}$. Here the lateral surface area is given by $\pi r s = \pi(s\alpha/2)s = (\pi s^2\alpha)/2$.

7. (a) $AC/AM = \sin AMC/\sin ACM = \sin 108°/\sin 36° = \sin 72°/\sin 36° = (2\sin 36° \cos 36°)/\sin 36° = 2\cos 36°$.

   (b) $\cos 18° = \sin 72° = 2\sin 36° \cos 36° = 2(2\sin 18° \cos 18°)(1 - 2\sin^2 18°) = 4\sin 18° \cos 18°(1 - 2\sin^2 18°)$.

   (c) Dividing by $\cos 18°$ in part (b), it follows that $1 = 4\sin 18°(1 - 2\sin^2 18°)$. So $0 = 8\sin^3 18° - 4\sin 18° + 1$, and we find that $\sin 18°$ is a root of the cubic equation $8x^3 - 4x + 1 = 0$. With $0 = 8x^3 - 4x + 1 = (2x - 1)(4x^2 + 2x - 1)$, we find that the roots of this cubic equation are $1/2$ and $(-1 \pm \sqrt{5})/4$. Since $0 < \sin 18° < \sin 30° = 1/2$, it follows that $\sin 18° = (-1 + \sqrt{5})/2$.

(d) $(1/2)(AC/AM) = \cos 36° = \alpha/2$. Consequently, $AC/AM = \alpha$.

9. **Proof:** Since $SR = ST$, it follows that $\angle SRT = \angle STR$. Likewise, $TQ = TS$ implies that $\angle Q = \angle QST$, and so $\angle R = \angle RTS = 2 \angle Q$. Therefore, $180° = \angle R + \angle QSR + \angle Q = 2 \angle Q + 2 \angle Q + \angle Q = 5 \angle Q$, so $\angle Q = 36°$. In Example 11.1, it was shown that $\cos 36° = \cos(\pi/5) = \alpha/2$.

Since $\triangle RQS$ is similar to $\triangle TSR$,

$$\frac{QR}{RS} = \frac{ST}{TR} = \frac{\sin 72°}{\sin 36°} = \frac{2 \sin 36° \cos 36°}{\sin 36°} = 2 \cos 36° = 2 \cos Q.$$

11. Let AB be one side of the pentagon. Then $\triangle AOB$ is isosceles with the measure of $\angle AOB$ equal to $360°/5 = 72°$. If M is the midpoint of AB, then OM bisects $\angle AOB$. Consequently, the measure of $\angle AOM$ is $36°$, so the radius of the inscribed circle is the length of OM, which is $\cos 36° = \cos(\pi/5) = \alpha/2$.

13. (a) Since the volume is 1 cubic foot, $1 = l \cdot w \cdot 1 = lw$. The diagonal of the solid is 2 feet and $2 = \sqrt{l^2 + w^2 + 1^2} \implies 4 = l^2 + w^2 + 1^2 \implies l^2 + w^2 = 3$. Consequently, $3 = l^2 + (1/l)^2$ and it follows that $l^4 - 3l^2 + 1 = 0$, a quadratic in l^2. So

$$l^2 = \frac{-(-3) \pm \sqrt{(-3)^2 - 4(1)(1)}}{2(1)} = \frac{3 \pm \sqrt{5}}{2}.$$

If $l^2 = (3 + \sqrt{5})/2 = ((1 + \sqrt{5})/2)^2$, then since $l > 0$, it follows that $l = (1 + \sqrt{5})/2 = \alpha$ and $w = 1/\alpha = -\beta = |\beta|$. If $l^2 = (3 - \sqrt{5})/2 = ((1 - \sqrt{5})/2)^2$, then $l = -\beta$ and $w = \alpha > l$. Consequently, the only solution is $l = (1 + \sqrt{5})/2 = \alpha$ and $w = 1/\alpha = -\beta = |\beta|$.

(b) The total surface area is $2 \cdot l \cdot w + 2 \cdot l \cdot 1 + 2 \cdot w \cdot 1 = 2 + 2\alpha + 2(-\beta) = 2 + 2(\alpha - \beta) = 2 + 2(\sqrt{5}) = 2(1 + \sqrt{5}) = 4((1 + \sqrt{5})/2) = 4\alpha$.

15. (a) $[\alpha/2 = \cos 36° = \cos 2(18°) = \cos^2 18° - \sin^2 18° = 1 - 2 \sin^2 18°]$ $\implies [2 \sin^2 18° = 1 - (\alpha/2) = (2 - \alpha)/2] \implies [\sin^2 18° = (2 - \alpha)/4]$ $\implies \sin 18° = (1/2)\sqrt{2 - \alpha}$, since $\sin 18° > 0$. Now $2 - \alpha = 2 - (1 + \sqrt{5})/2 = (3 - \sqrt{5})/2$ and $1/\alpha^2 = \dfrac{1}{\left((1+\sqrt{5})/2\right)^2} = \dfrac{4}{\left(1+\sqrt{5}\right)^2} = \dfrac{4}{6+2\sqrt{5}} = \dfrac{2}{3+\sqrt{5}} = \dfrac{2}{3+\sqrt{5}}\dfrac{3-\sqrt{5}}{3-\sqrt{5}} = \dfrac{6-2\sqrt{5}}{9-5} = \dfrac{6-2\sqrt{5}}{4} = \dfrac{3-\sqrt{5}}{2}$, so $2 - \alpha = 1/\alpha^2$, from which it follows that $\sqrt{2 - \alpha} = 1/\alpha$ and $\sin 18° = 1/2\alpha$.

(b) Let AB be one side of the regular decagon and let M be the midpoint of AB. For $\triangle BOM$, we find that $MB = s/2$, $OB = r$, and $\angle BOM = 18°$. Consequently, $1/(2\alpha) = \sin 18° = (s/2)/r$. From this it follows that $r = (2\alpha)(s/2) = s\alpha$.

(c) Consider $\triangle BOM$, as described in part (b) of this exercise. The area of $\triangle BOM = (1/2)r(s/2)\sin 72°$. In the solution of part (a), we learned that $\sin 18° = (1/2)\sqrt{2-\alpha}$, so $\sin 72° = \cos 18° = \sqrt{1 - \sin^2 18°} = \sqrt{1 - ((2-\alpha)/4)} = \sqrt{(2+\alpha)/4} = (1/2)\sqrt{2+\alpha}$. Therefore, the area of $\triangle BOM = (1/2)r(s/2)(1/2)\sqrt{2+\alpha} = (1/8)(s\alpha)s\sqrt{2+\alpha}$ and the area of the regular decagon is 20(the area of $\triangle BOM$) $= (5/2)s^2\alpha\sqrt{2+\alpha}$.

Exercises for Chapter 12

1. (a) From Example 12.2, the number of perfect matchings is $F_{11} = 89$.

(b) If a perfect matching contains the edge $x_4 y_4$, then we must remove the edges $x_3 x_4$, $x_4 x_5$, $x_4 y_4$, $y_3 y_4$, and $y_4 y_5$ from the graph. This leaves us with a ladder graph with three rungs and another with six rungs. There are F_4 perfect matchings for the ladder graph with three rungs and F_7 perfect matchings for the ladder graph with six rungs. So the number of perfect matchings that contain the edge $x_4 y_4$ is $F_4 F_7 = 3 \cdot 13 = 39$.

3. Let i_n count the number of independent sets of vertices for the comb graph with n teeth. There are two cases to consider: (i) The vertex y_n is not used. Then there are i_{n-1} independent sets that contain the vertex x_n, and another i_{n-1} such sets that do not contain the vertex x_n. (ii) The vertex y_n is included in the independent set. Now neither of the vertices x_n or y_{n-1} can be used. So, in this case, there are i_{n-2} independent sets that contain the vertex x_{n-1}, and another i_{n-2} independent sets that do not contain the vertex x_{n-1}. These two cases lead to the recurrence relation

$$i_n = 2i_{n-1} + 2i_{n-2}, \quad i_0 = 1, i_1 = 3.$$

(Here we determined that $i_0 = 1$ from $i_2 = 8$.)

To solve the recurrence relation, we consider the characteristic equation

$$r^2 - 2r - 2 = 0,$$

for which the roots are $1 \pm \sqrt{3}$. Therefore,

$$i_n = c_1(1 + \sqrt{3})^n + c_2(1 - \sqrt{3})^n.$$

To determine the constants c_1 and c_2, we use the initial conditions and find that

$$1 = i_0 = c_1 + c_2$$
$$3 = i_1 = c_1(1 + \sqrt{3}) + c_2(1 - \sqrt{3}).$$

This leads to $c_1 = (\sqrt{3} + 2)/(2\sqrt{3})$, $c_2 = (\sqrt{3} - 2)/(2\sqrt{3})$, and

$$i_n = \left[\frac{(\sqrt{3} + 2)}{2\sqrt{3}}\right](1 + \sqrt{3})^n + \left[\frac{(\sqrt{3} - 2)}{2\sqrt{3}}\right](1 - \sqrt{3})^n, \quad n \geq 0.$$

5. Let i_n count the number of independent sets of vertices for the given graph with $2n$ vertices. If the vertex y_n is included in the independent set, then we cannot include any of the vertices y_{n-1}, x_{n-1}, or x_n. There are i_{n-2} such sets of vertices—and an additional i_{n-2} independent sets for when we include the vertex x_n, thus excluding the vertices x_{n-1}, y_{n-1}, and y_n. Furthermore, there are i_{n-1} independent sets when both x_n and y_n are excluded. This leads us to the recurrence relation

$$i_n = i_{n-1} + 2i_{n-2}, \quad i_1 = 3, \ i_2 = 5.$$

From i_1 and i_2, we determine $i_0 = (1/2)(5 - 3) = 1$.

The characteristic equation for this recurrence relation is

$$r^2 - r - 2 = 0,$$

so the characteristic roots are -1 and 2. Therefore, $i_n = c_1(-1)^n + c_2(2^n)$. To determine the constants c_1 and c_2, we consider

$$1 = i_0 = c_1 + c_2$$
$$3 = i_1 = -c_1 + 2c_2,$$

from which it follows that $c_1 = -(1/3)$ and $c_2 = 4/3$. Consequently,

$$i_n = \left(\frac{-1}{3}\right)(-1)^n + \left(\frac{4}{3}\right)(2^n), \quad n \geq 0.$$

7. **Proof:** If the result is false, then there is a first case, say $n = r > 0$, where $\gcd(L_r, L_{r+1}) > 1$. However, $\gcd(L_{r-1}, L_r) = 1$. So there is a positive integer d such that $d > 1$ and d divides L_r and L_{r+1}. But we know that

$$L_{r+1} = L_r + L_{r-1}.$$

So if d divides L_r and L_{r+1}, it follows that d divides L_{r-1}. This then contradicts $\gcd(L_{r-1}, L_r) = 1$. Consequently, $\gcd(L_n, L_{n+1}) = 1$ for $n \geq 0$.

9. $\gcd(L_0, L_3) = \gcd(2, 4) = 2 \neq 1$.

11. **Proof (By Mathematical Induction):** For $n = 0$, we have $\sum_{r=0}^{0} L_r^2 = L_0^2 = 2^2 = 4 = (2 \times 1) + 2 = (L_0 L_1) + 2 = (L_0 L_{0+1}) + 2$. This demonstrates that

the conjecture is true for this first case and establishes the basis step for our inductive proof. So now we assume the conjecture true for some $n = k \ (\geq 0)$ and this gives us $\sum_{r=0}^{k} L_r^2 = L_k L_{k+1} + 2$. Turning to the case where $n = k + 1$ (≥ 1), we have

$$\sum_{r=0}^{k+1} L_r^2 = \left(\sum_{r=0}^{k} L_r^2 \right) + L_{k+1}^2 = L_k L_{k+1} + 2 + L_{k+1}^2$$
$$= L_{k+1}(L_k + L_{k+1}) + 2 = L_{k+1} L_{k+2} + 2.$$

Consequently, the truth of the case for $n = k + 1$ follows from the case for $n = k$. So our conjecture is true for all $n \geq 0$ by the Principle of Mathematical Induction.

As $L_0 = 2$, $\sum_{r=1}^{n} L_r^2 = \sum_{r=0}^{n} L_r^2 - 2^2 = L_n L_{n+1} + 2 - 4 = L_n L_{n+1} - 2$.

13. **Proof (By Mathematical Induction):** We see that for $n = 0$, $\sum_{r=0}^{0} L_{2r} = L_0 = 2 = 1 + 1 = L_1 + 1 = L_{2(0)+1} + 1$. So the result is true in this first case and this establishes the basis case for our inductive proof. So now we shall assume the result is true for some $n = k \ (\geq 0)$, that is, $\sum_{r=0}^{k} L_{2r} = L_{2k+1} + 1$. Then, for $n = k + 1 \ (\geq 1)$, we find that $\sum_{r=0}^{k+1} L_{2r} = \left(\sum_{r=0}^{k} L_{2r} \right) + L_{2(k+1)} = (L_{2k+1} + 1) + L_{2k+2} = (L_{2k+1} + L_{2k+2}) + 1 = L_{2k+3} + 1 = L_{2(k+1)+1} + 1$, so the result is true for $n = k + 1$. Consequently, the result follows for all $n \geq 1$ by the Principle of Mathematical Induction.

As $L_0 = 2$, $\sum_{r=1}^{n} L_{2r} = \sum_{r=0}^{n} L_{2r} - 2 = (L_{2n+1} + 1) - 2 = L_{2n+1} - 1$.

15. **Proof (By Mathematical Induction):** For $n = 1$, we see that $\sum_{r=1}^{1} r L_r = 1 L_1 = 1 \cdot 1 = 1$, while $n L_{n+2} - L_{n+3} + 4 = 1 L_3 - L_4 + 4 = 1(4) - 7 + 4 = 1$, so the result is true in this first case. This establishes the basis case for our inductive proof. Next we assume the result is true for some $n = k \ (\geq 1)$—that is, we assume that $\sum_{r=1}^{k} r L_r = k L_{k+2} - L_{k+3} + 4$. Now, for $n = k + 1 \ (\geq 2)$, we have

$$\sum_{r=1}^{k+1} r L_r = \sum_{r=1}^{k} r L_r + (k + 1) L_{k+1}$$
$$= (k L_{k+2} - L_{k+3} + 4) + (k + 1) L_{k+1}$$
$$= [(k + 1) L_{k+2} - L_{k+2} - L_{k+3} + 4] + (k + 1) L_{k+1}$$
$$= [(k + 1) L_{k+2} + (k + 1) L_{k+1}] - (L_{k+2} + L_{k+3}) + 4$$
$$= (k + 1)(L_{k+2} + L_{k+1}) - (L_{k+2} + L_{k+3}) + 4$$
$$= (k + 1) L_{k+3} - L_{k+4} + 4 = (k + 1) L_{(k+1)+2} - L_{(k+1)+3} + 4.$$

So the result for $n = k + 1$ follows from the result for $n = k$. Therefore, by the Principle of Mathematical Induction, this weighted sum formula (involving the Lucas numbers) holds for all $n \geq 1$.

Alternate Solution Using the Result in Exercise 10:

$$\sum_{r=1}^{n} rL_r = 1L_1 + 2L_2 + 3L_3 + \cdots + (n-1)L_{n-1} + nL_n$$

$$= \sum_{r=1}^{n} L_r + \sum_{r=2}^{n} L_r + \sum_{r=3}^{n} L_r + \cdots + \sum_{r=n-1}^{n} L_r + \sum_{r=n}^{n} L_r.$$

For this sum, note, for example, that L_2 occurs only in the first two summations while L_3 occurs only in the first three summations. Continuing we find that

$$\sum_{r=1}^{n} L_r + \sum_{r=2}^{n} L_r + \sum_{r=3}^{n} L_r + \cdots + \sum_{r=n-1}^{n} L_r + \sum_{r=n}^{n} L_r$$

$$= \sum_{r=1}^{n} L_r + \left(\sum_{r=1}^{n} L_r - \sum_{r=1}^{1} L_r\right) + \left(\sum_{r=1}^{n} L_r - \sum_{r=1}^{2} L_r\right) + \cdots$$

$$+ \left(\sum_{r=1}^{n} L_r - \sum_{r=1}^{n-2} L_r\right) + \left(\sum_{r=1}^{n} L_r - \sum_{r=1}^{n-1} L_r\right)$$

$$= n\sum_{r=1}^{n} L_r - \left(\sum_{r=1}^{1} L_r + \sum_{r=1}^{2} L_r + \cdots + \sum_{r=1}^{n-1} L_r\right) = n\sum_{r=1}^{n} L_r - \sum_{m=1}^{n-1}\left(\sum_{r=1}^{m} L_r\right)$$

$$= n(L_{n+2} - 3) - \sum_{m=1}^{n-1}(L_{m+2} - 3), \quad \text{by Exercise 10}$$

$$= nL_{n+2} - 3n - \sum_{m=1}^{n-1} L_{m+2} + \sum_{m=1}^{n-1} 3$$

$$= nL_{n+2} - 3n - \sum_{k=3}^{n+1} L_k + 3(n-1)$$

$$= nL_{n+2} - 3n - \left(\sum_{k=1}^{n+1} L_k - L_1 - L_2\right) + 3(n-1)$$

$$= nL_{n+2} - \sum_{k=1}^{n+1} L_k + 4 - 3 = nL_{n+2} - (L_{n+3} - 3) + 1, \quad \text{also by Exercise 10}$$

$$= nL_{n+2} - L_{n+3} + 4.$$

17. **Proof:** Using the Binet form for the Lucas numbers, we find that

$$\sum_{k=0}^{n} \binom{n}{k} L_k L_{n-k} = \sum_{k=0}^{n} \binom{n}{k} \left(\alpha^k + \beta^k\right)\left(\alpha^{n-k} + \beta^{n-k}\right)$$

$$= \sum_{k=0}^{n} \binom{n}{k} \left(\alpha^n + \alpha^{n-k}\beta^k + \alpha^k\beta^{n-k} + \beta^n\right)$$

$$= \sum_{k=0}^{n} \binom{n}{k}(\alpha^n + \beta^n) + \sum_{k=0}^{n} \binom{n}{k}\alpha^{n-k}\beta^k + \sum_{k=0}^{n} \binom{n}{k}\alpha^k\beta^{n-k}$$

$$= L_n \sum_{k=0}^{n} \binom{n}{k} + (\alpha + \beta)^n + (\alpha + \beta)^n$$

$$= 2^n L_n + 2, \quad \text{since } \alpha + \beta = 1.$$

19.
$$\lim_{n\to\infty}\frac{L_{n+1}}{L_n} = \lim_{n\to\infty}\frac{\alpha^{n+1}+\beta^{n+1}}{\alpha^n+\beta^n} = \lim_{n\to\infty}\frac{\alpha+\beta\,(\beta/\alpha)^n}{1+(\beta/\alpha)^n}.$$

Since $|\beta/\alpha| < 1$, $\lim_{n\to\infty}(\beta/\alpha)^n = 0$. Consequently,

$$\lim_{n\to\infty}\frac{L_{n+1}}{L_n} = \alpha.$$

21.
$$\lim_{n\to\infty}\frac{L_{n+k}}{F_n} = \lim_{n\to\infty}\sqrt{5}\,\frac{\alpha^{n+k}+\beta^{n+k}}{\alpha^n-\beta^n} = \sqrt{5}\lim_{n\to\infty}\frac{\alpha^k+\beta^k\,(\beta/\alpha)^n}{1-(\beta/\alpha)^n}.$$

Since $|\beta/\alpha| < 1$, $\lim_{n\to\infty}(\beta/\alpha)^n = 0$. Consequently,

$$\lim_{n\to\infty}\frac{L_{n+k}}{F_n} = \sqrt{5}\alpha^k.$$

23. **Proof:**

$$L_n^2 + L_{n+1}^2 = \left(\alpha^n+\beta^n\right)^2 + \left(\alpha^{n+1}+\beta^{n+1}\right)^2$$
$$= \alpha^{2n}+\beta^{2n}+2(\alpha\beta)^n+\alpha^{2n+2}+\beta^{2n+2}+2(\alpha\beta)^{n+1}$$
$$= (\alpha^{2n}+\beta^{2n})+(\alpha^{2n+2}+\beta^{2n+2})\quad[\text{Since } 2(\alpha\beta)^n+2(\alpha\beta)^{n+1}$$
$$= 2(-1)^n+2(-1)^{n+1}=0.]$$
$$= L_{2n}+L_{2n+2}.$$

25. **Proof:** This result follows from Candido's identity, which is given in part (a) of Exercise 24 for Chapter 4.

   Replace a by L_n and b by L_{n+1} in Candido's identity. The result then follows since $L_{n+2} = L_{n+1}+L_n = a+b$.

27. **Proof (By Mathematical Induction):** Since $L_{3\cdot0} = L_0 = 2$, the result is true for when $n = 0$. This establishes the basis step for our proof by induction. Now assume the result true for some $n = k\ (\geq 0)$. That is, assume that L_{3k} is even. For $n = k+1$, we have

$$L_{3(k+1)} = L_{3k+3} = L_{3k+2}+L_{3k+1}$$
$$= (L_{3k+1}+L_{3k})+L_{3k+1}$$
$$= 2L_{3k+1}+L_{3k}.$$

Since L_{3k} is even by the induction hypothesis, it now follows that $2L_{3k+1}+L_{3k}$ is even—that is, $L_{3(k+1)}$ is even. Consequently, the result follows for all $n \geq 0$ by the Principle of Mathematical Induction.

29. **Proof:** Using the Binet form for the Lucas numbers, we find that

$$\sum_{i=0}^{n} \binom{n}{i} L_{mi} L_{mn-mi} = \sum_{i=0}^{n} \binom{n}{i} \left(\alpha^{mi} + \beta^{mi} \right) \left(\alpha^{mn-mi} + \beta^{mn-mi} \right)$$

$$= \sum_{i=0}^{n} \binom{n}{i} \left(\alpha^{mn} + \left(\frac{\alpha}{\beta} \right)^{mi} \beta^{mn} + \left(\frac{\beta}{\alpha} \right)^{mi} \alpha^{mn} + \beta^{mn} \right)$$

$$= (\alpha^{mn} + \beta^{mn}) \sum_{i=0}^{n} \binom{n}{i} + \beta^{mn} \sum_{i=0}^{n} \binom{n}{i} \left(\frac{\alpha^m}{\beta^m} \right)^{i} 1^{n-i}$$

$$+ \alpha^{mn} \sum_{i=0}^{n} \binom{n}{i} \left(\frac{\beta^m}{\alpha^m} \right)^{i} 1^{n-i}$$

$$= 2^n L_{mn} + \beta^{mn} \left(1 + \frac{\alpha^m}{\beta^m} \right)^{n} + \alpha^{mn} \left(1 + \frac{\beta^m}{\alpha^m} \right)^{n}$$

$$= 2^n L_{mn} + \left(\beta^m \left(1 + \frac{\alpha^m}{\beta^m} \right) \right)^{n} + \left(\alpha^m \left(1 + \frac{\beta^m}{\alpha^m} \right) \right)^{n}$$

$$= 2^n L_{mn} + (\beta^m + \alpha^m)^n + (\alpha^m + \beta^m)^n$$

$$= 2^n L_{mn} + L_m^n + L_m^n = 2^n L_{mn} + 2L_m^n.$$

Exercises for Chapter 13

1. (a) If Cinda and Katelyn place a 1×2 curved rectangular tile on curved squares n and 1, then the remaining 23 curved squares can be covered in F_{24} ways. Otherwise, they can tile the 1×25 curved chessboard in F_{26} ways. Since $F_{n-1} + F_{n+1} = L_n$, there are $L_{25} (= 167, 761)$ ways in which Cinda and Katelyn can tile a 1×25 circular chessboard under these conditions.

   (b) Consider a $1 \times n$ circular chessboard, for $n \geq 4$. If we do not use the nth square, then we can place nontaking kings on the resulting $1 \times (n - 1)$ chessboard in F_{n+1} ways. [We observed this in Example 8.1.] If we use the nth square, then we cannot use the squares at positions 1 and $n - 1$. So now we can place nontaking kings on the resulting $1 \times (n - 3)$ chessboard in F_{n-1} ways. So nontaking kings can be placed on a $1 \times n$ circular chessboard in $F_{n-1} + F_{n+1} = L_n$ ways, from Property 13.1. (Note that this result is also true for $n = 0, 1, 2, 3$.) Consequently, Cinda and Katelyn can place nontaking kings on their 1×25 chessboard in $L_{25} (= 167, 761)$ ways.

3. **Proof:** Using the Binet forms for the Fibonacci and Lucas numbers, it follows that $F_n L_n = \left(\frac{\alpha^n - \beta^n}{\alpha - \beta} \right) (\alpha^n + \beta^n) = \left(\frac{\alpha^{2n} - \beta^{2n}}{\alpha - \beta} \right) = F_{2n}$.

5. **Proof:** Using the result from Exercise 4, we find that $L_{n-1} + L_{n+1} = (F_{n-2} + F_n) + (F_n + F_{n+2}) = 2F_n + (F_{n+2} + F_{n-2}) = 2F_n + (F_{n+1} + F_n + F_n - F_{n-1}) = 4F_n + (F_{n+1} - F_{n-1}) = 4F_n + F_n = 5F_n$.

7. **Proof:** Start with the result from Exercise 5 and add L_n to both sides of that equation. This results in $(L_n + L_{n-1}) + L_{n+1} = 5F_n + L_n$, or $L_{n+1} + L_{n+1} = 2L_{n+1} = 5F_n + L_n$.

   (*Note*: This result is the same as $2L_{n+1} - L_n = 5F_n$, which is given in Property 13.4.)

9. **Proof:** Using the Binet forms for the Fibonacci and Lucas numbers, we find that

$$
\begin{aligned}
L_{2m}L_{2n} &= (\alpha^{2m} + \beta^{2m})(\alpha^{2n} + \beta^{2n}) \\
&= \alpha^{2m+2n} + \alpha^{2m-2n}(\alpha^{2n}\beta^{2n}) + \beta^{2m-2n}(\alpha^{2n}\beta^{2n}) + \beta^{2m+2n} \\
&= \alpha^{2m+2n} + \alpha^{2m-2n}(\alpha\beta)^{2n} + \beta^{2m-2n}(\alpha\beta)^{2n} + \beta^{2m+2n} \\
&= \alpha^{2m+2n} + \alpha^{2m-2n}(-1)^{2n} + \beta^{2m-2n}(-1)^{2n} + \beta^{2m+2n} \\
&= \alpha^{2m+2n} + \alpha^{2m-2n} + \beta^{2m-2n} + \beta^{2m+2n}
\end{aligned}
$$

while

$$
\begin{aligned}
L_{m+n}^2 + 5F_{m-n}^2 &= \left(\alpha^{m+n} + \beta^{m+n}\right)^2 + 5\left[\frac{\alpha^{m-n} - \beta^{m-n}}{\alpha - \beta}\right]^2 \\
&= \left(\alpha^{m+n} + \beta^{m+n}\right)^2 + \left(\alpha^{m-n} - \beta^{m-n}\right)^2, \quad \text{since } \alpha - \beta = \sqrt{5} \\
&= \alpha^{2m+2n} + \beta^{2m+2n} + 2(\alpha\beta)^{m+n} + \alpha^{2m-2n} + \beta^{2m-2n} - 2(\alpha\beta)^{m-n} \\
&= \alpha^{2m+2n} + \beta^{2m+2n} + 2(-1)^{m+n} + \alpha^{2m-2n} + \beta^{2m-2n} - 2(-1)^{m-n} \\
&= \alpha^{2m+2n} + \beta^{2m+2n} + \alpha^{2m-2n} + \beta^{2m-2n} + 2[(-1)^{m+n} - (-1)^{m-n}] \\
&= \alpha^{2m+2n} + \beta^{2m+2n} + \alpha^{2m-2n} + \beta^{2m-2n} + 2[(-1)^{m+n} - (-1)^{m-n+2n}] \\
&= \alpha^{2m+2n} + \beta^{2m+2n} + \alpha^{2m-2n} + \beta^{2m-2n} + 2[(-1)^{m+n} - (-1)^{m+n}] \\
&= \alpha^{2m+2n} + \beta^{2m+2n} + \alpha^{2m-2n} + \beta^{2m-2n} + 2[0] \\
&= \alpha^{2m+2n} + \beta^{2m+2n} + \alpha^{2m-2n} + \beta^{2m-2n}.
\end{aligned}
$$

Consequently, $L_{2m}L_{2n} = L_{m+n}^2 + 5F_{m-n}^2$.

11. **Proof:** Using the Binet form for the Lucas numbers, the Binomial theorem, and the facts that $\sqrt{5} = \alpha - \beta = 1 - 2\beta$ and $-\sqrt{5} = \beta - \alpha = 1 - 2\alpha$, we have

$$
\begin{aligned}
\sum_{k=0}^{2n} (-1)^k \binom{2n}{k} 2^{k-1} L_k &= \frac{1}{2}\sum_{k=0}^{2n} (-2)^k \binom{2n}{k} \alpha^k + \frac{1}{2}\sum_{k=0}^{2n} (-2)^k \binom{2n}{k} \beta^k \\
&= \frac{1}{2}\sum_{k=0}^{2n} \binom{2n}{k}(-2\alpha)^k + \frac{1}{2}\sum_{k=0}^{2n} \binom{2n}{k}(-2\beta)^k \\
&= \frac{1}{2}\sum_{k=0}^{2n} \binom{2n}{k}(-2\alpha)^k 1^{2n-k} + \frac{1}{2}\sum_{k=0}^{2n} \binom{2n}{k}(-2\beta)^k 1^{2n-k} \\
&= \frac{1}{2}(1 - 2\alpha)^{2n} + \frac{1}{2}(1 - 2\beta)^{2n} = \frac{1}{2}(-\sqrt{5})^{2n} + \frac{1}{2}(\sqrt{5})^{2n} \\
&= \frac{1}{2}(5^n + 5^n) = 5^n.
\end{aligned}
$$

13. **Proof:** From the Binet forms for the Fibonacci and Lucas numbers and the Binomial theorem, we find that

$$\sum_{k=0}^{n} \binom{n}{k} F_k L_{n-k} = \sum_{k=0}^{n} \binom{n}{k} \left(\frac{\alpha^k - \beta^k}{\alpha - \beta}\right) \left(\alpha^{n-k} + \beta^{n-k}\right)$$

$$= \sum_{k=0}^{n} \binom{n}{k} \left(\frac{\alpha^n - \alpha^{n-k}\beta^k + \alpha^k\beta^{n-k} - \beta^n}{\alpha - \beta}\right)$$

$$= \sum_{k=0}^{n} \binom{n}{k} \left(\frac{\alpha^n - \beta^n}{\alpha - \beta}\right) + \sum_{k=0}^{n} \binom{n}{k} \frac{\alpha^k \beta^{n-k}}{\alpha - \beta} - \sum_{k=0}^{n} \binom{n}{k} \frac{\alpha^{n-k}\beta^k}{\alpha - \beta}$$

$$= \left(\frac{\alpha^n - \beta^n}{\alpha - \beta}\right) \sum_{k=0}^{n} \binom{n}{k} + \frac{1}{\alpha - \beta} \sum_{k=0}^{n} \binom{n}{k} \alpha^k \beta^{n-k}$$

$$- \frac{1}{\alpha - \beta} \sum_{k=0}^{n} \binom{n}{k} \alpha^{n-k}\beta^k$$

$$= F_n(2^n) + \frac{1}{\alpha - \beta}(\alpha + \beta)^n - \frac{1}{\alpha - \beta}(\alpha + \beta)^n = 2^n F_n.$$

15. **Proof:**

$$5F_{2n+3}F_{2n-3} = 5\left[\frac{\alpha^{2n+3} - \beta^{2n+3}}{\alpha - \beta} \cdot \frac{\alpha^{2n-3} - \beta^{2n-3}}{\alpha - \beta}\right]$$

$$= \alpha^{4n} - \beta^6(\alpha\beta)^{2n-3} - \alpha^6(\alpha\beta)^{2n-3} + \beta^{4n}$$

$$= (\alpha^{4n} + \beta^{4n}) - \alpha^6(-1)^{2n-3} - \beta^6(-1)^{2n-3}$$

$$= L_{4n} + (\alpha^6 + \beta^6) = L_{4n} + L_6 = L_{4n} + 18.$$

17. **Proof:** From the Binet forms for the Fibonacci and Lucas numbers, we find that

$$5F_m F_{m+n} = 5\frac{\alpha^m - \beta^m}{\alpha - \beta} \frac{\alpha^{m+n} - \beta^{m+n}}{\alpha - \beta}$$

$$= (\alpha^m - \beta^m)(\alpha^{m+n} - \beta^{m+n}), \quad \text{since } \alpha - \beta = \sqrt{5}$$

$$= \alpha^{2m+n} - \alpha^{m+n}\beta^m - \alpha^m\beta^{m+n} + \beta^{2m+n}$$

$$= \left(\alpha^{2m+n} + \beta^{2m+n}\right) - (\alpha\beta)^m(\alpha^n + \beta^n)$$

$$= L_{2m+n} - (-1)^m L_n, \quad \text{since } \alpha\beta = -1.$$

19. **Proof:** From the Binet forms for the Fibonacci and Lucas numbers, we find that

$$
F_m L_{m+n} = \left(\frac{\alpha^m - \beta^m}{\alpha - \beta} \right) (\alpha^{m+n} + \beta^{m+n})
$$

$$
= \frac{\alpha^{2m+n} - \alpha^{m+n}\beta^m + \alpha^m \beta^{m+n} - \beta^{2m+n}}{\alpha - \beta}
$$

$$
= \frac{\alpha^{2m+n} - \beta^{2m+n}}{\alpha - \beta} - (\alpha\beta)^m \frac{\alpha^n - \beta^n}{\alpha - \beta}
$$

$$
= F_{2m+n} - (-1)^m F_n, \quad \text{since } \alpha\beta = -1.
$$

21. (a) **Proof (By the Alternative, or Strong, Form of the Principle of Mathematical Induction):** When $n = 1$, $232L_1 + 144L_0 = 232 + 144(2) = 520$, which is divisible by 10. For $n = 2$, $232L_2 + 144L_1 = 232(3) + 144(1) = 840$, which is also divisible by 10. These two results constitute the basis step for our inductive proof. Now assume the result true for $n = 1, 2, \ldots, k - 1, k \ (\geq 2)$. That is, assume that $232L_n + 144L_{n-1}$ is divisible by 10 for $n = 1, 2, \ldots, k - 1, k \ (\geq 2)$. For $n = k + 1 \ (\geq 3)$, $232L_n + 144L_{n-1} = 232L_{k+1} + 144L_k = 232(L_k + L_{k-1}) + 144(L_{k-1} + L_{k-2}) = (232L_k + 144L_{k-1}) + (232L_{k-1} + 144L_{k-2})$, where both summands are divisible by 10, from the induction hypothesis. Consequently, $232L_n + 144L_{n-1}$ is divisible by 10 for all $n \geq 1$, by the Alternative, or Strong, form of the Principle of Mathematical Induction.

(b) Examining Table 12.1, we find that the units digit for L_0 and L_{12} is 2, the units digit for L_1 and L_{13} is 1, the units digit for L_2 and L_{14} is 3, $\ldots$, the units digit for L_{11} and L_{23} is 9, and the units digit for L_{12} and L_{24} is 2. Since no repetitive scheme appears for L_0 through L_{11}, we conjecture that the period here is 12. To prove this, we need to show that

for $n \geq 0$ and $r \geq 0$, with r fixed, $L_{12n+r} - L_r$ is divisible by 10.

Proof (By Mathematical Induction): For $n = 0$, $L_{12n+r} - L_r = L_r - L_r = 0$, which is divisible by 10. Consequently, the result is true for $n = 0$ and this establishes the basis step for our inductive proof. Assuming the result true for some arbitrary, but fixed, value $k \ (\geq 0)$, we have $L_{12k+r} - L_r$ divisible by 10. Now for $n = k + 1 \ (\geq 1)$, we find that

$$
\begin{aligned}
L_{12(k+1)+r} - L_r &= L_{(12k+r)+12} - L_r \\
&= F_{13}L_{12k+r} + F_{12}L_{12k+(r-1)} - L_r, \quad \text{by Exercise 20} \\
&= 233L_{12k+r} + 144L_{12k+(r-1)} - L_r \\
&= 233(L_{12k+r} - L_r) + 144(L_{12k+(r-1)} - L_{(r-1)}) \\
&\quad + 232L_r + 144L_{r-1}.
\end{aligned}
$$

By the induction hypothesis (for $n = k$ and r and, consequently, $r - 1$ fixed), $L_{12k+r} - L_r$ and $L_{12k+(r-1)} - L_{(r-1)}$ are both divisible by 10. From part (a), $232L_r + 144L_{r-1}$ is divisible by 10. Therefore, the result follows for all $n \geq 0$ and $r \geq 0$, with r fixed, by the Principle of Mathematical Induction.

23. **Proof: (By the Alternative, or Strong, Form of the Principle of Mathematical Induction):** For $n = 1$, $3^{n-1} - L_n = 3^0 - L_1 = 1 - 1 = 0 = 0 \cdot 5$. Likewise, for $n = 2$, $3^{n-1} - L_n = 3^1 - L_2 = 3 - 3 = 0 = 0 \cdot 5$. So the result is true in these first two cases and this establishes the basis for an inductive proof.

Assuming that $3^{n-1} - L_n$ is divisible by 5 for $n = 1, 2, \ldots, k \,(\geq 2)$, we consider the case for $n = k + 1$. We find that

$$
\begin{aligned}
3^{(k+1)-1} - L_{k+1} &= 3 \cdot 3^{k-1} - (L_k + L_{k-1}) \\
&= 3^{k-1} + 2 \cdot 3^{k-1} - L_k - L_{k-1} \\
&= (3^{k-1} - L_k) + 2 \cdot 3^{k-1} - L_{k-1} \\
&= (3^{k-1} - L_k) + 2 \cdot 3 \cdot 3^{k-2} - L_{k-1} \\
&= (3^{k-1} - L_k) + 5 \cdot 3^{k-2} + (3^{k-2} - L_{k-1}),
\end{aligned}
$$

where $(3^{k-1} - L_k)$ and $(3^{k-2} - L_{k-1})$ are divisible by 5 from the induction hypothesis. The result now follows for all $n \geq 1$ by the Alternative, or Strong, form of the Principle of Mathematical Induction.

25. **Proof:** For $n = 0$, $\sum_{r=0}^{n} G_r = \sum_{r=0}^{0} G_r = G_0 = b - a = G_2 - a =$ $G_{0+2} - a = G_{n+2} - a$. For $n = 1$, $\sum_{r=0}^{n} G_r = \sum_{r=0}^{1} G_r = G_0 + G_1 =$ $(b - a) + a = b = (a + b) - a = G_3 - a = G_{1+2} - a = G_{n+2} - a$. For $n \geq 2$,

$$
\begin{aligned}
\sum_{r=0}^{n} G_r &= G_0 + G_1 + \sum_{r=2}^{n} (aF_{r-2} + bF_{r-1}) = (b - a) + a \\
&\quad + a \sum_{r=2}^{n} F_{r-2} + b \sum_{r=2}^{n} F_{r-1} \\
&= b + a \sum_{r=0}^{n-2} F_r + b \sum_{r=0}^{n-2} F_{r+1} = b + a\,(F_n - 1) \\
&\quad + b\,(F_{n+1} - 1), \quad \text{by Property 4.5} \\
&= aF_n + bF_{n+1} + b - a - b = (aF_n + bF_{n+1}) - a = G_{n+2} - a.
\end{aligned}
$$

27. **Proof:**

$$
\begin{aligned}
\sum_{r=0}^{n} G_{2r} &= \sum_{r=0}^{2n} G_r - \sum_{r=1}^{n} G_{2r-1} \\
&= (G_{2n+2} - a) - [G_{2n} + (a - b)], \quad \text{from Exercises 25 and 26} \\
&= G_{2n+2} - G_{2n} - a - a + b \\
&= G_{2n+1} + b - 2a
\end{aligned}
$$

29. **Proof:**

$$\sum_{r=0}^{n} G_{m+r} = \sum_{r=0}^{n} (aF_{m+r-2} + bF_{m+r-1})$$

$$= a\sum_{r=0}^{n} F_{m+r-2} + b\sum_{r=0}^{n} F_{m+r-1}$$

$$= a(F_{m-2} + F_{m-1} + \cdots + F_{m+n-2}) + b(F_{m-1} + F_m + \cdots + F_{m+n-1})$$

$$= a\left[(F_{m+n} - 1) - (F_{m-1} - 1)\right] + b\left[(F_{m+n+1} - 1) - (F_m - 1)\right]$$

$$= a(F_{m+n} - F_{m-1}) + b(F_{m+n+1} - F_m)$$

$$= (aF_{m+n} + bF_{m+n+1}) - (aF_{m-1} + bF_m)$$

$$= G_{m+n+2} - G_{m+1}$$

31. (a) $\lim_{n\to\infty}(G_n/F_n) = c$.

(b) $\lim_{n\to\infty}(G_n/L_n) = c/(\alpha - \beta)$.

33. (a) (i) $\alpha = (1 + \sqrt{5})/2$, $\alpha^2 = (3 + \sqrt{5})/2$, $\alpha^3 = \alpha^2 + \alpha = (4 + 2\sqrt{5})/2$, and $\alpha^4 = \alpha^3 + \alpha^2 = (7 + 3\sqrt{5})/2 = (L_4 + F_4\sqrt{5})/2$.

(ii) In general, $\alpha^n = (L_n + F_n\sqrt{5})/2$. This follows from $L_n = \alpha^n + \beta^n$ and $F_n = (\alpha^n - \beta^n)/\sqrt{5}$.

(b) (i) $\beta = (1 - \sqrt{5})/2$, $\beta^2 = (3 - \sqrt{5})/2$, $\beta^3 = \beta^2 + \beta = (4 - 2\sqrt{5})/2$, and $\beta^4 = \beta^3 + \beta^2 = (7 - 3\sqrt{5})/2 = (L_4 - F_4\sqrt{5})/2$.

(ii) In general, $\beta^n = (L_n - F_n\sqrt{5})/2$.

Exercises for Chapter 14

1. **Proof (By Mathematical Induction):** For $n = 1$,

$$\mathbf{Q} = \mathbf{Q}^1 = \begin{bmatrix} 1 & 1 \\ 1 & 0 \end{bmatrix} = \begin{bmatrix} F_2 & F_1 \\ F_1 & F_0 \end{bmatrix},$$

so the result is true in this first case. This establishes the basis step for our inductive proof. Now assume the result true for some $n = k\ (\geq 1)$. That is, assume that for $n = k$,

$$\mathbf{Q}^k = \begin{bmatrix} F_{k+1} & F_k \\ F_k & F_{k-1} \end{bmatrix}.$$

Then for $n = k + 1\ (\geq 2)$, we have

$$\mathbf{Q}^n = \mathbf{Q}^{k+1} = \mathbf{Q}\mathbf{Q}^k = \begin{bmatrix} 1 & 1 \\ 1 & 0 \end{bmatrix}\begin{bmatrix} F_{k+1} & F_k \\ F_k & F_{k-1} \end{bmatrix}$$

$$= \begin{bmatrix} F_{k+1} + F_k & F_k + F_{k-1} \\ F_{k+1} & F_k \end{bmatrix} = \begin{bmatrix} F_{k+2} & F_{k+1} \\ F_{k+1} & F_k \end{bmatrix}.$$

Consequently, the result is true for all $n \geq 1$, by the Principle of Mathematical Induction.

3.

$$\lim_{n\to\infty} \frac{Q^n}{F_{n-1}} = \lim_{n\to\infty} \frac{1}{F_{n-1}} \begin{bmatrix} F_{n+1} & F_n \\ F_n & F_{n-1} \end{bmatrix} = \begin{bmatrix} \lim_{n\to\infty} \frac{F_{n+1}}{F_{n-1}} & \lim_{n\to\infty} \frac{F_n}{F_{n-1}} \\ \lim_{n\to\infty} \frac{F_n}{F_{n-1}} & \lim_{n\to\infty} \frac{F_{n-1}}{F_{n-1}} \end{bmatrix} = \begin{bmatrix} \alpha^2 & \alpha \\ \alpha & 1 \end{bmatrix}.$$

Since $\alpha^2 = \alpha + 1$, the answer can also be written as

$$\begin{bmatrix} \alpha + 1 & \alpha \\ \alpha & 1 \end{bmatrix}.$$

5. **Proof:** Take the tangent of both sides. On the left, we obtain $\tan(\arctan 2) = 2$. On the right, we find

$$\tan(2 \arctan |\beta|) = \frac{2 \tan(\arctan |\beta|)}{1 - \tan^2(\arctan |\beta|)}$$

$$= \frac{2 |\beta|}{1 - |\beta|^2} = \frac{2 |\beta|}{1 - \beta^2}$$

$$= \frac{2 |\beta|}{-\beta} = \frac{2 (-\beta)}{-\beta} = 2.$$

The result then follows because the steps shown here are reversible.

7. (a)

$$M^2 = \begin{bmatrix} 2 & 3 \\ 3 & 5 \end{bmatrix}, \quad M^3 = \begin{bmatrix} 5 & 8 \\ 8 & 13 \end{bmatrix}, \quad M^4 = \begin{bmatrix} 13 & 21 \\ 21 & 34 \end{bmatrix}, \quad M^5 = \begin{bmatrix} 34 & 55 \\ 55 & 89 \end{bmatrix}.$$

(b)

$$M = \begin{bmatrix} 1 & 1 \\ 1 & 2 \end{bmatrix} = \begin{bmatrix} F_1 & F_2 \\ F_2 & F_3 \end{bmatrix}, \quad M^2 = \begin{bmatrix} 2 & 3 \\ 3 & 5 \end{bmatrix} = \begin{bmatrix} F_3 & F_4 \\ F_4 & F_5 \end{bmatrix},$$

$$M^3 = \begin{bmatrix} 5 & 8 \\ 8 & 13 \end{bmatrix} = \begin{bmatrix} F_5 & F_6 \\ F_6 & F_7 \end{bmatrix}, \quad M^4 = \begin{bmatrix} 13 & 21 \\ 21 & 34 \end{bmatrix} = \begin{bmatrix} F_7 & F_8 \\ F_8 & F_9 \end{bmatrix},$$

$$M^5 = \begin{bmatrix} 34 & 55 \\ 55 & 89 \end{bmatrix} = \begin{bmatrix} F_9 & F_{10} \\ F_{10} & F_{11} \end{bmatrix}.$$

Based on these five results, we claim that for $n \geq 1$,

$$M^n = \begin{bmatrix} F_{2n-1} & F_{2n} \\ F_{2n} & F_{2n+1} \end{bmatrix}.$$

Proof (By Mathematical Induction): We see that the claim is true for $n = 1$ (as well as, 2, 3, 4, and 5). This establishes the basis step for our inductive proof. So assume the result true for some $n = k \ (\geq 1)$—that is, assume that

$$M^k = \begin{bmatrix} F_{2k-1} & F_{2k} \\ F_{2k} & F_{2k+1} \end{bmatrix}.$$

Now consider what happens when $n = k + 1 \ (\geq 2)$. We find that

$$
M^n = M^{k+1} = \begin{bmatrix} 1 & 1 \\ 1 & 2 \end{bmatrix}^{k+1} = \begin{bmatrix} 1 & 1 \\ 1 & 2 \end{bmatrix} \begin{bmatrix} 1 & 1 \\ 1 & 2 \end{bmatrix}^{k}
$$

$$
= \begin{bmatrix} 1 & 1 \\ 1 & 2 \end{bmatrix} \begin{bmatrix} F_{2k-1} & F_{2k} \\ F_{2k} & F_{2k+1} \end{bmatrix} = \begin{bmatrix} F_{2k-1}+F_{2k} & F_{2k}+F_{2k+1} \\ F_{2k-1}+2F_{2k} & F_{2k}+2F_{2k+1} \end{bmatrix}
$$

$$
= \begin{bmatrix} F_{2k+1} & F_{2k+2} \\ (F_{2k-1}+F_{2k})+F_{2k} & (F_{2k}+F_{2k+1})+F_{2k+1} \end{bmatrix}
$$

$$
= \begin{bmatrix} F_{2k+1} & F_{2k+2} \\ F_{2k+1}+F_{2k} & F_{2k+2}+F_{2k+1} \end{bmatrix}
$$

$$
= \begin{bmatrix} F_{2k+1} & F_{2k+2} \\ F_{2k+2} & F_{2k+3} \end{bmatrix} = \begin{bmatrix} F_{2n-1} & F_{2n} \\ F_{2n} & F_{2n+1} \end{bmatrix}.
$$

It follows from the Principle of Mathematical Induction that

$$
M^n = \begin{bmatrix} F_{2n-1} & F_{2n} \\ F_{2n} & F_{2n+1} \end{bmatrix},
$$

for all $n \geq 1$.

(c)

$$
\lim_{n\to\infty} \frac{M^n}{F_{2n-1}} = \lim_{n\to\infty} \frac{1}{F_{2n-1}} \begin{bmatrix} F_{2n-1} & F_{2n} \\ F_{2n} & F_{2n+1} \end{bmatrix}
$$

$$
= \lim_{n\to\infty} \begin{bmatrix} 1 & \frac{F_{2n}}{F_{2n-1}} \\ \frac{F_{2n}}{F_{2n-1}} & \frac{F_{2n+1}}{F_{2n-1}} \end{bmatrix} = \begin{bmatrix} 1 & \alpha \\ \alpha & \alpha^2 \end{bmatrix}
$$

Since $\alpha^2 = \alpha + 1$, the answer can also be expressed as

$$
\begin{bmatrix} 1 & \alpha \\ \alpha & \alpha+1 \end{bmatrix}.
$$

9. The value of the determinants in parts (a) and (b) is 0 because in both parts, the third row of the determinant is the sum of the first and second rows of the determinant. In general, for $n \geq 3$, the value of the following $n \times n$ determinant is 0.

$$\begin{vmatrix} F_1 & F_2 & F_3 & \cdots & F_n \\ F_2 & F_3 & F_4 & \cdots & F_{n+1} \\ F_3 & F_4 & F_5 & \cdots & F_{n+2} \\ \vdots & \vdots & \vdots & \vdots & \vdots \\ F_n & F_{n+1} & F_{n+2} & \cdots & F_{2n-1} \end{vmatrix}$$

11. (a) $\det(A_1) =$ the determinant of $A_1 = |3| = 3$

$$\det(A_2) = \begin{vmatrix} 3 & -1 \\ -1 & 3 \end{vmatrix} = 3 \cdot 3 - (-1)(-1) = 9 - 1 = 8$$

$$\det(A_3) = \begin{vmatrix} 3 & -1 & 0 \\ -1 & 3 & -1 \\ 0 & -1 & 3 \end{vmatrix} = 3 \begin{vmatrix} 3 & -1 \\ -1 & 3 \end{vmatrix} + (-1)(-1)^{1+2} \begin{vmatrix} -1 & -1 \\ 0 & 3 \end{vmatrix}$$

(Expanding by the top row) $= 3(8) + (-3) = 21$

$$\det(A_4) = \begin{vmatrix} 3 & -1 & 0 & 0 \\ -1 & 3 & -1 & 0 \\ 0 & -1 & 3 & -1 \\ 0 & 0 & -1 & 3 \end{vmatrix} = 3 \det(A_3) + (-1)(-1)^{1+2} \begin{vmatrix} -1 & -1 & 0 \\ 0 & 3 & -1 \\ 0 & -1 & 3 \end{vmatrix}$$

(Expanding by the top row) $= 3(21) - \begin{vmatrix} 3 & -1 \\ -1 & 3 \end{vmatrix}$

(Expanding by the first column) $= 3 \det(A_3) - \det(A_2) = 63 - 8 = 55$

(b) Since $\det(A_1) = 3 = F_4$, $\det(A_2) = 8 = F_6$, $\det(A_3) = 21 = F_8$, and $\det(A_4) = 55 = F_{10}$, it appears that for $n \geq 1$, $\det(A_n) = F_{2n+2}$.

Proof: The result is true for $n = 1$ and $n = 2$. This establishes the basis for a proof by the Alternative, or Strong, form of the Principle of Mathematical Induction. Assume the result true for $n = 1, 2, 3, \ldots, k - 2, k - 1$ (≥ 2). Then for the case where $n = k$ (≥ 3), we have

$$\det(A_k) = 3\det(A_{k-1}) + \begin{vmatrix} -1 & -1 & 0 & \cdots & & 0 \\ 0 & 3 & -1 & \cdots & & 0 \\ & & & \cdots & & \\ & & & \cdots & & -1 \\ 0 & 0 & 0 & \cdots & -1 & 3 \end{vmatrix}$$

(Expanding by the top row)

$$= 3\det(A_{k-1}) - \det(A_{k-2}) \text{ (Expanding by the first column)}$$
$$= 3F_{2(k-1)+2} - F_{2(k-2)+2} = 3F_{2k} - F_{2k-2}$$
$$= F_{2k} + 2F_{2k} - F_{2k-2} = F_{2k} + F_{2k} + (F_{2k} - F_{2k-2})$$
$$= F_{2k} + (F_{2k} + F_{2k-1}) = F_{2k} + F_{2k+1} = F_{2k+2}.$$

Consequently, the result is true for $n \geq 1$ by the Alternative, or Strong, form of the Principle of Mathematical Induction.

Exercises for Chapter 15

1. Since $57 = 3 \cdot 19$ and $75 = 3 \cdot 5^2$, we have $\gcd(57, 75) = 3$. Therefore, by the gcd Property for the Fibonacci numbers, we have

$$\gcd(F_{57}, F_{75}) = F_{\gcd(57,75)} = F_3 = 2.$$

Without the gcd Property, we would be trying to determine the greatest common divisor of $365, 435, 296, 162$ ($= F_{57}$) and $2, 111, 485, 077, 978, 050$ ($= F_{75}$). Since $365, 435, 296, 162 = 2 \cdot 37 \cdot 113 \cdot 797 \cdot 54833$ and

$$2, 111, 485, 077, 978, 050 = 2 \cdot 5^2 \cdot 61 \cdot 3001 \cdot 230686501,$$

we see that the greatest common divisor of these two integers is $2 = F_3$. But a great deal of calculation is needed to determine the prime factorizations of

$$365, 435, 296, 162 \text{ and } 2, 111, 485, 077, 978, 050.$$

3. No! Consider the following counterexample.

$$\gcd(L_3, L_6) = \gcd(4, 18) = 2, \text{ but } L_{\gcd(3,6)} = L_3 = 4.$$

Exercise for Chapter 16

1. To obtain the a_n ternary strings of length n for $n \geq 3$, consider the a_{n-1} strings of length $n - 1$. If we append 0 or 2 to the end of each of these strings, we obtain $2a_{n-1}$ of the a_n ternary strings of length n. If we append 1 to each of the a_{n-1} strings of length $n - 1$, we have included strings that end in 21. How many such strings are there? There are a_{n-2} strings of length $n - 1$ that terminate with 2. So there are $a_{n-1} - a_{n-2}$ ternary strings of length n that end in 1 but do not end in 21. Consequently,

$$a_n = 2a_{n-1} + (a_{n-1} - a_{n-2}) = 3a_{n-1} - a_{n-2}, \quad n \geq 3, \quad a_1 = 2, \quad a_2 = 5.$$

Solving this recurrence relation, we find that

$$a_n = \left(\frac{5 + \sqrt{5}}{10}\right)\alpha^{2n} + \left(\frac{5 - \sqrt{5}}{10}\right)\beta^{2n}$$

$$= \frac{1}{2}\left(\alpha^{2n} + \beta^{2n}\right) + \frac{1}{2}\left(\frac{1}{\sqrt{5}}\right)\left(\alpha^{2n} - \beta^{2n}\right)$$

$$= \frac{1}{2}L_{2n} + \frac{1}{2}F_{2n} = \frac{1}{2}(2F_{2n+1}), \quad \text{by Property 13.3}$$

$$= F_{2n+1}.$$

Exercise for Chapter 17

1. (a) Let a_n count the number of ways one can arrange n circles according to the four conditions stated in the exercise. If the circle at the right end of the bottom row has no circle in the upper row resting on it, then upon removing this circle (at the right end of the bottom row) we have a_{n-1} arrangements. If not, then there is a circle at the right end of the upper row and this circle rests on the circle at the right end of the bottom row. When both of these circles are removed, we have a_{n-2} arrangements. Since these two cases exhaust all possibilities and have nothing in common, it follows that

$$a_n = a_{n-1} + a_{n-2}, \quad n \geq 3, \ a_1 = 1, \ a_2 = 1.$$

Consequently, $a_n = F_n$, for $n \geq 1$.

(b) (i) $\binom{9}{0} = 1$, (ii) $\binom{8}{1} = 8$, (iii) $\binom{7}{2} = 21$, (iv) $\binom{6}{3} = 20$, (v) $\binom{5}{4} = 5$.

(c) $1 + 8 + 21 + 20 + 5 = 55 = F_{10}$.

(d) For $n \geq 1$, the number of arrangements of n circles, satisfying the four conditions, is

$$F_n = \sum_{i=0}^{\lfloor (n-1)/2 \rfloor} \binom{n-1-i}{i},$$

where $\lfloor (n-1)/2 \rfloor$ denotes the *floor* or *greatest integer* in $(n-1)/2$. (In general, for any real number x, (i) $\lfloor x \rfloor = x$, if x is an integer, and (ii) $\lfloor x \rfloor$ is the largest integer smaller than x, if x is not an integer. When x is not an integer, then $\lfloor x \rfloor$ is the integer immediately to the left of x on the real number line.)

(*Note*: The formula in part (d) provides another combinatorial proof for the result we established in parts (a) and (b) of Example 5.6.)

PART TWO: THE CATALAN NUMBERS

Exercises for Chapter 19

1. **Proof:**

$$\binom{2n}{n} - \binom{2n}{n-1} = \frac{(2n)!}{n!n!} - \frac{(2n)!}{(n-1)!(n+1)!}$$

$$= \frac{(2n)!(n+1)}{(n+1)!n!} - \frac{(2n)!n}{n!(n+1)!} = \frac{(2n)![(n+1)-n]}{(n+1)!n!}$$

$$= \left(\frac{1}{n+1} \right) \frac{(2n)!}{n!n!} = \left(\frac{1}{n+1} \right) \binom{2n}{n}.$$

(Note that the formula $C_n = (1/(n+1)) \binom{2n}{n}$ is valid for $n \geq 0$, not just $n \geq 1$.)

3. **Proof:**

$$\frac{1}{n}\binom{2n}{n-1} = \frac{1}{n}\frac{(2n)!}{(n-1)!(n+1)!} = \frac{(2n)!}{n!(n+1)!}$$

$$= \frac{1}{n+1}\frac{(2n)!}{n!n!} = \frac{1}{n+1}\binom{2n}{n} = C_n.$$

[This result also follows from the previous exercise because $\binom{2n}{n+1} = \binom{2n}{n-1}$.]

5. **Proof:**

$$4\binom{2n-1}{n} - \binom{2n+1}{n}$$

$$= \frac{4(2n-1)!}{n!(n-1)!} - \frac{(2n+1)!}{n!(n+1)!}$$

$$= \frac{4n(n+1)(2n-1)! - (2n+1)!}{n!(n+1)!}$$

$$= \frac{1}{n+1}\left[\frac{(2n)(2n+2)(2n-1)! - (2n+1)!}{n!n!}\right]$$

$$= \frac{1}{n+1}\left[\frac{(2n+2)(2n)! - (2n+1)(2n)!}{n!n!}\right]$$

$$= \frac{1}{n+1}\left[\frac{(2n)!}{n!n!}\right] = \left(\frac{1}{n+1}\right)\binom{2n}{n} = C_n.$$

7. **Proof:**

$$\frac{1}{2n+1}\binom{2n+1}{n} = \frac{1}{2n+1}\frac{(2n+1)!}{n!(n+1)!} = \frac{(2n)!}{(n+1)(n!)(n!)}$$

$$= \frac{1}{n+1}\frac{(2n)!}{(n!)(n!)} = \frac{1}{n+1}\binom{2n}{n} = C_n.$$

9. **Proof:**

$$\binom{2n+1}{n+1} - 2\binom{2n}{n+1} = \frac{(2n+1)!}{n!(n+1)!} - \frac{(2)(2n)!(n)}{(n+1)!(n-1)!(n)}$$

$$= \frac{(2n+1)! - 2n(2n)!}{(n!)(n+1)!} = \frac{(2n)![2n+1-2n]}{(n+1)(n!)(n!)}$$

$$= \frac{1}{n+1}\frac{(2n)!}{(n!)(n!)} = \frac{1}{n+1}\binom{2n}{n} = C_n.$$

11. **Proof:**

$$2C_n + \frac{2}{n}\binom{2n}{n-2} = 2\left(\frac{1}{n+1}\right)\binom{2n}{n} + \frac{2}{n}\binom{2n}{n-2}$$

$$= 2\left(\frac{1}{n+1}\right)\left(\frac{(2n)!}{n!n!}\right) + \frac{2}{n}\left(\frac{(2n)!}{(n-2)!(n+2)!}\right)$$

$$= 2\left(\frac{(2n)!}{(n+1)!n!}\right) + \frac{2(n-1)(2n)!}{n!(n+2)!}.$$

$$= \frac{2(n+2)(2n)! + 2(n-1)(2n)!}{n!(n+2)!}$$

$$= \left(\frac{1}{n+2}\right)\frac{(2n)![2n+4+2n-2]}{n!(n+1)!}$$

$$= \left(\frac{1}{n+2}\right)\frac{(2n)!(4n+2)}{n!(n+1)!}$$

$$= \left(\frac{1}{n+2}\right)\frac{(2n)!(2)(2n+1)}{n!(n+1)!}$$

$$= \left(\frac{1}{n+2}\right)\frac{(2n)!(2)(n+1)(2n+1)}{(n+1)(n!)(n+1)!}$$

$$= \left(\frac{1}{n+2}\right)\frac{(2n+2)(2n+1)(2n)!}{(n+1)!(n+1)!}$$

$$= \left(\frac{1}{n+2}\right)\frac{(2n+2)!}{(n+1)!(n+1)!}$$

$$= \frac{1}{(n+1)+1}\binom{2(n+1)}{n+1} = C_{n+1}.$$

13. **Proof:**

$$\binom{4n}{2n} - \binom{4n}{2n-1} = \binom{4n}{2n} - \frac{(4n)!}{(2n-1)!(2n+1)!}$$

$$= \binom{4n}{2n} - \frac{1}{2n+1}\frac{(4n)!(2n)}{(2n)!(2n)!}$$

$$= \binom{4n}{2n}\left[1 - \frac{2n}{2n+1}\right]$$

$$= \binom{4n}{2n}\left[\frac{2n+1}{2n+1} - \frac{2n}{2n+1}\right]$$

$$= \frac{1}{2n+1}\binom{2(2n)}{2n} = C_{2n}.$$

15. (a) **Proof:**

$$\Gamma\left(n+\frac{1}{2}\right) = \left(n-\frac{1}{2}\right)\Gamma\left(n-\frac{1}{2}\right)$$

$$= \left(n-\frac{1}{2}\right)\left(n-\frac{3}{2}\right)\Gamma\left(n-\frac{3}{2}\right)$$

$$= \left(n-\frac{1}{2}\right)\left(n-\frac{3}{2}\right)\left(n-\frac{5}{2}\right)\Gamma\left(n-\frac{5}{2}\right)$$

$$= \cdots$$

$$= \left(n-\frac{1}{2}\right)\left(n-\frac{3}{2}\right)\left(n-\frac{5}{2}\right)\cdots\left(n-\left(\frac{2n-1}{2}\right)\right)\Gamma\left(n-\left(\frac{2n-1}{2}\right)\right)$$

$$= \left(n-\frac{1}{2}\right)\left(n-\frac{3}{2}\right)\left(n-\frac{5}{2}\right)\cdots\left(n-\left(\frac{2n-1}{2}\right)\right)\Gamma\left(\frac{2n}{2}-\left(\frac{2n-1}{2}\right)\right)$$

$$= \left(n-\frac{1}{2}\right)\left(n-\frac{3}{2}\right)\left(n-\frac{5}{2}\right)\cdots\left(n-\left(\frac{2n-1}{2}\right)\right)\Gamma\left(\frac{1}{2}\right)$$

$$= \left(\frac{2n-1}{2}\right)\left(\frac{2n-3}{2}\right)\left(\frac{2n-5}{2}\right)\cdots\left(\frac{2n-(2n-1)}{2}\right)\sqrt{\pi}$$

$$= \left(\frac{2n-1}{2}\right)\left(\frac{2n-3}{2}\right)\left(\frac{2n-5}{2}\right)\cdots\left(\frac{1}{2}\right)\sqrt{\pi}$$

$$= \frac{(2n-1)(2n-3)(2n-5)\cdots 1}{2^n}\sqrt{\pi}$$

$$= \frac{(2n-1)(2n-3)(2n-5)\cdots 1}{2^n}\frac{2^n n!}{2^n n!}\sqrt{\pi}$$

$$= \frac{(2n-1)(2n-3)(2n-5)\cdots 1}{2^n}\frac{(2n)(2n-2)(2n-4)\cdots 2}{2^n n!}\sqrt{\pi}$$

$$= \frac{(2n)!}{2^n 2^n n!}\sqrt{\pi} = \frac{(2n)!\sqrt{\pi}}{4^n n!}.$$

(b) **Proof:**

$$\left(\frac{4^n}{\sqrt{\pi}}\right)\frac{\Gamma(n+(1/2))}{\Gamma(n+2)} = \left(\frac{4^n}{\sqrt{\pi}}\right)\left(\frac{(2n)!\sqrt{\pi}}{4^n n!}\right)\left(\frac{1}{(n+1)!}\right)$$

$$= \frac{(2n)!}{n!(n+1)!} = \frac{(2n)!}{(n+1)n!n!}$$

$$= \frac{1}{n+1}\frac{(2n)!}{n!n!} = \frac{1}{n+1}\binom{2n}{n} = C_n.$$

17. (a) Here we must travel five steps to the right and five steps up. This is similar to the result in Example 19.1 and the number of paths allowed here is $C_5 = 42$.

(b) In this case, the number of steps to the right is eight and the number of steps up is also eight. Therefore, the number of paths that are allowed here is $C_8 = 1430$.

19. (a) In total, there are $\binom{19}{12} = \binom{19}{7}$ paths from $(0, 0)$ to $(12, 7)$, each made up of 12 R's and seven U's. From these $\binom{19}{12}$ paths, we delete those that violate the stated condition—namely, those paths where the number of U's exceeds the number of R's (at some first position in the path). For instance, consider one such path:

$$R\,R\,R\,U\,R\,U\,U\,U\,U\,R\,R\,R\,R\,U\,R\,R\,R\,U.$$

Here the condition is violated, for the first time, after the fifth U. Transform the given path as follows:

$$R\,R\,R\,U\,R\,U\,U\,U\,U\,\mathbin{:}R\,R\,R\,R\,U\,R\,U\,R\,R\,U$$
$$\leftrightarrow R\,R\,R\,U\,R\,U\,U\,U\,U\,U\,U\,U\,U\,R\,U\,U\,U\,R.$$

Here the entries up to and including the first violation stay unchanged, while the entries following the first violation are changed: R's are replaced by U's and U's are replaced by R's. This correspondence shows us that the number of paths that violate the stated condition is the same as the number of paths made up of thirteen U's and six R's. So there are $\binom{19}{13}$ paths that violate the condition.

Therefore, the answer is

$$\binom{19}{12} - \binom{19}{13} = \frac{19!}{12!7!} - \frac{19!}{13!6!} = \frac{19!(13)}{13!7!} - \frac{19!(7)}{13!7!}$$

$$= \left(\frac{6}{13}\right)\frac{19!}{12!7!} = \left(\frac{12 + 1 - 7}{12 + 1}\right)\binom{19}{12}.$$

(b) Here the answer is

$$\binom{m + n}{n} - \binom{m + n}{n + 1} = \frac{(m + n)!}{n!m!} - \frac{(m + n)!}{(n + 1)!(m - 1)!}$$

$$= \frac{(m + n)!(n + 1) - (m + n)!m}{(n + 1)!m!}$$

$$= \left(\frac{n + 1 - m}{n + 1}\right)\left(\frac{(m + n)!}{n!m!}\right)$$

$$= \left(\frac{n + 1 - m}{n + 1}\right)\binom{m + n}{n}.$$

[Note that when $m = n$, this final result takes the form $\left(\frac{1}{n+1}\right)\binom{2n}{n}$, the formula for the nth Catalan number.]

21. (a) When she can touch the diagonal, the number of routes that Mary Lou can travel is $C_5 = (\frac{1}{6})(\binom{10}{5}) = 42$.

    (b) In this case the first step is R: $(0, 0) \rightarrow (1, 0)$ and the last step is U: $(5, 4) \uparrow (5, 5)$. The number of ways that Mary Lou can travel from $(0, 0)$ to $(5, 5)$ without crossing or touching the diagonal $(y = x)$ is the number of ways that Mary Lou can travel from $(1, 0)$ to $(5, 4)$ without crossing (but perhaps touching) the line $y = x - 1$. This is the same as the number of ways that Mary Lou can travel from $(0, 0)$ to $(4, 4)$ without crossing (but perhaps touching) the diagonal $(y = x)$, and this is

$$C_4 = \left(\frac{1}{5}\right)\binom{8}{4} = 14.$$

23. This is due to the fact that $\sum_{i=0}^{n} \binom{n}{i}^2 = \binom{2n}{n}$. To show this, consider the expansion of $(x + 1)^{2n}$. From the Binomial theorem, we know that the coefficient of x^n in the expansion of $(x + 1)^{2n}$ is $\binom{2n}{n}$. However, since $(x + 1)^{2n} = [(x + 1)^n]^2 = \left(\sum_{i=0}^{n} \binom{n}{i}x^i\right)\left(\sum_{i=0}^{n} \binom{n}{i}x^i\right)$, we may also write the coefficient of x^n as $\binom{n}{0}\binom{n}{n} + \binom{n}{1}\binom{n}{n-1} + \binom{n}{2}\binom{n}{n-2} + \cdots + \binom{n}{n-1}\binom{n}{1} + \binom{n}{n}\binom{n}{0} = \binom{n}{0}\binom{n}{0} + \binom{n}{1}\binom{n}{1} + \binom{n}{2}\binom{n}{2} + \cdots + \binom{n}{n-1}\binom{n}{n-1} + \binom{n}{n}\binom{n}{n} = \sum_{i=0}^{n} \binom{n}{i}^2$. This follows because $\binom{n}{r} = \binom{n}{n-r}$ for nonnegative integers n, r with $0 \leq r \leq n$.

Exercises for Chapter 20

1. The number of ways one can parenthesize the product $abcdefgh$ is $C_7 (= 429)$.

3. Consider the binary strings made up of four 1's and four 0's, where the number of 0's never exceeds the number of 1's, as the string is read from left to right. For each such string, replace the first 1 (from the left) with s_1, the second 1 with s_2, the third 1 with s_3, and the fourth 1 with s_4. Then, for the same string, replace the first 0 (from the left) with e_1, the second 0 with e_2, the third 0 with e_3, and the fourth 0 with e_4. From the result in Example 20.2, the number of such binary strings being considered here is $C_4 = 14$.

   So Samuel can order these eight courses in 14 different ways and still satisfy all of the prerequisites needed to complete the minor in engineering management in eight semesters.

Exercises for Chapter 21

1. The number of Dyck paths from $(0, 0)$ to $(8, 0)$ $[=(2\cdot4, 0)]$ is C_4. The number of Dyck paths from $(8, 0)$ to $(14, 0)$ is the same as the number of Dyck paths from $(0, 0)$ to $(6, 0)$ $[= (2\cdot3, 0)]$—and this is C_3. So there are $C_4C_3 = (14)(5) = 70$ Dyck paths from $(0, 0)$ to $(14, 0)$ that include the point $(8, 0)$.

3. (a) (i) D, D, D*, D, D   (ii) D, D*, D, D, D*, D*, D, D

   (b) (i) D, D, D, D*, D*, D, D*, D*   (ii) D, D*, D, D, D*, D*, D, D*

5.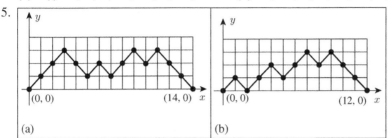

   (a)   (b)

7. The number of Dyck paths from $(0, 0)$ to $(16, 0)$ $[= (2 \cdot 8, 0)]$ is $C_8 = (1/9)\binom{16}{8} = 1430$. From Example 21.5, it follows that the number of times these Dyck paths meet the x-axis is $C_9 = (1/10)\binom{18}{9} = 4862$.

9. Replace each D step with a 1, each R step with $1, -1$, and each D* step with a -1. This results in the following sequences of 1's and -1's for parts (a)–(e) in Fig. 21.15: (a) $1, 1, 1, -1, -1, -1$; (b) $1, 1, -1, 1, -1, -1$; (c) $1, -1, 1, 1, -1, -1$; (d) $1, 1, -1, -1, 1, -1$; (e) $1, -1, 1, -1, 1, -1$. These are precisely the sequences that appear in Fig. 20.4 of Example 20.6. This correspondence follows upon replacing 3 by n, where $n \geq 0$. Consequently, the number of peakless Motzkin paths with a total of n D and R steps is C_n.

11. Consider the case for $n = 3$. For the paths in Fig. 21.17, replace each D with $1, 1$; each D* with $-1, -1$; each R_1 with $-1, 1$; and, each R_2 with $1, -1$. The result for the path in part (a) of Fig. 21.17 is $1, 1, 1, -1, -1, -1$, which is the sequence of three 1's and three -1's that appears in part (a) of Fig. 20.4 in Example 20.6. The paths in parts (b)–(e) of Fig. 21.17 likewise correspond, respectively, with parts (b)–(e) of Fig. 20.4. Since this correspondence follows upon replacing 3 by n, we find that for $n \geq 1$, these particular two-colored Motzkin paths from $(0, 0)$ to $(n, 0)$ are counted by the Catalan numbers.

13. (a) In this case, all of the two-colored Motzkin paths are made up of m R_1 steps and $10 - m$ R_2 steps, for $0 \leq m \leq 10$. The number of possible triples in this case is $\binom{10}{0} + \binom{10}{1} + \cdots + \binom{10}{10} = 2^{10}$.

    (b) For $n \geq 1$, the number of such triples is 2^{n-1}.

15. (a) There are 2^{10} possible outcomes for this situation with $\binom{10}{5}$ successes—that is, outcomes where there are five heads and five tails. So the probability is $\binom{10}{5}/2^{10} = 252/1024 = 63/256 \doteq 0.2461$.

    (b) Here the probability is $C_5/2^{10} = 42/1024 = 21/512 \doteq 0.0410$.

Exercise for Chapter 22

1. This is an application of the ideas set forth on 2 by n Young Tableaux in Example 22.1. Here $n = 6$, so the number of ways to make these arrangements is $C_6 = 132$.

Exercises for Chapter 23

1. (a) $(((ab)c)(de))$ (b) $((((ab)(cd))e)(fg))$

3. We know from Example 23.1 (and later from Example 30.1) that there are C_{n-2} ways to triangulate the interior of a convex polygon of n sides into $n-2$ triangles using $n-3$ diagonals, no two of which intersect within the interior of the polygon. When we remove any one of these $n-3$ diagonals, the result is a partition into $n-4$ triangles and one quadrilateral. Suppose the quadrilateral is $ABCD$ and that AC is the diagonal we removed. This partition, however, could have resulted from another triangulation where all the diagonals are the same except that now we have the diagonal BD instead of the diagonal AC. Consequently, for the C_{n-2} possible triangulations, each with $n-3$ diagonals, we arrive at twice the total number of partitions into $n-4$ triangles and one quadrilateral. Therefore, the number of possible partitions in this situation is

$$
\frac{1}{2}(n-3)\,C_{n-2} = \frac{1}{2}(n-3)\left(\frac{1}{(n-2)+1}\right)\binom{2(n-2)}{n-2}
$$
$$
= \frac{1}{2}\frac{(n-3)}{(n-1)}\binom{2n-4}{n-2} = \frac{1}{2}\frac{(n-3)}{(n-1)}\frac{(2n-4)!}{(n-2)!(n-2)!}
$$
$$
= \frac{1}{2}\frac{(n-3)}{(n-1)}\frac{(2n-4)(2n-5)!}{(n-2)!(n-2)!} = \frac{1}{2}(n-3)(2n-4)\frac{(2n-5)!}{(n-1)!(n-2)!}
$$
$$
= \frac{1}{2}\frac{(n-3)(2n-4)}{(n-3)(n-2)}\frac{(2n-5)!}{(n-1)!(n-4)!} = \frac{(2n-5)!}{(n-1)!(n-4)!} = \binom{2n-5}{n-4}.
$$

Exercises for Chapter 24

1. *Preorder Traversal:*

   1, 2, 6, 12, 16, 17, 3, 7, 8, 13, 4, 9, 5, 10, 11, 14, 15, 18, 19, 20

   Postorder Traversal:

   16, 17, 12, 6, 2, 7, 13, 8, 3, 9, 4, 10, 14, 18, 19, 20, 15, 11, 5, 1

3. (a) (i) This graph has four components, each an isolated vertex. (ii) This graph is connected. It has four vertices and six edges. It is the *complete* graph on four vertices and is denoted K_4. Here there is an edge for each pair of vertices. (iii) Here the graph is a path on four vertices. It is minimally connected in the sense that if any edge is removed (without removing the vertices for that edge), then the resulting graph is disconnected. (iv) This graph is connected. It has five vertices and seven edges. (v) This graph has three components. Two components consist of a single edge and the third component is an isolated vertex. (vi) This graph has two components. One is the complete graph K_4, described in part (ii), and the other component is a single isolated vertex.

   (b) A unit-interval graph is connected when the only time the number of 1's in its binary string equals the number of 0's is at the end of the string. To determine the number of components, consider the string in (v), for example. As we go from left to right, we find that the number of 1's equals the number of 0's for the substring 0011. Consequently, the two vertices for these two unit-intervals are the endpoints of an edge and this edge constitutes a component of the graph. Starting with the 0 in position 5 of the string, as we go from left to right once again, we find the substring 0011, which gives us a second component consisting of two vertices and the edge they determine. Finally, the remaining substring 01 provides the third component for the graph. This time it is an isolated vertex.

5. (a) For the binary string associated with a unit-interval graph G, start at $(0, 0)$ and replace each 0 with a D step from (x, y) to $(x + 1, y + 1)$ and each 1 with a D* step from (x, y) to $(x + 1, y - 1)$. The result is a Dyck path from $(0, 0)$ to $(2n, 0)$. Going in the reverse order, a Dyck path from $(0, 0)$ to $(2n, 0)$ yields a unit-interval graph on n vertices.

   (b) The unit-interval graphs on n vertices that are connected correspond with the Dyck paths from $(0, 0)$ to $(2n, 0)$ that never touch the x-axis. The number of these paths is C_{n-1}, for if we have a path from $(0, 0)$ to $(2(n - 1), 0)$, we can move that path one unit to the right and one unit upward so that it now starts at $(1, 1)$ and ends at $(2(n - 1) + 1, 1)$ [or $(2n - 1, 1)$] and never touches the x-axis. To this resulting path, we now attach a D step from $(0, 0)$ to $(1, 1)$ and a D* step from $(2n - 1, 1)$ to $(2n, 0)$. Furthermore, we note that this process is reversible.

Exercises for Chapter 25

1. **Proof:** Let $a_1 \in A$. Should a_1 be minimal, we are done. If not, there exists an element $a_2 \in A$ such that $a_2 \neq a_1$ and $a_2 \mathcal{R} a_1$. If a_2 is minimal, we are done. Otherwise, there exists an element $a_3 \in A$ such that $a_3 \neq a_2, a_3 \neq a_1$, and $a_3 \mathcal{R} a_2$. Now, fortunately, this cannot continue indefinitely because A is finite, so at some point we find that

$$a_k \mathcal{R} a_{k-1} \mathcal{R} \cdots \mathcal{R} a_3 \mathcal{R} a_2 \mathcal{R} a_1$$

with a_k minimal.

3. (a) $45 = 3^2 \cdot 5$. There are $3 \cdot 2 = 6$ divisors for this partial order and they can be totally ordered in $(1/4)\binom{6}{3} = 5$ ways.

   (b) $54 = 2 \cdot 3^3$. There are $2 \cdot 4 = 8$ divisors for this partial order and they can be totally ordered in $(1/5)\binom{8}{4} = 14$ ways.

   (c) $160 = 2^5 \cdot 5$. There are $6 \cdot 2 = 12$ divisors for this partial order and they can be totally ordered in $(1/7)\binom{12}{6} = 132$ ways.

Exercises for Chapter 26

1. For $n = 1$, there is only one possible distribution—namely, there is one object in box 1. In the case of $n = 2$, there are two distributions. We list these as (i) 1, 1— meaning there is one object in each of boxes 1 and 2; and (ii) 0, 2—indicating

that there is nothing in box 1 and two objects in box 2. The distributions for $n = 3$ can be listed as

(a) $0, 0, 3$ (b) $0, 1, 2$ (c) $0, 2, 1$ (d) $1, 0, 2$ (e) $1, 1, 1$.

If we subtract 1 from each entry in these distributions, we obtain the lists

(a) $-1, -1, 2$ (b) $-1, 0, 1$ (c) $-1, 1, 0$ (d) $0, -1, 1$ (e) $0, 0, 0$.

Reversing the order of the entries, we arrive at the lists

(a) $2, -1, -1$ (b) $1, 0, -1$ (c) $0, 1, -1$ (d) $1, -1, 0$ (e) $0, 0, 0$.

These are precisely the lists that appear in Fig. 26.5 of Example 26.5.

Consequently, for $n \geq 1$, the number of ways we can distribute n identical objects into n distinct boxes numbered $1, 2, 3, \ldots, n$ is the nth Catalan number C_n.

3. (a) It follows from Example 26.3 that there are $1430 (= C_8)$ monomials in the complete expansion.

   (b) The sum of the coefficients of all the monomials in the complete expansion can be obtained by assigning the value 1 to each of the variables $x_1, x_2, \ldots, x_8$. This is the same result we obtain when we assign the value 1 to each of the variables in $\prod_{i=1}^{8}(x_1 + x_2 + \cdots + x_i)$ and this is $1 \cdot 2 \cdot 3 \cdots 8 = 8! = 40{,}320$.

5. (i) The sequences are (a) $1, 2$ (b) $1, 3$ (c) $2, 3$ (d) $1, 4$ (e) $2, 4$.

   (ii) The sequences in part (i) can be placed in a one-to-one correspondence with the sequences in Fig. 26.1 of Example 26.1 as follows. For a sequence in Fig. 26.1, delete the leading 1 $(= a_1)$, then add 0 to the second term $(= a_2)$ and 1 to the third term $(= a_3)$. (In general, we add $i - 2$ to a_i.) In this way, the sequences in parts (a)–(e) of Fig. 26.1 generate the respective sequences in parts (a)–(e) of part (i). To go in the reverse order, start a sequence with first term 1. Now consider a sequence b_1, b_2 as in parts (a)–(e) of part (i). Follow the leading 1 with b_1 and then add the term $b_2 - 1$. [In general, we add the term $b_i - (i - 1)$.] This is how we can go from the sequences in parts (a)–(e) in part (i) to the respective sequences in parts (a)–(e) of Fig. 26.1 in Example 26.1. Consequently, for $n \geq 2$, there are C_n sequences of positive integers $b_1, b_2, b_3, \ldots, b_{n-1}$ where $b_i \leq 2i$, for $1 \leq i \leq n - 1$, and $b_1 < b_2 < b_3 < \cdots < b_{n-1}$.

Exercise for Chapter 27

1. (a) (i) Applying this process to the balls labeled 3 and 4, we obtain $C_2 (= 2)$ possible sets with just one tennis ball. To each of these sets, we add the ball with the label 2. (ii) When the process is applied to the balls labeled 1

and 2 and then 3 and 4, there are $C_3 (= 5)$ possible sets of two tennis balls on the lawn. Of these, there are $C_3 - C_2 = 5 - 2 = 3$ that contain the ball with the label 1.

(b) (i) Applying this process to the balls labeled 3 and 4 and then 5 and 6, we obtain $C_3 (= 5)$ possible sets of two tennis balls. To each of these sets, we add the ball with the label 2. (ii) When the process is applied to the balls labeled 1 and 2, and then 3 and 4, and finally 5 and 6, there are $C_4 (= 14)$ possible sets of three tennis balls on the lawn. Of these, there are $C_4 - C_3 = 14 - 5 = 9$ that contain the ball with the label 1.

(c) (i) C_n (ii) $C_{n+1} - C_n$

Exercise for Chapter 28

1. Here $2n - 1 = 11$, so $n = 6$ and the expected number of games is

$$E_6 = 6\sum_{k=0}^{5} C_k p^k q^k \doteq 9.224.$$

Exercises for Chapter 29

1. **Proof (By Mathematical Induction):** If $n = 4$, we find that $C_4 = 14 > 4 + 2$, so the result is true for this first case and this establishes the basis step for our inductive proof. Now assume the result true for some $k\ (\geq 4)$—that is, assume that $C_k > k + 2$. For $n = k + 1\ (\geq 5)$, we find that

$$C_{k+1} = \frac{1}{(k+1)+1}\binom{2(k+1)}{k+1} = \frac{1}{k+2}\frac{(2k+2)!}{(k+1)!(k+1)!}$$

$$= \frac{1}{k+2}\frac{2k+2}{k+1}(2k+1)\left[\frac{1}{k+1}\frac{(2k)!}{k!k!}\right] = \frac{4k+2}{k+2}C_k$$

$$> \left(\frac{4k+2}{k+2}\right)(k+2), \quad \text{by the induction hypothesis}$$

$$= 4k+2 = 3k + (k-1) + 3$$

$$> 3k+3, \quad \text{since } k \geq 4$$

$$> k+3 = (k+1)+2.$$

Since the truth of the result at $n = k$ implies the truth at $n = k + 1$, it follows that the result is true for all $n \geq 4$, by the Principle of Mathematical Induction.

3. (a) $C_3 C_7 = 5 \cdot 429 = 2145$.   (b) $C_7 C_3 = 429 \cdot 5 = 2145$.
   (c) $C_3 C_4 C_3 = 5 \cdot 14 \cdot 5 = 350$.   (d) $C_3 C_4 C_2 C_1 = 5 \cdot 14 \cdot 2 \cdot 1 = 140$

5. Since $2940/(C_3 C_4) = 2940/(5 \cdot 14) = 42 = C_5$, we find that $n = 3 + 4 + 5 = 12$.

Exercises for Chapter 30

1. By drawing in the diagonal from v_1 to v_7, we are now looking at two convex polygons, one with seven sides and the other with five sides. We can triangulate the polygon with seven sides in C_5 ways and the polygon with five sides in C_3 ways. Consequently, we can triangulate the given decagon, using the diagonal from v_1 to v_7, in $C_5 C_3 = 42 \cdot 5 = 210$ ways.

3. The number of ways one can triangulate the interior of the convex hexagon with vertices $v_1, v_2, v_3, v_4, v_5, v_6$ is $C_4 = 14$. Consequently, the number of ways one can triangulate the interior of the convex $(n-4)$-gon with vertices $v_1, v_6, v_7, \ldots, v_{n-1}, v_n$ is $(823,004)/14 = 58,787 = C_{11}$. So the number of sides in Tim's convex n-gon is $n = 6 + 13 - 2 = 17$. (Here we subtract 2 to account for the fact that the diagonal from v_1 to v_6 is counted among the sides of both the convex hexagon and the convex $(n-4)$-gon, but does not appear among the sides of the original convex n-gon.)

Exercises for Chapter 31

1. (a) The ith component of $\overline{p}$ is $n + 1 - p_i$, so the ith component of $\overline{\overline{p}}$ is $n + 1 - (n + 1 - p_i) = p_i$, the ith component of p. Consequently, $\overline{\overline{p}} = p$.

   (b) The permutation $r = 132456$ is the only one of the four permutations that avoids the 312 pattern.

   (c) $\overline{p} = 625134, \overline{q} = 241365, \overline{r} = 645321, \overline{s} = 426315$
   The permutation $\overline{r} = 645321$ is the only one of these four permutations that avoids the 132 pattern.

   (d) Of the 120 permutations of $\{1, 2, 3, 4, 5\}$, there are $C_5 (= 42)$ permutations that avoid the 132 pattern.

   (e) Of the $n!$ permutations of $\{1, 2, 3, \ldots, n\}$, there are C_n permutations that avoid the 132 pattern.

   (f) Since the 231 pattern is the reverse of the 132 pattern, for $n \geq 3$, there are C_n permutations of $\{1, 2, 3, \ldots, n\}$ that avoid the 231 pattern.

3. (a) No   (b) Yes   (c) No   (d) Yes

5. (a) Not stack-sortable—this permutation contains the 231 pattern—as exhibited by $3, 4, 2$.

   (b) This permutation avoids the 231 pattern. Consequently, it is stack-sortable.

   (c) Not stack-sortable—this permutation contains the 231 pattern—as exhibited by $5, 7, 4$.

   (d) Not stack-sortable—this permutation contains the 231 pattern—as exhibited, for example, by $4, 7, 1$.

7. The permutation p can be sorted in this way if and only if p avoids the pattern 123.

Exercises for Chapter 32

1. (a) There are $C_{12} (= 208,012)$ ways for these 24 superheroes to all simultaneously shake hands with no pairs of arms crossing.

(b) There are C_3 ($= 5$) ways for the six superheroes at seats $2, 3, \ldots, 7$ to all shake hands simultaneously with no pairs of arms crossing. The superheroes at seats $9, 10, \ldots, 24$ can all shake hands simultaneously with no pairs of arms crossing in C_8 ($= 1430$) ways. Therefore, the 24 superheroes can all shake hands with no pairs of arms crossing and with Michael shaking hands with Rebecca in $C_3 C_8 = 5 \cdot 1430 = 7150$ ways.

(c) In this case, there are $C_3 C_3 C_3 C_1 = 5 \cdot 5 \cdot 5 \cdot 1 = 125$ ways in which the 24 superheroes can all shake hands simultaneously with Michael shaking hands with Rebecca, Alberto shaking hands with Cara, and no pairs of arms crossing.

3. We need to find n so that $C_4 \cdot C_n = 235, 144$. Since $C_4 = 14$, it follows that $C_n = (235, 144)/14 = 16, 796$, so $n = 10$. Consequently, there are $8 + 2 + 20 = 30$ superheroes seated at this convention table.

5. (1) If the four points are collinear, then the convex hull is the line segment they determine—including whichever two points are the endpoints of the segment.

(2) If only three of the points are collinear, then the convex hull is the perimeter and interior of a triangle. This triangle has the three collinear points along one side—namely, the line segment they determine, including whichever two points are the endpoints of the segment. These two points are now two of the vertices of the triangle. The remaining point is the other vertex for the triangle.

(3) If no three of the points are collinear, then there are two possibilities: (i) If the triangle determined by three of the points is such that the fourth point is on the perimeter of the triangle or in the interior of the triangle, then the triangle is the convex hull of the four points. (ii) Otherwise, the convex hull is the perimeter and interior of the quadrilateral determined by the four points.

7. (1) Draw a rectangle with corner points $(0, 0)$, $(1, 0)$, $(1, 4)$, and $(0, 4)$. Then place any of the five staircases for $n = 3$ to the right of this staircase so that its lower left endpoint is at the point $(1, 0)$, and its lower right endpoint is at the point $(4, 0)$. This accounts for five of the staircases.

(2) Draw a rectangle with corner points $(0, 0)$, $(2, 0)$, $(2, 3)$, and $(0, 3)$. Then place the one staircase for $n = 1$ on top of this staircase so that its lower left endpoint is at the point $(0, 3)$ and its lower right endpoint is at the point $(1, 3)$. Finally, place either of the staircases for $n = 2$ on the right (of the rectangle drawn), so that its lower left endpoint is at the point $(2, 0)$ and its lower right endpoint is at the point $(4, 0)$. This accounts for $(1)(2) = 2$ of the staircases.

(3) Draw a rectangle with corner points $(0, 0)$, $(3, 0)$, $(3, 2)$, and $(0, 2)$. Place either of the staircases for $n = 2$ on top of this rectangle so that its lower left endpoint is at the point $(0, 2)$ and its lower right endpoint is at the point $(2, 2)$. Now place the one staircase for $n = 1$ on the right (of the rectangle drawn), so that its lower left endpoint is at the point $(3, 0)$, and its lower

right endpoint is at the point $(4, 0)$. This accounts for $(2)(1) = 2$ of the staircases.

(4) Finally, draw a rectangle with corner points $(0, 0)$, $(4, 0)$, $(4, 1)$, and $(0, 1)$. Then place any of the five staircases for $n = 3$ on top of this rectangle so that its lower left endpoint is at $(0, 1)$ and its lower right endpoint is at $(3, 1)$. This provides the final five staircases.

So in total, there are $5 + 2 + 2 + 5 = 14 = C_4$ staircases.

9. (a) If the bottom row is made up of seven pennies, then there are six spaces available to place additional pennies to form the part of the arrangement above the bottom row. (i) One space can be selected in $\binom{6}{1} = 6$ ways, so there are six ways to have an arrangement with eight pennies, and (ii) $\binom{6}{2} = 15$ ways to have an arrangement with nine pennies. (iii) There are $\binom{6}{3} = 20$ ways to select three of the six available spaces and they provide 20 of the arrangements with ten pennies. However, in this case, it is also possible to place two pennies in consecutive spaces (in the row above the bottom row) in five ways, and then place the final penny above these two at the row two levels above the bottom row. In total, there are $\binom{6}{3} + 5 = 20 + 5 = 25$ arrangements where there are 10 pennies.

(b) (i) $\binom{n-1}{1} = n - 1$ (ii) $\binom{n-1}{2}$ (iii) $\binom{n-1}{3} + (n - 2)$. [*Note:* For $n = 3$, $\binom{n-1}{3} = \binom{2}{3} = 0$.]

Exercises for Chapter 33

1. (a) From $n + 1$ 1's and n 0's, one can make up $(2n + 1)!/((n + 1)!n!) = \binom{2n+1}{n+1} = \binom{2n+1}{n}$ binary strings.

(b) Each string in part (a) accounts for $2n + 1$ cyclic shifts, including itself. Consequently, the number of distinguishable binary strings, where one such string cannot be obtained from another by a cyclic shift, is
$$\frac{1}{2n+1}\binom{2n+1}{n} = \frac{1}{2n+1}\frac{(2n+1)!}{(n+1)!n!} = \frac{1}{n+1}\frac{(2n)!}{n!n!} = \frac{1}{n+1}\binom{2n}{n} = C_n.$$

(c) For the C_n strings made up from $n + 1$ 1's and n 0's, where no string is a cyclic shift of another, the Narayana number $N(n, k)$ counts the number of strings where there are k occurrences of the substring "10." For example, $N(4, 2) = 6$ accounts for the strings

100011011, 100100111, 100110011, 101000111, 101100011, 101111000.

3. The number of terms that contain (exactly) six of the variables $x_1, x_2, x_3, x_4, x_5, x_6, x_7, x_8$ is the Narayana number $N(8, 6) = (1/8)\binom{8}{6}\binom{8}{5} = (1/8)(28)(56) = 196$. (This can also be obtained from $N(8, 6) = (1/6)\binom{8}{5}\binom{7}{5} = (1/6)(56)(21) = 196$, using the result in Exercise 6 for this chapter.]

5. The number of Dyck paths from $(0, 0)$ to $(20, 0)$ is $16,796 = C_{10}$. So the number with seven peaks is $N(n, k)$, where $n = 10$ and $k = 7$. Consequently, there are $N(10, 7) = (1/10)\binom{10}{7}\binom{10}{6} = (1/10)(120)(210) = 2520$ Dyck paths from $(0, 0)$ to $(20, 0)$ with seven peaks.

7. **Proof:**

$$N(n, n + 1 - k) = \frac{1}{n}\binom{n}{n+1-k}\binom{n}{n+1-k-1}$$

$$= \frac{1}{n}\binom{n}{n-k+1}\binom{n}{n-k}$$

$$= \frac{1}{n}\binom{n}{k-1}\binom{n}{n-k} = \frac{1}{n}\binom{n}{n-k}\binom{n}{k-1}$$

$$= \frac{1}{n}\binom{n}{k}\binom{n}{k-1} = N(n, k).$$

9. **Proof:**

$$\binom{n}{k-1}N(n, k+1) = \frac{n!}{(k-1)!(n-k+1)!}\frac{1}{n}\binom{n}{k+1}\binom{n}{k}$$

$$= \frac{n!}{(k-1)!(n-k+1)!}\frac{1}{n}\frac{n!}{(k+1)!(n-k-1)!}\frac{n!}{k!(n-k)!}$$

$$= \frac{n!}{(k+1)!(n-k-1)!}\frac{1}{n}\frac{n!}{k!(n-k)!}\frac{n!}{(k-1)!(n-k+1)!}$$

$$= \binom{n}{k+1}\frac{1}{n}\binom{n}{k}\binom{n}{k-1} = \binom{n}{k+1}N(n, k).$$

11. (a) $N(k, k) = \frac{1}{k}\binom{k}{k}\binom{k}{k-1} = \frac{1}{k}(1)(k) = 1$

$$N(k + 1, k) = \frac{1}{k+1}\binom{k+1}{k}\binom{k+1}{k-1} = \frac{1}{k+1}(k+1)\binom{k+1}{k-1} = \binom{k+1}{k-1} =$$

$$\frac{1}{2}(k)(k+1) = \frac{1}{2}(k^2 + k)$$

(b) **Proof:**

$$N(k + p, k) < N(p + k + 1, k)$$

$$\implies \frac{1}{k+p}\binom{k+p}{k}\binom{k+p}{k-1} < \frac{1}{k+p+1}\binom{k+p+1}{k}\binom{k+p+1}{k-1}$$

$$\implies \frac{1}{k+p}\frac{(k+p)!}{k!p!}\frac{(k+p)!}{(k-1)!(p+1)!} < \frac{1}{k+p+1}\frac{(k+p+1)!}{k!(p+1)!}\frac{(k+p+1)!}{(k-1)!(p+2)!}$$

$$\implies \frac{1}{k+p}\frac{(k+p)!}{p!}\frac{(k+p)!}{(p+1)!} < \frac{1}{k+p+1}\frac{(k+p+1)!}{(p+1)!}\frac{(k+p+1)!}{(p+2)!}$$

$$\implies \frac{1}{k+p} < \frac{1}{k+p+1}\frac{(k+p+1)}{(p+1)}\frac{(k+p+1)}{(p+2)}$$

$$\implies \frac{1}{k+p} < \frac{(k+p+1)}{(p+1)(p+2)}$$

$$\implies (p+1)(p+2) < (k+p)(k+p+1)$$

The last inequality is true because $k \geq 2$ and the result follows because each of the above steps is reversible—that is, each logical implication is actually a logical equivalence.

Exercises for Chapter 34

1. *Method 1*

$$M_{11} = \sum_{k=0}^{5} \binom{11}{2k} C_k$$

$$= \binom{11}{0} C_0 + \binom{11}{2} C_1 + \binom{11}{4} C_2 + \binom{11}{6} C_3 + \binom{11}{8} C_4 + \binom{11}{10} C_5$$

$$= 1 \cdot 1 + 55 \cdot 1 + 330 \cdot 2 + 462 \cdot 5 + 165 \cdot 14 + 11 \cdot 42 = 5798$$

Method 2

$$M_{11} = M_{11-1} + \sum_{k=0}^{11-2} M_k M_{11-2-k}$$

$$= M_{10} + \sum_{k=0}^{9} M_k M_{9-k}$$

$$= M_{10} + M_9 M_0 + M_8 M_1 + M_7 M_2 + M_6 M_3 + M_5 M_4$$

$$+ M_4 M_5 + M_3 M_6 + M_2 M_7 + M_1 M_8 + M_0 M_9$$

$$= 2188 + 835 \cdot 1 + 323 \cdot 1 + 127 \cdot 2 + 51 \cdot 4 + 21 \cdot 9$$

$$+ 9 \cdot 21 + 4 \cdot 51 + 2 \cdot 127 + 1 \cdot 323 + 1 \cdot 835$$

$$= 2188 + 835 + 323 + 254 + 204 + 189$$

$$+ 189 + 204 + 254 + 323 + 835$$

$$= 5798$$

Method 3

$$M_{11} = \frac{3(11-1)M_{11-2} + (2 \cdot 11 + 1)M_{11-1}}{11 + 2}$$

$$= \frac{3(10)M_9 + 23M_{10}}{13}$$

$$= \frac{30 \cdot 835 + 23 \cdot 2188}{13} = \frac{25050 + 50324}{13}$$

$$= \frac{75374}{13} = 5798$$

3. (a)   (i) $(-1)^0 |C_0| = 1 \cdot 1 = 1$

(ii) $(-1)^1 \begin{vmatrix} C_0 & C_1 \\ 1 & C_0 \end{vmatrix} = (-1) \left[C_0^2 - C_1 \cdot 1 \right] = (-1)[1 - 1] = 0$

(iii) $(-1)^2 \begin{vmatrix} C_0 & C_1 & C_2 \\ 1 & C_0 & C_1 \\ 0 & 1 & C_0 \end{vmatrix} = C_0 \begin{vmatrix} C_0 & C_1 \\ 1 & C_0 \end{vmatrix} - 1 \begin{vmatrix} C_1 & C_2 \\ 1 & C_0 \end{vmatrix}$

$$= C_0 \cdot 0 - [C_1 C_0 - C_2] = -[1 \cdot 1 - 2] = 1$$

(iv) $(-1)^3 \begin{vmatrix} C_0 & C_1 & C_2 & C_3 \\ 1 & C_0 & C_1 & C_2 \\ 0 & 1 & C_0 & C_1 \\ 0 & 0 & 1 & C_0 \end{vmatrix} = -\left(C_0 \begin{vmatrix} C_0 & C_1 & C_2 \\ 1 & C_0 & C_1 \\ 0 & 1 & C_0 \end{vmatrix} - 1 \begin{vmatrix} C_1 & C_2 & C_3 \\ 1 & C_0 & C_1 \\ 0 & 1 & C_0 \end{vmatrix} \right)$

$$= -\left(1 - \left(C_1 \begin{vmatrix} C_0 & C_1 \\ 1 & C_0 \end{vmatrix} - 1 \begin{vmatrix} C_2 & C_3 \\ 1 & C_0 \end{vmatrix} \right) \right)$$

$$= -1 + C_1 \cdot 0 - (C_2 C_0 - C_3) = -1 - (2 \cdot 1 - 5) =$$
$$-1 + 3 = 2.$$

(b) *Conjecture*: (This is actually true for $n \geq 0$.)

$$(-1)^n \begin{vmatrix} C_0 & C_1 & C_2 & \ldots & C_{n-1} & C_n \\ 1 & C_0 & C_1 & \ldots & C_{n-2} & C_{n-1} \\ 0 & 1 & C_0 & \ldots & C_{n-3} & C_{n-2} \\ \vdots & \vdots & \vdots & \vdots & \vdots & \vdots \\ 0 & 0 & 0 & \ldots & 1 & C_0 \end{vmatrix} = F_n.$$

5. $(n + 2)s_{n+2} - 3(2n + 1)s_{n+1} + (n - 1)s_n = 0$

$$8s_8 - 3(13)s_7 + 5s_6 = 0$$
$$8s_8 = 39(903) - 5(197)$$
$$= 34232$$
$$s_8 = 4279 = \frac{1}{2}R_7 = \frac{1}{2}(8558)$$

$(n + 2)s_{n+2} - 3(2n + 1)s_{n+1} + (n - 1)s_n = 0$

$$9s_9 - 3(15)s_8 + 6s_7 = 0$$
$$9s_9 = 45(4279) - 6(903)$$
$$= 187137$$
$$s_9 = 20793 = \frac{1}{2}R_8 = \frac{1}{2}(41586)$$

Exercises for Chapter 35

1. **Proof:**

$$\frac{1}{n}\binom{kn}{n-1} = \frac{1}{n}\frac{(kn)!}{(n-1)!(kn-n+1)!}$$

$$= \frac{(kn)!}{n![(k-1)n+1]!}$$

$$= \frac{1}{(k-1)n+1}\frac{(kn)!}{n!(kn-n)!}$$

$$= \frac{1}{(k-1)n+1}\binom{kn}{n} = C_k(n).$$

3. (a) There are three possible labels: 1, 2, and 3.

   (b) There are 12 possible sets of two labels:

$$\{1,2\} \quad \{1,3\} \quad \{1,4\} \quad \{1,5\} \quad \{1,6\}$$
$$\{2,3\} \quad \{2,4\} \quad \{2,5\} \quad \{2,6\}$$
$$\{3,4\} \quad \{3,5\} \quad \{3,6\}$$

   (c)

Set of Two Labels	Number of Corresponding Sets with Three Labels
$\{1,2\}$	7
$\{1,3\}$	6
$\{1,4\}$	5
$\{1,5\}$	4
$\{1,6\}$	3
$\{2,3\}$	6
$\{2,4\}$	5
$\{2,5\}$	4
$\{2,6\}$	3
$\{3,4\}$	5
$\{3,5\}$	4
$\{3,6\}$	3

Here, for example, the seven sets of three labels that contain the labels 1 and 2 are

$$\{1,2,3\} \quad \{1,2,4\} \quad \{1,2,5\} \quad \{1,2,6\} \quad \{1,2,7\} \quad \{1,2,8\} \quad \{1,2,9\}$$

The total number of sets of three labels is 55.

   (d) Note that $12 = C_3(3)$ and $55 = C_3(4)$. Based on these results, we conjecture that the number of sets of n labels that would appear on the lawn after the first $3n$ labeled tennis balls have been tossed into the box is $C_3(n+1)$.

5. (a) There are four possible labels: 1, 2, 3, and 4.

   (b) There are 22 possible sets of two labels:

$$\{1, 2\} \ \{1, 3\} \ \{1, 4\} \ \{1, 5\} \ \{1, 6\} \ \{1, 7\} \ \{1, 8\}$$
$$\{2, 3\} \ \{2, 4\} \ \{2, 5\} \ \{2, 6\} \ \{2, 7\} \ \{2, 8\}$$
$$\{3, 4\} \ \{3, 5\} \ \{3, 6\} \ \{3, 7\} \ \{3, 8\}$$
$$\{4, 5\} \ \{4, 6\} \ \{4, 7\} \ \{4, 8\}$$

   (c)

Set of Two Labels	Number of Corresponding Sets with Three Labels
$\{1, 2\}$	10
$\{1, 3\}$	9
$\{1, 4\}$	8
$\{1, 5\}$	7
$\{1, 6\}$	6
$\{1, 7\}$	5
$\{1, 8\}$	4
$\{2, 3\}$	9
$\{2, 4\}$	8
$\{2, 5\}$	7
$\{2, 6\}$	6
$\{2, 7\}$	5
$\{2, 8\}$	4
$\{3, 4\}$	8
$\{3, 5\}$	7
$\{3, 6\}$	6
$\{3, 7\}$	5
$\{3, 8\}$	4
$\{4, 5\}$	7
$\{4, 6\}$	6
$\{4, 7\}$	5
$\{4, 8\}$	4

Here, for example, the ten sets of three labels that contain the labels 1 and 2 are

$$\{1, 2, 3\} \ \{1, 2, 4\} \ \{1, 2, 5\} \ \{1, 2, 6\} \ \{1, 2, 7\}$$
$$\{1, 2, 8\} \ \{1, 2, 9\} \ \{1, 2, 10\} \ \{1, 2, 11\} \ \{1, 2, 12\}$$

The total number of sets of three labels is $(10 + 9 + 8 + 7 + 6 + 5 + 4) + (9 + 8 + 7 + 6 + 5 + 4) + (8 + 7 + 6 + 5 + 4) + (7 + 6 + 5 + 4) = 49 + 39 + 30 + 22 = 140$.

   (d) Note that $22 = C_4(3)$ and $140 = C_4(4)$. Based on these results, we conjecture that the number of sets of n labels that would appear on the lawn after the first $4n$ labeled tennis balls have been tossed into the box is $C_4(n + 1)$.

INDEX

Fibonacci and Catalan Numbers: An Introduction, First Edition. Ralph P. Grimaldi.
© 2012 John Wiley & Sons, Inc. Published 2012 by John Wiley & Sons, Inc.

Printed and bound by CPI Group (UK) Ltd, Croydon, CR0 4YY

16/04/2025

14658368-0004